NOTABLE WOMEN
IN THE
LIFE SCIENCES

NOTABLE WOMEN IN THE LIFE SCIENCES

A Biographical Dictionary

EDITED BY
Benjamin F. Shearer and Barbara S. Shearer

Greenwood Press
Westport, Connecticut • London

Library of Congress Cataloging-in-Publication Data

Notable women in the life sciences : a biographical dictionary /
   edited by Benjamin F. Shearer and Barbara S. Shearer.
     p. cm.
   Includes bibliographical references (p.) and index.
   ISBN 0–313–29302–3 (alk. paper)
   1. Women life scientists—Biography—Dictionaries. I. Shearer,
Benjamin F. II. Shearer, Barbara Smith.
QH26.N68   1996
574'.092'2—dc20   95–25603

British Library Cataloguing in Publication Data is available.

Library of Congress Catalog Card Number: 95–25603
ISBN: 0–313–29302–3

First published in 1996

Greenwood Press, 88 Post Road West, Westport, CT 06881
An imprint of Greenwood Publishing Group, Inc.

Printed in the United States of America

The paper used in this book complies with the
Permanent Paper Standard issued by the National
Information Standards Organization (Z39.48–1984).

10 9 8 7 6 5 4 3 2

Copyright Acknowledgments

The editors and publisher are grateful for permission to reprint excerpts from
the following copyrighted material:

Ruth Hubbard, "Reflections on My Life as a Scientist," *Radical Teacher* (Jan.
1986): 3–7.

On the cover: Roberta Bondar, Alice Evans, Mimi Koehl, and
Rebecca Lancefield

Contents

Introduction

This volume provides biographical entries about 97 women who have made significant contributions to the life sciences from antiquity to the present. Twenty-nine photographs accompany the entries. From long-recognized historical figures and women whose names were "starred" in the first editions of *American Men of Science,* to contemporary MacArthur Foundation "genius" award winners, Blackwell Medal winners, National Medal of Science winners, Nobel Prize winners, and winners of other extraordinary awards such as the Lasker prize and Watermann prize, the unique stories of these prominent women can inspire other talented young women to follow in their footsteps.

Virtually everyone working in women's biography has bemoaned the lack of sources. Research into the lives of women in science can be especially frustrating, because biographical information beyond standard listings in *American Men of Science* is often nonexistent. Such listings only tempt the researcher; they often do not answer important questions about the subjects' lives. In fact, the many dead ends might be partially explained by the irony that American women scientists found themselves listed for years in a work entitled *American Men of Science.*

Taken as a whole, the biographical entries in this volume provide a broad, historical sweep of the work of women in the life sciences. However, the emphasis is on twentieth-century women, most notably on those whose work continues. Disciplines represented in this volume include anatomy, bacteriology, biology, botany, embryology, entomology, genetics, horticulture, medicine, ornithology, pathology, pharmacology, physiology, and zoology. Whenever extant sources allow, the entries go beyond the basic facts found in standard biographical dictionaries. This is most certainly true for those subjects who contributed their own words, through interviews or autobiographical contributions, to the endeavor. The cooperation of these living scientists adds a vital and unique element to their biographical profiles.

Selection of subjects for inclusion in this volume began with a review of a number of excellent and widely known biographical sources, including Marilyn Bailey Ogilvie's *Women in Science: Antiquity Through the Nineteenth Century*, H.J. Mozans's *Woman in Science*, Beatrice Levin's *Women and Medicine*, and Margaret Rossiter's pioneering *Women Scientists in America: Struggles and Strategies to 1940*. Caroline L. Herzenberg's *Women Scientists from Antiquity to the Present: An Index* and *Women in the Scientific Search: An American Bio-Bibliography*, by Patricia Joan Siegel and Kay Thomas Finley, were indispensable not only in building research sources, but also in understanding the tremendous scope of women's achievements in science. The Howard Hughes Medical Institute, the American Medical Women's Association, and the MacArthur Foundation kindly supplied lists of award-winning scientists. The annual "Women in Science" sections of the journal *Science* were helpful in identifying noted contemporary scientists; the personal assistance of the staff of the institute for Scientific Information in helping to identify "high-impact" working scientists was most appreciated.

From hundreds of possible subjects, a draft list appropriate to the size of this volume was created with the intent to represent the disciplines of the life sciences in a broad historical sweep but with an emphasis on the contemporary; to present the stories of women scientists who have been recognized by their peers for the importance of their contributions; and to ensure an international perspective. Decisions for inclusion necessarily had to take into account the existence of sufficient biographical sources for a chosen subject, a sad fact of working in the biography of women. The call for contributors brought an incredible response and the assistance of members of the Association for Women in Science, many of whom are contributors to this volume. Their perusal of the draft list of subjects resulted in further revisions and a nearly final list. Some living scientists who had been selected chose not to participate in the project or could not be contacted.

This volume presents the work of 66 contributors whose boundless enthusiasm for the project was born of the desire to recognize outstanding women in their own fields as well as an awareness of the paucity of biographical resources concerning women in science that has hampered scholarship in the past. Among the contributors—all active participants in their own fields—are educators, biologists, entomologists, pathologists, immunologists, physicians, biochemists, physiologists, zoologists, anthropologists, librarians, archivists, historians, and science writers.

Special effort has been made to explain the work of each scientist clearly in terms familiar to general readers as well as to high school students, to whom this volume is aimed. The biographical entries, which profile the lives of the scientists, make reference to significant developmental influences, obstacles, and achievements. They are accompanied by lists of key dates in the scientists' lives and bibliographies of sources for further information. Cross-references to entries on other scientists appear in **bold** type. The index listings assist in identifying the scientists by ethnic group and nationality.

NOTABLE WOMEN
IN THE
LIFE SCIENCES

HATTIE ELIZABETH ALEXANDER

(1901–1968)

Physician

Birth	April 5, 1901
1923	A.B., Goucher College
1923–26	Bacteriologist, Public Health Services of the United States and Maryland
1930	M.D., Johns Hopkins School of Medicine
1932–66	Professor, Dept. of Pediatrics, Columbia–Presbyterian Medical Center
1954	Stevens Triennial Prize
1961	First woman to receive the Oscar B. Hunter Memorial Award
1964	First woman president of American Pediatric Society
Death	June 24, 1968

Hattie Elizabeth Alexander became the leading authority on the treatment of bacterial meningitis. Her research saved countless children's lives. As a result of this work, she became interested in DNA research and was instrumental in persuading the scientific community to accept genetic research.

Hattie Alexander was born the second of eight children. She was raised in the center of Baltimore, Maryland, and was educated at the all-girl Western High School, where she was known for her spirit. Young Hattie enjoyed many social and athletic activities there, participating in baseball, basketball, and hockey.

She won a partial scholarship to Goucher College, graduating with her bachelor's degree in 1923. As an undergraduate, Hattie was a mediocre student. In fact, some faculty at Goucher—Dr. Jessie L. King, Dr. Gallagher, and Dr. Lilian Welsh—despaired at her lack of application. No one suspected that this "C" student would eventually become a pediatrician of world renown, the discoverer of a cure for bacterial meningitis.

After graduation she worked for three years as a bacteriologist in the Public Health Services of the United States and Maryland to earn money to put herself through medical school. At night she studied physics at George Washington University. Motivated to pursue her dream of becoming a doctor, she finally entered the Johns Hopkins University School of Medicine and received her M.D. degree in 1930. While at Johns Hopkins, she established an outstanding academic record. She chose pediatrics as her specialty and undertook two internships. The first was at the Harriet Lane Home in Baltimore, where she developed her keen interest in *Hemophilus* influenza meningitis. Then she traveled to New York City and interned at the famed Babies Hospital of the Columbia–Presbyterian Medical Center. Following her internship, she was granted an appointment there to the Department of Pediatrics. In 1958 she became a full professor, and in 1966 she was given emerita status until her death in 1968.

Dr. Alexander shouldered many responsibilities at the Columbia–Presbyterian Medical Center. These included administrative duties, research, and teaching. She did not like the traditional lecture approach to teaching; rather, she preferred to challenge her students to explore and always present convincing data to support their decisions. "What is your evidence?" and "What makes you think so?" were frequently recurring questions. She was famous for the examples she set in her warm bedside manner, and her pursuit of excellence and dedication to service in all her endeavors was outstanding and inspirational.

The early treatment for influenza meningitis offered no hope to its young victims. This common disease was 100 percent fatal. Dr. Alexander tried a number of techniques that were unsuccessful, even including horse antiserum. She and an immunochemist, Dr. Michael Heidelberger, combined medical experience with biological theories to discover a cure. A massive amount of bacilli from the spinal fluid of children with influenza meningitis was injected intravenously into healthy rabbits. The antibody produced in the rabbits was therapeutically efficient. Alexander and Heidelberger successfully developed a rabbit antiserum that was a complete cure for the fatal disease, saving the lives of countless children. In just two short years, fatalities dropped to a low of 20 percent. Then sulfanilamide and other sulfonamides were used in combination with the serum. Dr. Alexander continued her work in antibiotic therapy, eventually perfecting the therapy and decreasing the fatality rate to 10 percent.

Dr. Alexander was among the first scientists to determine that bacterial resistance to antibiotics was linked to genetic mutations. Her early work on *Hemophilus* influenza meningitis encouraged her later research in bacterial genetics. Dr. Alexander and Grace Leidy, her assistant, worked together for ten years. Armed with the early reports from the Rockefeller Institute, they explored how deoxyribonucleic acid (DNA) changed genetic characteristics and streptomycin treatment, identifying the transformation of *Hemophilus* influenza attributed to DNA. Their reports and experiments were vital to the acceptance of DNA research by the medical and scientific communities.

In 1953, Dr. Alexander organized a team of scientists to determine if the gene in viruses is similar biochemically to those of bacteria. This research led her to believe that "the principles controlling the inheritance of genetic traits in microorganisms may prove applicable to control of genetic traits in human cells. Such control may make it possible to regulate the kinds of growth that are responsible for neoplasms."[1]

Her colleagues and students admired her courage and scientific objectivity. "Hattie was quite serious minded, intent on and dedicated to her work not only with her mind but with her heart."[2] For her outstanding scientific contributions, she won numerous honors and awards; she found time to contribute to numerous pediatric textbooks and to write 65 research papers as an individual and co-author; and she was a trustee of Goucher College. Her first award was the $500 E. Mead Johnson Award for Research in Pediatrics in 1942. This was for her accomplishment in the treatment of *Hemophilus* influenza. For her outstanding essay on a medical subject, she won the Stevens Triennial prize in 1954 from the Trustees of Columbia University; and in 1961 she was the first woman to receive the Oscar B. Hunter Memorial Award of the American Therapeutic Society. (Jonas Salk had been the first winner of the Oscar B. Hunter Memorial Award.) In 1964 she was the first woman to be elected president of the American Pediatric Society.

Dr. Alexander also found time to enjoy life. She traveled a great deal, enjoyed listening to classical music and speedboating around Long Island Sound, and cultivated exotic plants in her own greenhouse. During the last years of her life, she served as a pediatric consultant at Presbyterian Hospital and emerita professor of pediatrics at the College of Physicians and Surgeons of Columbia University. She died from cancer on June 24, 1968, at age 67. A colleague memorialized Dr. Alexander after her death:

The death of Hattie Alexander, on June 24 last, took from pediatrics one of its most loyal and effective participants. Her skill as a clinician and as an investigator of communicable disease won her wide recognition. . . . Several years before she retired, she suffered a spontaneous subarachnoid hemorrhage which rendered her unconscious for weeks. The possibility of its pathogenesis in an intracranial aneurysm was recognized, but arteriography had shown nothing decisive and her neurosurgeons opted for conservative management. This left her with the knowledge of the probability of recurrence. Despite this knowledge, she resumed her work as though nothing had happened; and, so far as I know, she never alluded to the matter and never lost a trace of élan or of her enjoyment of life. Similarly, in her last illness when it became clear almost to the point of certainty that her abdominal pain was caused by rapid enlargement of her liver with metastatic carcinoma (she had undergone a radical mastectomy some 20 months before), she would not be content with the indirect evidence of disturbed liver function but insisted on the confirmation of an abdominal exploration. The results, at her request, were told to her in full. She knew what she was up against.

Her courage never wavered, and she was her cheerful self right to the end. Her life is a happy inspiration to all who knew her.[3]

Dr. Hattie Elizabeth Alexander was an extraordinary role model. Throughout her career she displayed a steadfast devotion to scientific research, a sheer love of work, brilliance, humility, and warmth. Perhaps one of the children who was saved by her antiserum will dedicate her life to others as she did.

Notes

1. Lenore Turner, "From C Student to Winning Scientist," *Goucher Alumnae Quarterly* (1962): 18–20.

2. Charlotte Stout Singewald, "Hattie E. Alexander. Died June 24, 1968." *Goucher Alumnae Quarterly* (1968): 35–36.

3. Rustin McIntosh, M.D., "Hattie Alexander," *Pediatrics* 42 (Sept. 1968): 544.

Bibliography

"Dr. Hattie Alexander, 67, Dies; Columbia Research Pediatrician." *New York Times* 117 (June 25, 1968): 41.

McIntosh, Rustin. "Hattie Alexander." *Pediatrics* 42 (Sept. 1968): 544.

National Cyclopaedia of American Biography. New York: J. T. White, 1898–   .

Notable American Women: The Modern Period. Cambridge and London: Belknap Press, 1980.

O'Neill, Lois Decker, ed. *The Woman's Book of World Records and Achievements.* Garden City, N.Y.: Anchor Press/Doubleday, 1979.

Siegel, Patricia Joan, and Kay Thomas Finley. *Women in the Scientific Search.* Me-
    tuchen, N.J., and London: Scarecrow Press, 1985.
Singewald, Charlotte S. "Hattie E. Alexander. Died June 24, 1968." *Goucher Alum-
    nae Quarterly* (Fall 1968): 35–36.
Turner, Lenore. "From C Student to Winning Scientist." *Goucher Alumnae Quar-
    terly* (Winter 1962): 18–20.

MARGARET A. IRWIN

DOROTHY HANSINE ANDERSEN
(1901–1963)
Pathologist, Physician

Birth	May 15, 1901
1922	A.B., Mount Holyoke College
1926	M.D., Johns Hopkins University
1929	Accepted appointment in pathology at Columbia University
1935	M.S.D., Columbia University
1938	Described cystic fibrosis; E. Mead Johnson Award
1948	Borden Award for research in nutrition
1954	Elizabeth Blackwell Award
1959	Described cystic fibrosis in young adults
Death	March 3, 1963

Dorothy Hansine Andersen discovered and named the disease cystic fibrosis. She also developed a diagnostic test to detect the disease. Although she was considered "a character" by some, her contribution to
medicine and to the victims of cystic fibrosis cannot be ignored.

Dorothy Andersen was born in Asheville, North Carolina, into the
quintessentially American family of Hans Peter Andersen of the Danish
island of Bornholm and Mary Louise Mason of Chicago. Mr. Andersen
had come to the United States at age 8 and grew up on a farm in Dan-
ville, Vermont. Mrs. Andersen was a descendent of Benning Wentworth
(for whom Bennington, Vermont, was named) and Sir John Wentworth,
a colonial governor of New Hampshire.

When Dorothy's father died in 1914, Andersen and her invalid mother moved from Asheville to Vermont. Andersen attended St. Johnsbury Academy, graduating in 1918, and considered Vermont her home state. Admitted to Mount Holyoke College following her high school graduation, Dorothy was studying there when her mother died in 1920, leaving her with no close relatives. Alone, Andersen completed her bachelor's degree at Mount Holyoke in 1922 and immediately began medical school. She graduated from Johns Hopkins in 1926. There she worked in the laboratory of **Dr. Florence Sabin**, who was the first woman to enroll, graduate, become a faculty member, and work as a full professor at Johns Hopkins. Andersen wrote two papers under Sabin's direction even before completing her M.D. degree. Her thesis, on the lymph system and circulation of the ovary in the sow, was accepted for publication by *Contributions to Embryology*.

Andersen interned at the University of Rochester in 1926 and 1927, teaching anatomy. After a stellar academic career, her rise stalled following her entry into the workplace. In 1928 she completed a surgical internship at Strong Memorial Hospital but was denied a residency in surgery. She was also refused an appointment in pathology at the University of Rochester because she was female. Instead, she became an assistant in the pathology department of the College of Physicians and Surgeons at Columbia University in 1929. There she began research into the relationship of endocrine glands to the female reproductive cycle. In 1930, Andersen was appointed instructor at the medical school, a position that allowed her to teach. She continued her research in endocrinology and reproduction, which led to a medical research degree from Columbia in 1935.

In that same year, at age 34, Andersen moved to Babies Hospital at the Columbia–Presbyterian Medical Center, where she remained until her death. Martha Wollstein had just retired as pathologist there. Andersen joined Beryl Paige as assistant pathologist, later succeeding her as pathologist. Andersen also held an appointment as pediatrician. She began studying heart defects in infants, amassing a collection of specimens that she carefully catalogued.

In December of that momentous year, Andersen performed a postmortem examination of a child who had presented with celiac disease. She found a lesion on the child's pancreas, an anomaly that triggered her curiosity as a researcher. She looked for similar cases in autopsy files and case reports in the literature. Andersen slowly uncovered the pattern of a previously unnamed disorder. On May 5, 1938, in Cincinnati, Ohio, she presented her research on the disease—which she called cystic fibrosis—to a joint meeting of the American Pediatric Society and the Society for Pediatric Research. The publication of her classic work, "Cystic Fibrosis of the Pancreas and Its Relation to Celiac Disease," won her the E. Mead Johnson Award in 1938.

Institutions are always pleased to have internationally recognized researchers on their staff, but adoration of Andersen was far from universal. Perhaps because of being left alone so young, or perhaps because of the difficulty she faced as a woman when she began her career as a researcher, Andersen was extremely independent. On her own, she learned the chemistry and clinical skills needed to continue her research into cystic fibrosis. She was unconditionally generous and friendly to students and residents, but she rarely collaborated with peers. Unwilling simply to describe cystic fibrosis, she looked for a means of diagnosing and even curing it—unusual work for a pathologist. She developed a way to obtain duodenal fluid and analyze its enzyme content, leading to a diagnostic test. Ultimately her research group's discovery that sufferers had an increase in sweat salinity led to the development of the simple "sweat test," still used to diagnose cystic fibrosis today. She also published results of chemotherapy for respiratory tract infections in the hereditary disease.

The clinical skills she learned led to her appointment as assistant attending pediatrician in 1945. She was described as always being aware of her limitations as a clinician, "though the awareness usually exceeded the limitations."[1] During World War II she served as consultant to the Armed Forces Institute of Pathology. During the war, open-heart surgery emerged as a new field; Andersen's knowledge and collection of heart defects served as an important resource. In fact, cardiologists at Babies Hospital were not allowed to operate until having completed a course Andersen developed in the anatomy and embryology of the heart.

Andersen's research interests included the genetics of cystic fibrosis and glycogen storage disease. In 1948 she won the Borden Award for research in nutrition. Throughout this busy period of research she continued her patient care, kept up a full teaching load, conducted seminars, and presented at clinical and pathological conferences. With this schedule, it is perhaps understandable that she was not the best-groomed person, with cigarette ashes often falling on her lab coat. Not wearing the American 1950s female uniform of pearls, a dress, and high heels at a time when women were headed out of the factories of World War II and back into the home, Andersen stood out. Critics denounced her untidy appearance and that of her laboratory, her purposeful disregard of convention, and her unfeminine hobbies of canoeing, hiking, and carpentry. Less criticized were (1) her open acknowledgment that she had indeed made an important discovery, and (2) the way her research into cystic fibrosis crossed the lines of pathology into diagnosis and treatment. Her professional record outweighed all these criticisms, and she was appointed chief of pathology in 1952 following Dr. Paige's retirement. In 1958 she became a full professor. Andersen's last major paper was published in 1959, on the occurrence of cystic fibrosis in young adults. Previously, all who inherited the disease had died in childhood.

Andersen received many honors, including the Citation of Mount Holyoke College (1952), the Elizabeth Blackwell Award (1954), the Great Heart Award of the Variety Club of Philadelphia (1963), and the Distinguished Service Medal of the Columbia–Presbyterian Medical Center (1963). She was a Diplomate of the American Board of Pathology, an Honorary Fellow of the American Academy of Pediatrics, and an Honorary Chair of the National Cystic Fibrosis Research Foundation.

Shy and perhaps lonely, Andersen kept problems to herself, even the lung cancer that she developed in the early 1960s. She often invited young researchers in her lab to her retreat, a farm in the Kittatinny mountains of northern New Jersey. She worked her visitors as vigorously and as pleasantly there as she did in her lab: "My guests find some of the entertainment strenuous, for they shared in building the fireplace, the kitchen chimney and in doing carpentry and other manual labor around the place. It is also a good place for sketching, photography, birds, flowers, cooking, eating, and conversation."[2] A dry, New England sense of humor enlivened her quiet reserve, and her glüg parties always kicked off the Christmas season at Babies Hospital (glüg is a hot concoction of burgundy, cognac, cinnamon, cloves, and other cheerful ingredients). Although she was not a typical feminist of the age, she spoke out against sex discrimination, fought for professional equality, and led her life as she pleased. This practical disregard for convention allowed her to become an expert carpenter, stone mason, canoeist, and world-class medical researcher. Andersen died of lung cancer in New York on March 3, 1963, after an operation in 1962.

Notes

1. Douglas S. Damrosch, "Dorothy Hansine Andersen," *Journal of Pediatrics* 65 (Oct. 1964): 478.
2. *Ibid.*, pp. 478–79.

Bibliography

Andersen, Dorothy Hansine. "Cystic Fibrosis of the Pancreas and Its Relation to Celiac Disease: Clinical and Pathologic Study." *American Journal of Diseases in Children* 56 (Aug. 1938): 344–99.
Damrosch, Douglas S. "Dorothy Hansine Andersen." *Journal of Pediatrics* 65 (Oct. 1964): 477–79.
Di Santágnese, P.A., and Dorothy H. Andersen. "Cystic Fibrosis of the Pancreas in Young Adults." *Annals of Internal Medicine* (1959): 1321–30.
"Dorothy H. Andersen." *Journal of the American Medical Association* 184 (May 25, 1963): 670.
"In Memoriam. Dr. Dorothy Andersen." *Stethoscope*, April 1963.

Machol, Libby. "Andersen, Dorothy Hansine." In *Notable American Women: The Modern Period*, eds. Barbara Sicherman and Carol Hurd Green, pp. 18–20. Cambridge, Mass.: Belknap Press of Harvard University, 1980.

Siegel, Patricia Joan, and Kay Thomas Finley. *Women in the Scientific Search: An American Bio-Bibliography 1724–1979*. Metuchen, N.J.: Scarecrow Press, 1985.

KELLY HENSLEY

ELIZABETH GARRETT ANDERSON

(1836–1917)

Physician

Birth	June 9, 1836
1861	Began a five-year apprenticeship prescribed by the Society of Apothecaries
1865	Received Licentiate of the Society of Apothecaries (LSA) allowing her to practice medicine after passing a licensing examination by the Society of Apothecaries (the first woman to do so)
1866	Became the first English woman to be entered in the Medical Register
1870	M.D., University of Paris
1871	Married J.G. Anderson
1883	Dean, London School of Medicine for Women
1903	President, London School of Medicine for Women
Death	December 17, 1917

Elizabeth Garrett Anderson was a pioneer: the first English woman to be certified as a physician, the first woman to receive a medical degree from the Sorbonne, the first woman to hold the office of mayor in any city of England. She changed history.

Elizabeth Garrett was born in London in 1836, but her family moved to Aldeburgh early in her life. Her father, businessman Newson Garrett, believed in education for his children, boys and girls alike. Her mother, Louisa Dunnell Garrett, was more educated than her husband; besides managing a household with six daughters and four sons, she helped in his office. Elizabeth was the second child. When she was 13 she went to boarding school with her older sister, Louise. The school was typical of its day. Students spoke French but learned little else of practical value. Elizabeth did, however, make many lifelong friends at the school.

In March 1859, Elizabeth Garrett met **Dr. Elizabeth Blackwell**, who was giving a series of lectures on "Medicine as a Profession for Ladies." Garrett's father arranged a letter of introduction to Dr. Blackwell, who liked the younger woman and believed she would also become a doctor. Aware that she could study in America or Switzerland, Garrett was discouraged because foreign medical degrees were not recognized in England. The Medical Act of 1858 stated that to be on the Medical Register, a person must be licensed by a qualified examining board of a British university. At the time, no British university accepted female medical students.

Elizabeth's father accompanied her (proper young ladies had to be accompanied) to local physicians in search of advice. These doctors pointed out that she could never be on the Medical Register. Her father protested that Blackwell was already on it, but he was told it had happened because of loopholes that had since been closed. These sorts of roadblocks challenged Mr. Garrett. When asked why she didn't want to become a nurse, Elizabeth replied, "Because I prefer to earn a thousand rather than twenty pounds a year."[1] Through her refusal to quit fighting for her education, Elizabeth won her father's complete support.

In 1860 Elizabeth began medical training as a nurse at the greatest teaching hospital of the time—Middlesex. This appointment was arranged by a friend of Dr. Blackwell's who suggested that Elizabeth try a six-month trial as a nurse. The young woman had no medical experience, and this would be a crash course in the unsanitary, bloody nature of medicine in the mid-1800s.

Elizabeth remained detached and undaunted in this atmosphere. After three months, she abandoned her pretense of being a nurse and made her rounds as an unofficial medical student. Barred from lectures at first, she eventually wore down resistance from doctors and students with her self-effacing manner. The arrangement lasted until June 1861, when she was the only student who could answer a visiting professor's question. This prompted some male students to appeal for her dismissal from Middlesex.

On August 17, 1861, Elizabeth applied to join the Society of Apothecaries. Although they were not medical doctors, apothecaries could practice medicine in England. The Board said it would allow her to take the licensing examination after the required five-year apprenticeship, perhaps thinking that the pretty young woman would marry, breed, and give up her foolish notions long before then.

At age 25, Elizabeth began a wearing season of studying with tutors and attempting to learn at formal institutions. She studied with individual tutors such as Dr. Stevenson Macadam, one of the best-known doctors of the time, and Sir James Simpson, previously tutor to Elizabeth Blackwell. She learned midwifery from Dr. Alexander Keiller and surgery from Dr. David Murray, who was experimenting with vaccination. During this hodgepodge credit collecting, she was required to carry parts of cadavers back to her rooms because she was not allowed to work in the dissecting labs. She was refused admission to formal classes at Middlesex, Oxford, Cambridge, London University, and St. Andrews in Scotland. The notoriety from these rejections placed her under a spotlight, but that light also shone on the institutions that spurned her.

In 1865 Elizabeth brought all her certificates to the Society of Apothecaries. Following the advice of their lawyers, the Governing Board allowed her to undertake the examination. There was no legal way to deny her, because there was no sex clause in their charter. They enacted one after Elizabeth successfully passed the test and became a licentiate of the Society. The LSA degree was not as prestigious as an M.D. degree, but it allowed her to practice medicine in England.

In 1866 Elizabeth Garrett entered the Medical Register, the first woman after Blackwell and the first female citizen of England to do so. No other female name would appear for twelve years. Her proud father rented Elizabeth an office and hung her shingle himself. Subsequently she established St. Mary's Dispensary for Women. (A success beyond anyone's speculation, St. Mary's became the New Hospital for Women in 1872. After her death, it was renamed the Elizabeth Garrett Anderson Hospital.) Needing a staff position at a regular hospital in order to have colleagues with whom to consult, she adopted her most self-effacing manner and applied for a position at the Shadwell Hospital for Children. At least one member of the Board, a bachelor, was struck by her beauty, if not her credentials, and she received the position.

Despite these successes as a doctor, Garrett did not think her medical education would be complete until she had a university degree. Now 33 years old, she was not willing to wait for English schools of medicine to change their minds about admitting females. She decided to follow Blackwell's example and go to France for her degree. The British ambassador to France went all the way to Empress Eugenie and obtained permission for Garrett to begin taking exams from the University of Paris in March 1869. She wrote her thesis on headaches and defended it in June 1870, becoming the first woman to receive an M.D. degree from the Sorbonne. The British Medical Registry refused to recognize her French degree, but she was a Doctor of Medicine to her own satisfaction.

Now with her own practice and dispensary and all the credentials she could obtain, Elizabeth had done everything right. The weight of being a medical pioneer lessened, and she settled with delight into the life of

a practitioner. When she first opened St. Mary's, a Dr. Nathaniel Heck-
ford had helped during a cholera epidemic. Now he was opening the
East London Hospital and asked her to become a medical officer. There
was some opposition from that hospital's board, including member
James George Skelton Anderson. He prepared to vote against her appointment but reversed himself after listening to her presentation. Elizabeth began her duties there in March 1870. She and James began to see
a lot of each other. When he asked her to marry him, she agreed. Elizabeth had no intention of giving up her work, but James had no intention
of asking her to do so either. They were wed on February 9, 1871. Elizabeth wrote to a friend of the marriage: "I am sure that the woman
question will never be solved in any complete way so long as marriage
is thought to be incompatible with freedom and an independent career,
and I think there is a very good chance that we may be able to do
something to discourage this notion."[2]

Elizabeth and her husband had three children. The oldest, Louisa, became a physician and her mother's biographer. The second, Margaret,
died of an infectious disease at 15 months of age, a hazard in that time
when not even the child of a medical trailblazer was safe. At age 40,
Elizabeth delivered a healthy boy, Alan, her last child.

In 1873, Elizabeth was inadvertently admitted to the British Medical
Association. For nineteen years she was the only female member, holding
offices and giving addresses, because an action at the 1878 meeting excluded all other women. In the years 1874–1878, medical education for
women was legally secured; but the right to a residency was not won
until 1896. A living example of what female doctors could do, Elizabeth
performed what is believed to be the first surgical removal of an ovary
by a woman in 1878.

In 1877, **Dr. Sophia Jex-Blake** asked Elizabeth to help organize the
London School of Medicine for Women. Education for women—especially medical education—was an irresistible campaign for Dr. Garrett,
and the school was founded the next year. She and James raised
thousands of pounds to support it. In 1883 she became dean of the London School, an office she held for twenty years. She then became president, holding the largely ceremonial position until her death. One of her
first duties as dean was to present two students for graduation to the
chancellor of London University. This man had been instrumental in
rejecting her application to his college years earlier. Elizabeth so enjoyed
the situation that she invited her father to come to London for the graduation ceremonies.

In 1902 the Andersons retired to Elizabeth's hometown of Aldeburgh.
James died in 1907. In 1908 Elizabeth succeeded him as mayor, becoming
the first female mayor in England, and was returned to office in her own
right in the next election. In 1914, at the start of World War I, she went

to London to watch her daughter, Dr. Louisa Garrett Anderson, depart for France as part of the Women's Hospital Corps, wishing she was young enough to go along as well. Elizabeth Garrett Anderson died at age 81 on December 17, 1917. Said Dr. Lauder Brunton, "There is no question that she did more for the cause of women in medicine in England than any other person."[3]

Notes

1. Geoffrey Marks and Willis K. Beatty, *Women in White* (New York: Charles Scribner's Sons, 1972), p. 97.
2. Olive M. Anderson, "Elizabeth Garrett Anderson and Her Contemporaries," *Ulster Medical Journal* 27 (1957): 104.
3. Lauder Brunton, "Some Women in Medicine," *Canadian Medical Association Journal*, 48 (1943): 64.

Bibliography

Alic, Margaret. *Hypatia's Heritage*. Boston, Mass.: Beacon Press, 1986.
Anderson, Louisa Garrett. *Elizabeth Garrett Anderson: 1836–1917*. London: Faber and Faber, Ltd., 1939.
Anderson, Olive M. "Elizabeth Garrett Anderson and Her Contemporaries." *Journal of the Medical Women's Federation* 40 (July 1958): 167–79. [Article also appeared in *Ulster Medical Journal* 27 (1957): 97–107.]
Brunton, Lauder. "Some Women in Medicine." *Canadian Medical Association Journal*, 48 (1943): 60–65.
Bynum, W.F., and Roy Porter. *Companion Encyclopedia of the History of Medicine*, Vol. 2, pp. 898–901. London: Routledge, 1993.
"Dr. Elizabeth Garrett Anderson—An English Pioneer." *Women in Medicine* 54 (Oct. 1936): 22.
"Elizabeth Garret [sic] Anderson, M.D." *Medical Woman's Journal* 48 (Dec. 1941): 384–85.
Fancourt, Mary St. J. *They Dared to Be Doctors: Elizabeth Blackwell, Elizabeth Garrett Anderson*. London: Longmans Green and Co., 1965.
Levin, Beatrice. *Women and Medicine*. Lincoln, Neb.: Media Publishing, 1988.
Lovejoy, Esther Pohl. *Women Doctors of the World*. New York: Macmillan, 1957.
Manton, Jo. *Elizabeth Garrett Anderson*. New York: E.P. Dutton and Co., 1965.
Marks, Geoffrey, and Willis K. Beatty. *Women in White*. New York: Charles Scribner's Sons, 1972.

KELLY HENSLEY

VIRGINIA APGAR

(1909–1974)

Anesthesiologist

Birth	June 7, 1909
1933	M.D., Columbia University, College of Physicians and Surgeons
1949	Full Professor, Columbia Medical School
1952	Presented Apgar Score System
1959	M.P.H., Johns Hopkins University School of Hygiene and Public Health; joined National Foundation—March of Dimes
1966	Member, Board of Trustees, Mount Holyoke College
1967	Director, Basic Research Division, National Foundation
1973	Ralph M. Waters Award, American Society of Anesthesiology
Death	August 7, 1974

Virginia Apgar is perhaps best remembered for developing the Apgar Score System, which is used to evaluate newborns in the first minutes after birth. A closer look at her life and accomplishments reveals much more that merits recognition. Colleagues and friends have described her as a person with unmeasurable energy and enthusiasm who loved to teach, to learn, and to care for people. Both personally and professionally, she dedicated herself to improving the quality of human life. She was an intensely independent woman who was able to overcome the traditional boundaries of her time to achieve prominence as a scientist and a humanitarian.

Virginia was born in 1909 in Westfield, New Jersey. She was one of two surviving children of Helen May (Clarke) and Charles Emory Apgar. Her father was a businessman who spent much of his spare time pursuing hobbies in astronomy, wireless telegraphy, and music. The family was a busy one, and music was a favorite activity. Lawrence Clarke Apgar, her older brother, went on to a career in musicology. Virginia played the violin, viola, and cello as an amateur throughout her life.

Her ambition of becoming a doctor most likely derived from her early childhood years when she was influenced by her father's interest in science. As a student in the Westfield public schools, Virginia earned excellent grades in the sciences, participated in athletics, and nearly failed her cooking classes. After graduating from Westfield High School in 1925, she entered Mount Holyoke College. Because her family was not wealthy she financed her own education, supporting herself through scholarships and by working in the college library, waiting on tables, and working as a laboratory aide. At the same time, she reported for the college newspaper, played violin in the orchestra, acted in dramatic productions, and played on seven varsity teams. She majored in zoology with minors in chemistry and physiology. In 1929 she received her B.A. degree.

Virginia went on to study medicine at Columbia University's College of Physicians and Surgeons, where she was one of only a few women students. She was elected to Alpha Omega Alpha, the medical honor society, and graduated fourth in her class. Her financial difficulties continued during these years; despite scholarships, she had to borrow money from family members to continue her studies.

Immediately after receiving her M.D. degree in 1933, she was granted a prestigious surgical internship at the Columbia–Presbyterian Medical Center. Dr. Apgar was only the fifth woman to have earned this distinction. Two years and over 200 operations later, she turned to a career in anesthesiology. A professor of surgery had convinced her that earning a living as a surgeon would be very difficult for a woman. She had loans to repay. This was the time of the Great Depression, when even male surgeons were having difficulty establishing themselves in the field. Because anesthesiology was still in its early stages of development, Dr. Apgar saw the opportunity of doing pioneering work.

She sought the best training available at the time. Nurses had previously been responsible for the administration of anesthetics, so she began her instruction with the nurse-anesthetists at Columbia. This was followed by residencies at the University of Wisconsin and Bellevue Hospital in New York City. In 1937, Dr. Apgar became the fiftieth physician to be certified as an anesthesiologist by the American Board of Anesthesiology.

After initially serving as instructor and then assistant professor at Columbia, she was appointed clinical director of the division of anesthesiology at the Columbia–Presbyterian Medical Center. This was the first time a woman had been chosen to head a department there, and under her leadership it became one of the best in the country. By 1949, Dr. Apgar had advanced to the first full professorship of anesthesiology at Columbia, thereby becoming the first woman to hold a full professorship at that university.

Although she spent her medical career as a teacher and researcher, she remained a student throughout her life. She never became rigid, being always ready to accept new information and make necessary changes. She believed one should seek knowledge and keep abreast of developments in order to provide the best care possible. In fact, she considered it negligent and irresponsible not to do so.

Dr. Apgar was well known for her sense of humor and openness, qualities that made her an excellent teacher. She believed in doing what was right and was never too proud to admit her own mistakes. In any emergency, inside or outside the hospital, she was always prepared to help. Before the advent of Good Samaritan laws, when medical students were being taught to protect themselves first when treating emergencies, she was teaching the opposite. Those who came in contact with her remember how she made each one of them feel special and important.

The delivery room was always one of the most interesting and enjoyable places for Dr. Apgar, and she devoted part of her career to the study of anesthesia in childbirth. The anesthesiologist was responsible for administering anesthesia to the mother and also for resuscitating the newborn if necessary. After years of clinical observation, Dr. Apgar felt that too little attention was being paid to the condition of the baby. She believed that serious disorders, later diagnosed in the nursery, could be prevented or treated sooner if someone examined the infant in the delivery room. It is for her work in this area that she first achieved public prominence.

Her goal was to have someone examine the baby during the first few minutes after birth. The system she ultimately developed, the Apgar Score System, consists of five observations to be made during the first 60 seconds: checks on the heart rate, respiration, muscle tone, reflexes, and color. Each of the five signs is rated on a scale of zero through two, with a maximum score of ten. This system, presented in 1952, was later modified to include an additional check at five minutes. It continues as a standard procedure in hospitals in the United States and many countries throughout the world.

In 1959, after 30 years at Columbia, Dr. Apgar took a year off and earned a Master of Public Health degree from the Johns Hopkins University School of Hygiene and Public Health. Instead of returning to Columbia, she accepted a position with the National Foundation—March of Dimes. Here she became head of the division on congenital malformations. The rest of her life was devoted to the prevention and early detection of birth defects, resulting in lasting and significant contributions to the welfare of mothers and children worldwide. She taught, lectured, and wrote articles on the subject, traveling internationally to educate the public and to raise funds.

During these years she also held the first faculty position in the United States to include birth defects as a subspecialty. This was at Cornell University Medical College, where she served as lecturer and then as clinical professor of pediatrics. In 1973 she was appointed lecturer in the Department of Medical Genetics at the Johns Hopkins School of Public Health. During the same year she became the first woman to receive the Gold Medal for Distinguished Achievement in Medicine from the Alumni Association of the Columbia College of Physicians and Surgeons.

Dr. Apgar lived in Tenafly, New Jersey, and enjoyed gardening, fishing, photography, and golf. Her love of music never dimmed. She especially enjoyed performing chamber music with and for her friends. She was an active member of the Amateur Music Players and the Catgut Acoustical Society, whose members built their own stringed instruments. She was also an ardent stamp collector and belonged to the American Philatelic Society. A few years before her death, her adventuresome spirit and love of learning led her to take flying lessons. Her hope was to one day fly under the George Washington Bridge. Cooking was one of the few things she disliked; she jokingly remarked that she had never married because she hadn't found a man who could cook.

Dr. Apgar always maintained a strong dedication and affection for Mount Holyoke, and she served as an alumna trustee from 1966 to 1971. The college honored her with its Alumnae Award in 1954 and with an honorary Doctor of Science degree in 1965. These were only two of the many awards and honors she received in recognition of her efforts and contributions.

Although Dr. Apgar admitted to being disappointed in not becoming a surgeon, she did not believe that being a woman was an obstacle to her career. She believed that women could achieve as individuals. Although they might have to work harder to remain even with men, she felt that women were liberated as soon as they were born. In conversation she was known to express anger at the restrictions placed on her as a woman, but publicly she did not speak out against them. This was typical of women at the time.

The last years of her life were plagued by progressive liver disease. She died in her sleep on August 7, 1974, at age 65. Despite obstacles, Dr. Apgar achieved recognition as one of the most outstanding women in anesthesiology. Her way of practicing medicine, her teaching, and her research continue to benefit mothers and children today. She believed in doing good on earth; and her humility, thoughtfulness, honesty, and enthusiasm contributed to her personal success as a physician, teacher, and human being.

Bibliography

Apgar, Virginia, and Joan Beck. *Is My Baby All Right? A Guide to Birth Defects.* New York: Trident Press, 1972.

Calmes, Selma Harrison. "Virginia Apgar: A Woman Physician's Career in a

Developing Specialty." *Journal of the American Medical Women's Association* 39 (1984): 184–88.

Current Biography, 1968, s.v. "Apgar, Virginia."

James, L. Stanley. "Fond Memories of Virginia Apgar." *Pediatrics* 55 (1975): 1–4.

———. "Memories of Virginia Apgar." *Teratology* 10 (1974): 213–15.

Medovy, H. "The Apgar Score: A Living, Working Memorial (Dr. Virginia Apgar)." *Canadian Medical Association Journal* 111 (1974): 1023, 1025.

Smith, Christianna. "In Memoriam: Dr. Virginia Apgar '29." *Mount Holyoke Alumnae Quarterly* 58 (1974): 178–79.

Waldinger, Robert J. "Apgar, Virginia," in *Notable American Women: The Modern Period*. Cambridge and London: Belknap Press, 1980.

ELEANOR L. LOMAX

ASPASIA

(ca. 200s)

Physician

Aspasia was a second-century Greco-Roman physician whose writings in obstetrics, gynecology, and surgery were accepted and practiced for over one thousand years. She should not be confused with the Aspasia of the fifth century B.C., a learned writer and teacher who became the wife of Pericles. We know little about the life of the later Aspasia. Some historians speculate that the later Aspasia may have been a beautiful Phoenician, mistress of the kings of Persia, or that "Aspasia" was simply the title of a lost text on diseases of women written by a man.

Although biographical knowledge of her is sparse, we do have access to fragments of her medical writings quoted in a four-volume encyclopedia written in the sixth century. Aetios of Amida, who was a court physician to the Byzantine emperor Justinian I, wrote the *Tetrabiblon*, which covered all medical knowledge up to the beginning of the sixth century. The last section of the encyclopedia deals specifically with obstetrics and diseases of women. Of the 38 chapters of this section that Aetios assigned to quotes from physicians of the past, ten were written by Aspasia and one by Aspasia and Rufus. Later writers have concluded on the basis of Aspasia's quotations that she must have been considered a significant contributor to obstetric and gynecological knowledge. In fact, Aspasia's work survived for a thousand years only to be replaced by those of **Trotula**.[1]

The extracts of Aspasia's writing included in the *Tetrabiblon* deal mostly with managing pregnancy, diagnosing fetal positions, treating painful menstruation, and performing difficult childbirth procedures. Her important technique of rotating the fetus in a breech presentation was a procedural advance. Taking a commonsense approach, she stressed the importance of preventive medicine during pregnancy.

In most of her treatments she suggested the use of herbs, other plants, seeds, or oils—even devising treatments to prevent conception for women for whom it was hazardous. Aspasia performed other surgical procedures such as operations for hernias, blood vessel repair, and stitching to close wounds; and she provided post–operative advice.

Aspasia advised newly pregnant women to guard against "fear, sadness, and all other kinds of mental disturbances" and to avoid "violent exercises, retention of the breath, and sudden blows to the hip bone." She also advised against lifting heavy objects, sitting on hard chairs, and jumping. On the positive side, however, Aspasia recommended "moderate amounts of food, trips in a sedan chair, short walks, gentle massage, and wool spinning." But her advice for the last month of pregnancy was to "curtail the food intake and the more vigorous activities."[2]

Notes

1. James Ricci, *Aetios of Amida, The Gynaecology and Obstetrics of the VIth Century, A.D.* (Philadelphia: Blakiston Company, 1950), p. 5; Kate Campbell Hurd-Mead, *A History of Women in Medicine, from the Earliest Times to the Beginning of the Nineteenth Century* (Haddam, Conn.: Haddam Press, 1938), p. 66; Margaret Alic, *Hypatia's Heritage: A History of Women in Science from Antiquity through the Nineteenth Century* (Boston: Beacon Press, 1986), p. 33.

2. Ricci, *Aetios of Amida*, p. 22.

Bibliography

Alic, Margaret. *Hypatia's Heritage: A History of Women in Science from Antiquity through the Nineteenth Century.* Boston: Beacon Press, 1986.

Eron, Carol. "Women in Medicine and Health Care," in *The Women's Book of World Records and Achievements.* Garden City, N.Y.: Anchor Press/Doubleday, 1979.

Hurd-Mead, Kate Campbell. *A History of Women in Medicine, from the Earliest Times to the Beginning of the Nineteenth Century.* Haddam, Conn.: Haddam Press, 1938.

Mozans, H.J. *Woman in Science.* New York: D. Appleton and Co., 1913. (Reprinted: Cambridge, Mass.: MIT Press, 1974.)

Ogilvie, Marilyn Bailey. *Women in Science: Antiquity through the Nineteenth Century.* Cambridge, Mass.: MIT Press, 1986.

Ricci, James. *Aetios of Amida: The Gynaecology and Obstetrics of the VIth Century, A.D.* Philadelphia: Blakiston Co., 1950.

Schacher, Susan, coordinator. *Hypatia's Sisters: Biographies of Women Scientists, Past and Present*. Seattle, Wash.: Feminists Northwest, 1976.

CAROL BROOKS NORRIS

MARY ELLEN AVERY

(1927–)

Pediatrician

Birth	May 6, 1927
1947	Phi Beta Kappa
1948	A.B., *summa cum laude*, Wheaton College, Norton, MA
1952	M.D., Johns Hopkins University School of Medicine
1954	Internship in Pediatrics, Johns Hopkins
1954–57	Residency in Pediatrics, Johns Hopkins
1957–59	Research Fellow, Pediatrics, Harvard Medical School
1959–60	Fellow, Pediatrics, Johns Hopkins Medical School
1960–64	Assistant Professor of Pediatrics, Johns Hopkins
1964–65	Associate Professor of Pediatrics, Johns Hopkins
1969–74	Professor and Chairman, Dept. of Pediatrics, McGill University
1971	Fellow, American Academy of Arts and Sciences
1974–	Thomas Morgan Rotch Professor of Pediatrics, Harvard Medical School
1974–85	Physician-in-Chief, Children's Hospital, Boston
1985–	Physician-in-Chief, Emerita, Children's Hospital, Boston
1988	Alpha Omega Alpha (Medical Honorary Society), Johns Hopkins University
1990–91	President, American Pediatric Society
1991	Virginia Apgar Award, American Academy of Pediatrics; National Medal of Science
1994	Member, National Academy of Science

Dr. Mary Ellen Avery, a pediatrician, discovered the cause of infant respiratory distress syndrome (RDS) and helped develop treatments for this

Mary Ellen Avery. Photo courtesy of Mary Ellen Avery.

condition. Through her research, Avery found that RDS results from the lack of a fluid called pulmonary surfactant, which normally coats the internal surface of the lungs. This soap-like fluid enables the lungs to retain air after exhalation, making it easier for them to expand with the next breath. Development of the cells that secrete surfactant occurs late in pregnancy, so infants born prematurely are often unable to produce enough of the fluid to breathe easily. Until recently, RDS was responsible for the deaths of an estimated 10,000 premature newborns every year in the United States alone. As a result of Avery's work, the lives of countless infants have been saved—both in the United States and worldwide.

The inspiration that led Avery to a career in medicine came from three sources. The earliest was her neighbor, Dr. Emily Bacon, a professor of pediatrics at Woman's Medical College of Pennsylvania. As a child, Mary Ellen visited the hospital with her and watched as Bacon cared for sick children. "In the 1930s, there was a perception that there wasn't a place for women in the world of science," Avery said, "but my neighbor was proof that it wasn't so." "Then," she continued, "I went to a Dr. Kildare movie and became enamored by what a physician can do." Finally, her own pediatrician with his caring devotion, "which in those days included making house calls," was a person whom she wanted to emulate. "I was also influenced by the general respect with which people held their physicians." In addition, she had "the encouragement of both parents, who more or less made me feel like I could do anything."

Personal experience also motivated Avery's decision to focus on diseases of the lungs. In 1952, just months after completing medical school at Johns Hopkins, Avery was diagnosed with tuberculosis. She was sent to the Trudeau Sanitarium, where she was to have complete bed rest for a year, the recommended treatment at a time when drugs to cure the disease were still being developed. Having been given this prescription, she began to question the logic behind the treatment. Why, she wondered, would lying on one's back for a year be at all beneficial in treating pulmonary tuberculosis? Failing to receive an adequate explanation, she checked out of the sanitarium after two days and recovered at home over the course of a year. Being laid up with a pulmonary disease for this period of time, Avery naturally began to think and to read about lungs. In the end, she decided to become a pediatrician with a subspecialty in pulmonary disease. Her questioning of standard practice and her willingness to challenge conventional wisdom served her well as she embarked upon her research career.

After finishing her internship and residency in pediatrics, Avery began a research fellowship at Harvard Medical School. During this time she learned firsthand about RDS. "I was in the autopsy room a lot," she said. The scene became familiar—premature infants who had been born alive and breathed air for several hours died in respiratory distress. On autopsy, the lungs appeared "liver-like," lacking the normal spongy texture of healthy lung tissue; significantly, there was "no foam" or soapy fluid normally seen in the lungs. "This observation was well known, but the significance was not," Avery said.

Pathologists had many theories to explain these observations. At the time that Avery began her work, RDS was called hyaline membrane disease, named for the presence of a layer of cellular debris seen in the lungs of affected infants. Despite the fact that these membranes were never seen in stillborn infants and were seen only in infants who had breathed air for at least a couple of hours, many researchers believed that the presence of this material was the cause of the respiratory distress. It was later shown that the hyaline membranes formed as a result of the damage to the tissues that occurred when insufficient surfactant was present. The early preoccupation with the *presence* of the hyaline membranes may well have delayed the discovery of the *absence* of surfactant.

After making her early observations, Avery decided that she could not do any further work on the problem until she had developed a deeper understanding of the physiology of the lung. She decided to work with Dr. Jere Mead, a professor at the Harvard School of Public Health (HSPH) who was interested in fundamental aspects of pulmonary mechanics and physiology. With Mead, Avery began to explore the characteristics of the fluid that lines the lung, the foam that was lacking in the infants who died of RDS. At one point in her work Mead posed the question, "Where does the foam come from?" This led her to the scientific literature, where she read about foams and surface tension, the cohesive force between water molecules at the interface between air and liquid.

One day at lunch Dr. James Whittenberger, a professor at HSPH, asked Avery what she was working on. She told him about her work on foams. He suggested that she contact Dr. John A. Clements, a colleague of his at the U.S. Army Chemical Center at Edgewood, Maryland. Over the Christmas holidays she went to Edgewood to meet with him. Clements had constructed a Wilhelmy balance, which he used to measure surface tension in relation to surface area. He had noted the unusually low surface tension achieved by lung extracts. He also recognized the role of the foamy material of the lung in preventing atelectasis, the collapse of the individual alveoli where gas exchange takes place. Clements named this anti-atelectasis factor "pulmonary surfactant."

When Avery saw how crucial this material was in keeping the lungs inflated, she began to wonder whether the lungs of the infants who died might be missing pulmonary surfactant. She then tested her hypothesis. Avery and Mead constructed their own balance and measured the surface tension in lung extracts both from infants who had died of RDS and from those who had died of other causes. They found that the minimum surface tension in lung extracts from babies who had died of RDS was four times higher than that of babies who had died of other causes. The high surface tension caused the babies' lungs to collapse between each breath. Based on the results of her studies, she concluded that surfactant is absolutely essential to keeping air in the lungs and that its deficiency is responsible for RDS in infants.

Avery's work also launched a discipline that she describes as the "study of the metabolism of the lung." When she first began work in this field, "the prevailing view of the lung was that it was simply a huge surface area that let O_2 go one way and CO_2 go the other." Her work on surfactant provided insight into the nature of the lung tissue. "Then, molecular biologists got involved," she said, leading to a better understanding of the function of the proteins that make up surfactant. Now that the nature of surfactant is understood, she said, "hyaline membrane disease could really be classified as surfactant protein A deficiency disease." She continued, "Now we know where the genes are, and that there is a genetic timer of lung maturation."

She notes, however, that "another whole chapter in this story is getting this information to the bedside." The first application of her discovery of the importance of surfactant and its absence in premature infants was to encourage the use of glucocorticoids just before delivery to accelerate lung development in babies thought to be at risk of RDS. During development, glucocorticoids, hormones produced by the fetal adrenal glands, act as a switch that tells cells to "stop multiplying and differentiate." Glucocorticoid treatment just before birth promotes the maturation of the cells that produce surfactant. Avery's work has also led to the development of artificial surfactant, which can be used in combination with glucocorticoid treatment to increase surfactant availability.

Beyond saving the lives of premature infants in this country, Avery's work has involved child healthcare worldwide. She has helped facilitate the delivery of basic medical care to children as a visiting professor in several Third World countries and through her work with UNICEF. "Part of our living has to be a citizen," she said. Of her laboratory work, she said that the process of research involves "creating new knowledge, and once created, it is there forever; it is your own little taste of immortality." However, she emphasizes the importance of the practical aspect of her work: "The knowledge exists; now it's a matter of making it universally available."[1]

Note

1. Material for this entry came primarily from a personal interview with Dr. Mary Ellen Avery and from Julius H. Comroe, *Retrospectroscope: Insights into Medical Discovery* (Menlo Park, Calif.: Von Gehr Press, 1977).

Bibliography

Avery, M.E., and J. Mead. "Surface Properties in Relation to Atelectasis and Hyaline Membrane Disease." *American Journal of Diseases of Children* 97 (1959): 517–23.

Avery, M.E., and T.A. Merritt. "Surfactant Replacement Therapy." *New England Journal of Medicine* 324 (1991): 910.

Comroe, Julius H. *Retrospectroscope Insights into Medical Discovery.* Menlo Park, Calif.: Von Gehr Press, 1977.

NANCY C. LONG

FLORENCE MERRIAM BAILEY

(1863–1948)

Ornithologist, Nature Writer

Birth	August 8, 1863
1882–86	Special student, Smith College; co-founded the first Smith College Audubon Society
1889	Published first book of local nature observations, *Birds through an Opera Glass*
1896	Published *A-Birding on a Bronco* about wild birds in California
1899–1900	Married naturalist Vernon Bailey; joined her husband's expedition to study southwestern wildlife
1902	Published *Handbook of Birds of the Western United States*
1928	Published *Birds of New Mexico*
1929	Elected as first woman fellow of the American Ornithologists' Union (AOU)
1931	Brewster Medal, the AOU's highest award
Death	September 22, 1948

Florence Merriam Bailey was one of the most renowned and influential ornithologists, or bird researchers, in the early twentieth century. She began her career as a nature writer, publishing popular articles and books that encouraged public concern for conservation and interest in nature study. After marrying field naturalist Vernon Bailey, she joined government expeditions to the western United States, gathering information about bird populations. Her later publications grew increasingly scientific, and her success inspired other women to seek careers in field research.

Florence Augusta Merriam was born in 1863. Her father had retired from business to manage his country estate in Locust Grove, New York, and she grew up surrounded by a family of naturalists. Florence's mother taught her astronomy, an aunt taught her botany, and her father took her camping and exploring. One older brother, C. Hart Merriam, invited her along when he trapped and collected mammals, but she preferred watching birds. In an era when women were expected to lead quiet lives at home, Florence's family encouraged her lively interest in outdoor science.

Florence Merriam Bailey. Photo reprinted by permission of
the Sophia Smith Collection and College Archives, Smith
College.

Because her early education consisted primarily of home tutoring, she
could enroll only as a nondegree student at Smith College in 1882. There
were few science courses at Smith, so her studies were concentrated in
literature and languages. Outside of class, she read with alarm about the
slaughter of wild birds to decorate ladies' hats. Many Smith students
followed the fashion craze, but Florence believed they would stop wearing the plumage of dead birds if they knew more about live ones. She
organized the first Audubon Society chapter at Smith and recruited
famed naturalist John Burroughs to lead student nature walks. Returning
home after college, she expanded her efforts to encourage bird study in
articles for *Audubon Magazine* and in her first book, *Birds through an Opera
Glass*. She wrote to brother Hart that her aim in life was "to leave the
world better for my having lived," and she believed the best way to
attain that goal was through her writing.[1]

Also concerned about social problems, Florence spent winters assisting at Working Girls' Clubs in New York City. Thoughts of a social work career were dashed, however, by declining health. Ill with tuberculosis, she began traveling to find healthy climates. In 1894 she spent the spring at an uncle's ranch in San Diego County, California. Each morning she rode the range searching for migrating species, and after lunch she sat for hours watching nesting birds. The book about her experiences, *A-Birding on a Bronco*, is filled with personal stories about her discoveries in "a new bird world."[2]

Apparently cured of tuberculosis, Florence Merriam decided to join her brother in Washington, D.C., where he headed the U.S. Biological Survey. Hart Merriam introduced her to the nation's leading naturalists, helping his sister avoid the intellectual isolation suffered by many women scientists. Frank Chapman and other prominent ornithologists helped with her next book, *Birds of Village and Field*. Most professional ornithologists of the period focused on bird classification, using specimens shot for museum collections. In contrast, her book trained amateurs how to identify common birds without a gun. She wrote that bird identification required only four things, "a scrupulous conscience, unlimited patience, a notebook, and an opera glass."[3] Applauded by critics, the book attracted many enthusiasts to the bird-watching hobby. She reached even more people by helping to found the Audubon Society of the District of Columbia and by conducting bird-watching classes for school-teachers.

The work in Washington also brought her closer to a childhood acquaintance, Vernon Bailey, who worked as a field agent for the Biological Survey. The couple married in 1899. That spring, the Baileys embarked on a camping expedition to collect information about southwestern wildlife. Their rugged trip to Texas and the New Mexico Territory set a pattern for 30 years of exploration. Vernon Bailey, a mammal specialist, trapped mice, moles, and other specimens for the Survey collection, while Florence scoured the brush for birds. The Survey budget necessitated primitive camping, so Florence gamely rode in pack trains, slept under the stars, and laughed at finding rattlesnakes near the sleeping bags. At night the couple relaxed by the fire, writing notes on the day's observations.

Winters in Washington were devoted to turning notes into articles, books, and government documents. Florence Bailey began a huge task of combining her field observations with those of other scientists to produce a comprehensive guidebook, *Handbook of Birds of the Western United States*. Unlike her earlier writings, the *Handbook* employs scientific terminology and precise descriptions of bird distribution, appearance, and reproduction. One chapter, written by her husband, discusses methods of collecting birds for taxonomic study. She still wrote warm, personal bird essays for nature magazines, but publication of the *Handbook* in 1902 signaled her maturation as a rigorous scientist.

Research travels also broadened her concerns about wildlife conservation. Hunting for the plumage market had dropped off as new laws restricted the feather trade, but bird populations still faced numerous pressures. Hunting for meat and sport threatened some species, and Florence Bailey had seen the destructive effects of habitat loss even in the remote areas she visited. She wrote and spoke frequently about conservation problems and intensified her education work with the D.C. Audubon Society. Both Baileys led nature classes for young people, especially Boy Scouts.

The couple's unique scientific partnership continued through the years. At home between expeditions, they hosted dinner parties for the nation's elite scientists. Their desks stood back-to-back in their study, although they usually published their findings independently. Fieldwork in the Pacific Northwest led to a joint government publication in 1918, *Wild Animals of Glacier National Park* (Florence described bird life, and Vernon described the mammals). Both scientists still enjoyed arduous expeditions despite advancing age.

In the 1920s the Baileys began preparing complementary reports on New Mexico wildlife based on several years of fieldwork. Florence Bailey's book, *Birds of New Mexico*, appeared in 1928. Her preface described it as "a statement of complete range, descriptions of the birds, their nests, eggs, and food, together with accounts of their general habits."[4] Over 800 pages long, with maps and color illustrations, it became her most admired and honored work. In 1929, Florence Merriam Bailey became the first woman elected as a fellow of the American Ornithologists' Union (AOU). Two years later the AOU awarded her its most prestigious prize, the Brewster Medal, for *Birds of New Mexico*. Bailey's accomplishments were again acknowledged in 1933 when the University of New Mexico awarded her an honorary LL.D. (Doctor of Laws) degree.

Dr. Bailey continued to study, write, and travel. Her last book, *Among the Birds in the Grand Canyon Country*, was published when she was 76 years old. After her husband's death in 1942, she lived quietly in their Washington home and wrote letters to young relatives about local birds. Her work had not only inspired a generation of women ornithologists to pursue scientific careers but also shaped public attitudes toward wildlife and conservation. After Florence Bailey's death in 1948, one friend wrote in a eulogy, "She was the keenest of observers, and she was able to describe what she saw and heard in words and phrases as sprightly and graceful as the birds themselves."[5]

Notes

1. Harriet Kofalk, *No Woman Tenderfoot: Florence Merriam Bailey, Pioneer Naturalist* (College Station: Texas A&M University Press, 1989), p. 43.

2. Deborah Strom, ed., *Birdwatching with American Women: A Selection of Nature Writings* (New York: Norton, 1986), p. 36.

3. Kofalk, *No Woman Tenderfoot*, p. 80.

4. Florence Merriam Bailey, *Birds of New Mexico* (Santa Fe: New Mexico Department of Game and Fish, 1928), p. 2.

5. Paul H. Oehser, "Florence Merriam Bailey: Friend of Birds," *Nature Magazine* 35 (Mar. 1950): 154.

Bibliography

Bailey, Florence Merriam. *Birds of New Mexico.* Santa Fe: New Mexico Department of Game and Fish, 1928.

Bonta, Marcia Myers. *Women in the Field: America's Pioneering Women Naturalists.* College Station: Texas A&M University Press, 1991.

Gibbon, Felton, and Deborah Strom. *Neighbors to the Birds: A History of Birdwatching in America.* New York: Norton, 1988.

Kofalk, Harriet. *No Woman Tenderfoot: Florence Merriam Bailey, Pioneer Naturalist.* College Station: Texas A&M University Press, 1989.

Oehser, Paul H. "Florence Merriam Bailey: Friend of Birds." *Nature Magazine* 35 (March 1950): 153–54.

Strom, Deborah, ed. *Birdwatching with American Women: A Selection of Nature Writings.* New York: Norton, 1986.

JULIE DUNLAP

ELIZABETH BLACKWELL
(1821–1910)
Physician

Birth	February 3, 1821
1832	Moved to New York City with family
1838	Moved to Cincinnati; father died, leaving family impoverished
1838–44	Worked as a teacher in a family school
1845–46	Studied medicine with John Dickson of Asheville, NC

Elizabeth Blackwell. Photo reprinted by permission of The
Schlesinger Library, Radcliffe College.

1846–47	Studied medicine with Samuel Dickson of Charleston, SC
1847	Began studies at Geneva Medical College
1848	Worked and studied at Philadelphia Hospital
1849	Published medical thesis on combating disease; graduated from Geneva Medical College
1852	Published *The Laws of Life, with Special Reference to the Physical Education of Girls*
1853	Opened a dispensary in New York City
1857	Established the New York Infirmary for Women and Children
1859	First woman to be entered in the Medical Register of the United Kingdom
1868	Established the Woman's Medical College of the New York Infirmary
1871	Charter Member, National Health Society of England

1876	Retired from medicine but continued to write and publish
1878	Published *Counsel to Parents on the Moral Education of Their Children*
1884	Published *The Human Element in Sex*
Death	May 31, 1910

Elizabeth Blackwell, the first woman physician admitted to practice in the United States and Great Britain, was born in Counterslip, near Bristol, England, in 1821. She was the fourth child of Samuel and Hannah (Lane) Blackwell, whose surviving children numbered nine. The children were given tutoring, boys and girls alike, owing to the liberal attitude of their father. Elizabeth spent a happy childhood with her family in England, summering in the country and thriving in the busy household where social reform was the topic of dinner conversation. However, fire ruined her father's sugar refinery, and in 1832 Samuel Blackwell moved the family to America to start anew.

Life in America was not easy for the Blackwells, who settled in New York City. Although Samuel had a house and refinery, he did not believe in using cane sugar harvested by slaves; and his efforts at refining beet sugar failed. Despite this, the Blackwells took full advantage of the offerings of New York City, often attending lectures and abolitionist meetings. Elizabeth was influenced by the open-minded attitude with which her parents viewed religion. She attended the services of various churches, comparing what she experienced at each. Religion was an important part of her life, but she was constantly open to the flow of new ideas and would join churches or study doctrines that suited her mindset at the time. The Blackwells were also socially aware, engendering in Elizabeth a passion for the abolitionist cause and what would be a lifelong concern with social and moral issues.

The Blackwells moved to Cincinnati in 1838 so that Samuel could pursue further business ventures, but there he fell ill and died, devastating the close-knit family and leaving them impoverished. Elizabeth was concerned that she contribute to the family purse, musing, "I wish I could devise some good way of maintaining myself, but the restrictions which confine my dear sex render all my aspirations useless."[1] This is perhaps the first inkling that Elizabeth would become a career woman, for her determination to be independent would carry her through the difficulties of forging a life separate from any man. In a note from this time, she described herself as

distinguished for independence and decision of character—her will very strong—high minded and a great regard for her own opinion—not disposed to trifle or will she be trifled with . . . can fix her

mind with great intenseness and connectedness...unites rich
thought with strong imagination—when interested very enthusiastic, but takes much to excite...intellect large—more talent than
she displays for the want of language—can understand and com-
prehend better than communicate...poor to recite and not very
quick to commit to memory but never forgets when acquired—her
learning disciplines her mind—always investigating principles and
wishes to know whys and wherefores.[2]

In the interest of family solvency, however, Elizabeth and her mother
and sisters formed a school in Cincinnati at which they all taught in
order to provide income and education for the younger siblings.

After six years of teaching, including a position in Henderson, Ken-
tucky, boredom and ambition propelled Elizabeth away from the family.
She set out to study medicine with John Dickson of Asheville, North
Carolina. There she was able to teach in a small academy run by Dickson
while having access to his medical library. When the academy closed
owing to Dickson's ill health, she moved to Charleston, South Carolina,
where she continued in a similar arrangement with Dickson's brother,
Samuel.

The exact reasons for Elizabeth's decision to pursue medicine have not
survived in her writings. She does mention, however, that the urging of
a close friend who was very ill first gave her the notion. The friend
would have liked to have a lady doctor, she said, for "the delicate nature
of the disease made the methods of treatment a constant suffering for
her."[3] At first Elizabeth stated that the idea was impossible, because she
"hated everything connected with the body and could not bear the sight
of a medical book."[4] However, she was horrified at her attraction for
men and was determined to be independent, stating:

> At this very time when the medical career was suggested to me, I
> was experiencing an unusually strong struggle between attraction
> towards a highly educated man with whom I had been very inti-
> mately thrown....I grew indignant with...a struggle that weak-
> ened me, and resolved to take a step that I hoped might cut the
> knot I could not untie, and so recover full mental freedom! I finally
> made up my mind to devote myself to medical study.[5]

After two years of medical reading, Elizabeth applied to medical colleges in New York and Philadelphia but was rejected from all of them.
Finally, in 1847 she was accepted at the Geneva Medical College in upstate New York and began her studies there that year. She discovered
later that her acceptance, which had been the unanimous vote of the
student body, had been a riotous joke among her classmates. The effect

of her presence, however, was "the most perfect order and quiet" in the previously rowdy country classrooms.[6] Even the anatomy course, previously taught with many bawdy jokes, improved after Elizabeth refused to be excused.

> The Professor adhered closely to his text throughout, without having occasion to amuse the class with his usual vulgar anecdotes, and the class on its part observed the most perfect decorum. The older students declared that this was the first course of lectures on that subject, by the Professor, that they had ever been able to follow connectedly, and the Professor himself acknowledged that this was the first course in which he was thoroughly interested.[7]

Following her graduation in 1849 Elizabeth traveled to Europe, where she furthered her training at La Maternité in Paris. While giving an injection there for purulent ophthalmia, her own eye became infected; although she wrote to her uncle during her recovery that "I still mean to be at no very distant day *the first lady surgeon in the world*," the resulting blindness in one eye dashed all such hopes.[8] Despite this setback she continued her studies in England, returning to the United States in 1851 to open a practice in New York.

Although it is for her admission to medical practice that she is best known, Elizabeth Blackwell's enduring legacy is her formation of institutions dedicated both to the training of women for medical practice and to the care of women and children. She made a name for herself in New York by delivering a series of lectures and publishing, in 1852, *The Laws of Life, with Special Reference to the Physical Education of Girls.* In 1853 she opened a dispensary that became the New York Dispensary for Women and Children in 1854. Blackwell felt strongly that care should be provided for the poor, especially women and children, and that her facility should also train women doctors and nurses. In keeping with this philosophy she adopted 7-year-old Katherine Barry, an orphan from Randalls Island. In 1857 she established the New York Infirmary for Women and Children with her sister, Emily, who had also become a physician, and Dr. Marie Zakrzewska, who later founded the New England Hospital for Women and Children. Although the trustees of the new hospital were men, it was the first medical facility in which the clinical work was conducted entirely by women.

Feeling that progress had been made in the United States, Blackwell traveled to England in 1859, as she was now "desirous of learning what openings existed in England for the entrance of women into the medical profession."[9] There she promoted her ideas for a year, during which time she became the first woman enrolled as a recognized physician in the Medical Register of the United Kingdom.

Although she returned later that year to work at the New York Infirmary, progress was interrupted by the Civil War. Elizabeth Blackwell became active in training war nurses, who were sorely needed to sustain the troops. "It was not until this great national rebellion was ended that the next step in the growth of the infirmary could be taken."[10] By this time, women were being trained at women's medical colleges in Boston and Philadelphia, but Blackwell believed they were not being held to sufficiently high standards and thus were not able to do the job later, proving arguments against them. To combat this, she encouraged the trustees of the infirmary to apply for a college charter, which was granted in 1865.

In 1869 Elizabeth Blackwell returned to England, where she spent the rest of her life lecturing, writing, and crusading for the end of the double standard of sexual morality that condemned women, especially prostitutes. In her willingness to speak of such subjects, she was ahead of her time and shocked many. Her focus on hygiene, morality, and sexual issues contributed to the scope of medical study. Elizabeth Blackwell died on May 31, 1910, in Kilmun, Scotland, where she and her adopted daughter had been spending their summers for many years. The efforts of her later years can be summarized by the motto of the National Health Society, of which she was a charter member: "Prevention is better than cure."

Writing in 1895, Elizabeth Blackwell looked back on her life's work:

In 1869 the early pioneer work in America was ended. During the twenty years which followed the graduation of the first woman physician, the public recognition of the justice and advantage of such a measure had steadily grown. Throughout the Northern States the free and equal entrance of women into the profession of medicine was secured. In Boston, New York and Philadelphia special medical schools for women were sanctioned by the Legislatures, and in some long established colleges women were received as students in the ordinary classes.[11]

Owing to Elizabeth Blackwell's intellect, determination, and spirit, the medical profession was opened to women.

Notes

1. Elizabeth Blackwell, Diary. Quoted in Nancy Sahli, *Elizabeth Blackwell, M.D., 1821–1910* (New York: Arno Press, 1982), p. 17.

2. Elizabeth Blackwell to unknown (fragment), January 28, 1836. Blackwell Family Papers, Schlesinger Library, Radcliffe College.

3. Elizabeth Blackwell, *Pioneer Work in Opening the Medical Profession to Women* (New York: E.P. Dutton, 1914).

4. *Ibid.*, p. 22.

5. Elizabeth Blackwell, MS Autobiography. Blackwell Family Papers, Schlesinger Library, Radcliffe College.

6. Stephen Smith, "A Woman Student in a Medical College," in *Memory of Dr. Elizabeth Blackwell and Dr. Emily Blackwell* (New York: Knickerbocker Press, 1946), p. 9.

7. *Ibid.*, p. 12.

8. Blackwell, *Pioneer Work.*

9. *Ibid.*, p. 172.

10. *Ibid.*, p. 191.

11. *Ibid.*, p. 194.

Bibliography

Baker, Rachel. *The First Woman Doctor.* New York: Julian Messner, 1944.

Blackwell, Elizabeth. *Counsel to Parents on the Moral Education of Their Children.* New York: Brentano's Literary Emporium, 1880.

———. *Essays in Medical Sociology.* London: Ernest Bell, 1902.

———. *The Human Element in Sex.* London: J & A Churchill, 1894.

———. *Lectures on the Laws of Life, with Special Reference to Girls.* London: Sampson Low & Co., 1871.

———. *Pioneer Work in Opening the Medical Profession to Women.* New York: E. P. Dutton, 1914.

———. "Ship Fever: An Inaugural Thesis, submitted for the degree of M.D. at Geneva Medical College, January 1849." *Buffalo Medical Journal and Monthly Review* 4, no. 9 (Feb. 1849): 523–31.

Notable American Women, 1607–1950. Cambridge and London: Belknap Press, 1971.

Ross, Ishbel. *Child of Destiny: The Life Story of the First Woman Doctor.* New York: Harper & Brothers, 1949.

Sahli, Nancy. *Elizabeth Blackwell, M.D. (1821–1910): A Biography.* New York: Arno Press, 1982.

Wilson, Dorothy Clarke. *I Will Be a Doctor! The Story of America's First Woman Physician.* Nashville: Abington Press, 1983.

ELIZABETH LUNT

RACHEL LITTLER BODLEY

(1831–1888)

Botanist

Birth   December 7, 1831

1849–60   Assistant Teacher, Wesleyan Female College

1862–65	Chair and Professor of Natural Studies, Cincinnati Female Seminary
1865–77	First woman to be Professor and Chair of Chemistry and Toxicology, Female Medical College (later Woman's Medical College of Pennsylvania)
1871	A.M., Wesleyan Female College
1876	Corresponding Member, New York Academy of Sciences; Charter Member, American Chemical Society
1877–88	Dean of College, Woman's Medical College
1879	M.D., Woman's Medical College
1883	Board of Public Charities of the State of Pennsylvania
Death	June 15, 1888

Rachel Littler Bodley was the first woman professor and chair of chemistry and toxicology at the Female Medical College in Philadelphia as well as dean of the college for over a decade. As a scientist, she was a charter member of the American Chemical Society and the teacher of countless young women, whom she sent all over the world as physicians. As a citizen, she devoted herself to the education of children at home and in missions abroad. Her influence in the education of women in science is inestimable.

Rachel Bodley was born in Cincinnati, Ohio, in 1831 to Anthony and Rebecca Bodley. Rachel was the eldest daughter and the third of five children. The Bodleys were a deeply religious family, holding the qualities of obedience and reverence in high regard.

The Bodley family also held a deep respect for education. Mrs. Rebecca Bodley opened a private school in Cincinnati, which was known then as the "Athens of the West," where its citizens had the opportunity to listen to speakers from the women's movement, abolitionists, and members of New England Reformers at the Lyceum. Rachel's first teacher was her mother, and she attended her mother's school until her twelfth year. In 1844, Rachel entered the Wesleyan Female College of Cincinnati. The College had been founded in 1842 and was the first chartered college for women in the world. The quality of education offered at this institution of higher learning was excellent, surpassing any offered at existing seminaries and schools. Rev. P.B. Wilber, president of the school from 1842 to 1859, cultivated the moral character as well as the intellect of the women attending Wesleyan Female College. With their Christian faith and solid training, the graduates performed fine works for the church and communities.

Rachel was an outstanding student. During her five years at Wesleyan Female College, she achieved incredible academic success. She was friendly as well as studious, kind, and encouraging to her homesick and less talented classmates.

In 1849, Rachel graduated and accepted a post as assistant teacher at Wesleyan Female College. She remained there until 1860, having achieved the post of Preceptress in the Higher Collegiate Studies. She proved to be a skilled teacher who was devoted to her students. Rachel challenged her students, strived to make each lesson interesting, and endeavored to whet their thirst for knowledge. She thoroughly loved learning and teaching.

Rachel was so devoted to the complete development of her students that she gave weekly religious instruction to them. Her dedication provided a true model of Christian character that left a great and lasting impression on her students.

Rachel left teaching to further her own education in the hard sciences. During the fall of 1860, she traveled to Philadelphia to become a special student at the Polytechnic College of Pennsylvania in the areas of advanced chemistry and physics. The Polytechnic College of Pennsylvania was, at that time, the leading institution of higher education in the field of applied sciences, but it closed during her second year there. Rachel finally decided to study practical anatomy and physiology at the Woman's Medical College of Pennsylvania. Following completion of her studies in February 1862, she returned to Ohio. The Cincinnati Female Seminary appointed her professor of natural sciences, a post she held for three years.

As professor of natural sciences she made valuable contributions in the area of botanical science. At this time she wrote the *Catalogue of Plants Contained in the Herbarium of Joseph Clark: Arranged According to Natural System.* Joseph Clark was an avid nature lover who had emigrated from Scotland. He lived in Cincinnati for 35 years and died in 1858. The Cincinnati Female Seminary was the recipient of his extensive collections.

Rachel described the condition of Clark's collections at the seminary as "chaos reigning in the domain of science." She decided to enter "single-handed upon the Herculean task of making these collections available to science" by classifying them "according to the natural system." She described her long task and her feelings about it:

> The mass was carefully opened, the plants identified and finally arranged in labelled sheets of uniform size, and the whole placed in a convenient herbarium case, where it is now in complete readiness for reference study. During these years I have labored patiently and faithfully upon it in my leisure hours, and it is only now in my fourth summer vacation that I have finished the classification and arrangement of this herbarium. I have found my work womanly, secluded, ennobling; and I submit to educated

women of this vicinity whether, since these pursuits fail through lack of patronage, they may not enter upon them, and as they find opportunity, become workers in, or patronesses of science. Only the will is lacking; cultivated talent, wealth and opportunity are abundant.[1]

The preparation and publication of the catalogue received praise from Dr. Asa Gray, the renowned American botanist.

In 1865 the Female Medical College of Pennsylvania offered Rachel the chair of the chemistry and toxicology department. She knew the Corporator of the Female Medical College (later the Woman's Medical College of Pennsylvania), Isaac Barton. They met in 1854 when Rachel and her brother were traveling in the Great Lakes region. Barton introduced her to Ann Preston. With Rachel's acceptance of the post, she became the first woman professor of chemistry. At the time there were three women on the faculty at the college: Ann Preston, Emeline Horton Cleveland, and Mary Scarlett-Dixon.

Chemistry and medicine were forming greater bonds. Medicines such as painkillers and sedatives were being developed. Anna E. Broomall, a former student of Professor Rachel Bodley, discovered the use of chloroform in obstetrics. With new equipment coming into the laboratory, Professor Bodley studied the use of morphia and quinine. Her primary interest was in human health and its relationship to the laws of chemistry. She continued in her endeavor to serve God and inspire students as well as to secure the recognition and respect women deserved. Professor Bodley used her time and talents for the benefit of the college, her students, and humanity. She loved scientific studies, but she was also dedicated to the missionary cause. Emeline Horton Cleveland, a young woman friend of like mind, studied at the Woman's Hospital of the Woman's Medical College of Pennsylvania to become a missionary doctor.

Professor Bodley graduated but did not complete her medical course of study. Her regime was the classical curriculum. She also studied mathematics, French, German, art, phonography, microscopy, and elocution and took private music lessons. Her students and colleagues admired her pursuits. She found that the courses stimulated her and improved her teaching. In 1860, while serving as chair at the Woman's Medical College of Pennsylvania, she began and completed the regular course of medical study.

Professor Bodley was an adventurer. Her only recreation occurred during summer vacations, when she took long journeys to the Atlantic coast, the Rocky Mountains, the shores of the Great Lakes, and the Gulf of Mexico. The explorer in her was kindled in her youth by family outings. Her brothers were often her traveling companions. Rachel thoroughly enjoyed the great outdoors. In later years she journeyed alone, always accompanied by the manuals of Gray and Chapman and a botanical trunk. The plants she collected over the last 20 years of her life supplied abundant companionship. Professor Bodley was also a popular lecturer and visiting faculty member. She traveled during the summers of 1866–1869 to Long Island, Philadelphia, and Cincinnati. She taught at the Howland School in Cayuga Lake, New York, for five seasons from 1870 to 1874.

Serving as dean placed even more demands on Professor Bodley. She was busy with construction of a new building, moving, and mundane tasks such as correspondence, but she also made significant changes that expanded the curriculum. The college grew in size and prominence under her deanship. With her achievements and kindness, she developed contacts around the world. She stayed in constant touch with graduates and groomed relationships with scientists and physicians. Her enthusiasm inspired many former students: Amy S. Barton, the first woman professor of ophthalmology; Lillian Welsh, the college physician at Goucher College; Anna E. Broomall; and many others. Professor Bodley also taught chemistry to many future missionaries, including Lucinda Coombs, Mary Seeyle, Clara Swain, and Sarah Seward. These women went on to found hospitals and clinics overseas.

Mrs. **Anandibai Joshee** was the college's first foreign student. She stayed with Dean Bodley while she studied; when she graduated in 1886, Bodley brought to the attention of Queen Victoria that this woman from India had just become a physician. In such ways, Bodley encouraged her students as physicians and inspired them as scientists. In 1874 she admonished the graduates in a valedictory address that

> Another duty incumbent upon you, after the faithful ministration to the sick and afflicted, is the continued cultivation of scientific knowledge. Woman may yet be "numbered in the vanguard of original investigators." You enter upon your active duties at a time when beyond all the ages that have preceded it, it is glorious for a woman to live and toil. This great nation is astir.[2]

Rachel Littler Bodley's contributions to science, education, literature, and the lives of women were considerable. She was a distinguished scientist. Many honors were bestowed on her from learned societies in the historical and the natural sciences. These include: election as a corresponding member to the State Historical Society of Wisconsin; election as a member of the Academy of Natural Sciences of Philadelphia in 1871; an A.M. degree during the same year from her alma mater (she was the first of three alumnae to receive such a degree). The Cincinnati Society of Natural History elected her a corresponding member in 1873; in 1876 she became a corresponding member of the New York Academy of Sciences and a member of the American Chemical Society. Bodley was instrumental in arranging the centennial celebration of chemistry honoring Dr. Joseph Priestly, who discovered oxygen. The Woman's Medical College of Pennsylvania conferred the degree of M.D. to Professor Rachel Littler Bodley in 1879. In 1880 the esteemed Franklin Institute of Philadelphia elected her as a member. As such she was invited to lecture, giving six lectures on "Household Chemistry." Rachel Bodley contributed to the education of the youth of Philadelphia by serving on a number of boards, including the Public Educational Society of Philadelphia. The Board of Public Charities of the State of Pennsylvania appointed her and six other women to help inspect institutions in the county of Philadelphia. Bodley also enriched the lives of the citizens of Philadelphia. Her influence and delight in scientific research spread worldwide. She was intelligent yet modest, bold yet patient, always kind and generous. Her organizational skills, attention to detail, and superb use of time allowed her to accomplish her goals. Serving the Lord and bettering the lives of women were the credos she followed.

Notes

1. Rachel Littler Bodley, *Catalogue of Plants Contained in the Herbarium of Joseph Clark: Arranged According to Natural System* (Cincinnati: R.P. Thompson, Printer, 1865), Preface.

2. Rachel L. Bodley, "Valedictory Address to the Twenty-Second Graduating Class of the Woman's Medical College of Pennsylvania, March 13, 1874" (Philadelphia: Rogers, 1874), p. 13.

Bibliography

Alsop, Gulielma F. "Rachel Bodley, 1831–1888." *Journal of the American Medical Women's Association* 4 (Dec. 1949): 534–36.

Bodley, Rachel L. *Introductory Lecture to the Class of the Woman's Medical Center of Pennsylvania Delivered at the Opening of the Nineteenth Annual Session, Oct. 15, 1868.* Philadelphia: Merrihew, 1868.

Bolton, Sarah K. *Successful Women.* Boston: D. Lothrop, 1880.

Miles, Wyndham D. "Rachel L. Bodley, 1831–1888." In *American Chemists and Chemical Engineers.* Washington, D.C.: American Chemical Society, 1976.

Notable American Women, 1607–1950. Cambridge and London: Belknap Press, 1971.

Papers Read at the Memorial Hour: Commemorative of the Late Prof. Rachel L. Bodley, M.D. Philadelphia: Woman's Medical College of Pennsylvania, 1888.

                                                          MARGARET A. IRWIN

MARIE ANNE VICTOIRE GALLAIN BOIVIN

(1773–1841)

Physician

Birth	April 9, 1773
1797	Married Louis Boivin
1800	Received diploma in midwifery, Hospice de la Maternité, Paris
1814	Received gold medal of civic merit from King of Prussia
1819	Won silver medal from the Society of Medicine of Paris in an open competition for writings on internal hemorrhage of the uterus
1827	M.D., *honoris causa*, University of Marburg, Germany
1828	Received a commendation from the Royal Society of Medicine at Bordeaux for her research on miscarriage
Death	May 16, 1841

"Mme. Boivin . . . has an eye at the tip of each finger!" exclaimed Guillaume Dupuytren, the best trained French surgeon of the early nineteenth century.[1] It was to Mme. Boivin's care that he had entrusted his only daughter, the Countess of Beaumont, because he knew the birth would be difficult. His confidence was well placed, and his grandchild was delivered successfully.

American obstetrician Charles D. Meigs wrote in *Females and Their Diseases* in 1848 that "her [Boivin's] writings prove her to have been a most learned physician and as she enjoyed a very large practice, her science and her clinical expertise, as well as her own personal knowledge, are more to be relied on than all male physicians together."[2]

Professor Daniel Wyttenbach, a leading Greek scholar and a contemporary of Boivin, called her "the French Agnodice." In fourth-century B.C. Athens, Agnodice had attracted a large following of women whose modesty would not permit them to seek the care of male physicians.

Mme. Boivin wrote more than any other medical woman of her time, and more than most of her male colleagues. Through her writings, midwifery attained scientific status. She was one of France's foremost midwives and has been rated by some as the most outstanding obstetrician of her time.

Mme. Boivin was greatly disappointed to be refused admission to the

Royal Academy of Medicine. Its rules did not permit the admission of women, and it would not change the rules to allow her to be honored. She wryly remarked that *"les sage-femmes de l'Académie n'ont pas voulu du moi"* (the midwives in the Academy didn't need me).[3] Despite having carried out her greatest research at the University of Paris, she was denied a degree by the Medical Faculty there. Seven years before her death, the French writer Alexis Delacoux said that to disallow women from competing with men for elevated positions in science was an anachronism and a denial of reason.

Marie Anne Victoire Gallain was born in 1773 in the town of Montreuil, near Versailles. As a child she was educated by the nuns of the Visitation of Marie Leszcznska, where she developed an interest and ability in arts and sciences. The French Revolution began when she was 16 years old. She took refuge with her sister, who was the supervisor of hospitals in Etampes, near Paris. The surgeon there taught her anatomy and midwifery.

In 1797, at age 24, she returned to live with her family and married Louis Boivin, an assistant in the Bureau of National Domains. She gave birth to a daughter and was widowed in 1798. To support herself and her daughter, she entered the Hospice de la Maternité (the maternity hospital in Paris) as a student midwife and became assistant to **Marie-Louise Lachapelle**, who had established the hospital and was its first director. Mme. Lachapelle had perhaps greater obstetrical experience than anyone in Paris up to that time.

Mme. Boivin received her midwifery diploma in 1800, and it was due to her efforts that a school of midwifery was established at the Maternité. After receiving her diploma, she practiced in Versailles. Her daughter died tragically during this time. In 1801 Mme. Boivin returned to the Maternité, where she was named supervisor-in-chief. She and Mme. Lachapelle worked together until 1811, when Mme. Boivin's first work, a handbook written strictly for midwives, was published.

François Chaussier, surgeon, anatomist, and physician-in-chief of the Maternité, had seen Boivin's illustrations. Although Mme. Boivin had modestly thought they should not be published, he convinced her otherwise. Her book contained over one hundred exquisitely detailed woodcuts showing various positions of the fetus in the uterus. She knew her audience and wrote in a straightforward style without descriptions of pathological conditions or surgical treatments. The popularity of this book, which eventually went into four editions and was translated into Italian and German, foreshadowed her brilliant professional future.

Her handbook fulfilled a pressing need, because no book like it had been published since 1668. It was used to teach student midwives in Berlin, and distinguished German midwives followed her precepts unwaveringly. French-born Brazilian obstetrician Marie Durocher, who was one of the first woman doctors in Latin America, was greatly influenced by this and other works by Boivin.

Whether some petty jealousies alone or the success of her book in itself caused an irreparable rift between her and Mme. Lachapelle is unknown, but Mme. Boivin's position at the Maternité was abolished after her book was published and she was dismissed. Boivin went on to direct various general and maternity hospitals in the Paris area while continuing to write prolifically. Because she did not want to leave France, she turned down prestigious offers by the Empress of Russia; she also refused opportunities to return to the Maternité following Mme. Lachapelle's death, because she thought she would not be accepted in her predecessor's place.

In 1814 the King of Prussia conferred on her the Order of Civic Merit. The years following saw the publication of numerous books describing scientific aspects of obstetrics. Throughout her career she published case reports in medical journals, but these were less important than her books. In 1818 she translated Edward Rigby's *Treatise on Hemorrhages of the Uterus* because she could not find satisfactory studies on this subject by French physicians. The book was widely known in the United States and England. She followed this with a translation of Duncan Stewart's *Treatise on Uterine Hemorrhage*, which had been written in England two years previously. She also wrote a historical account of the subject, considering uterine hemorrhage author-by-author from ancient times to 1818, and added that to her translations. Her extensive research enabled her own book on internal hemorrhage of the uterus to garner the open prize of the Medical Society of Paris in 1819. The Society had assumed the writer was a man, referring to Mme. Boivin as "he" in discussing her work.

In 1827 she published what was arguably her most original work (*Nouvelles recherches sur l'origine, la nature et le traitement de la mole vesiculaire ou grossesse hydatique*), her classic description of the hydatiform mole (an abnormal pregnancy resulting in a mass of cysts), which became her principal independent contribution to pathology. Also in 1827 the University of Marburg awarded her the degree of Doctor of Medicine, *honoris causa*, an honor rarely extended to a woman.

Mme. Boivin was among the first practitioners to use the stethoscope, invented in 1816, to listen to the fetal heartbeat. Although she was practicing at a time when sterility was unknown, she insisted on strict cleanliness and condemned the too-frequent use of forceps. She was one of a dozen or so inventors of a vaginal speculum. She designed a smaller, cone-shaped speculum to cause as little pain as possible in comparison to the larger instruments then commonly used. In 1828 she invented a pelvimeter for internal measurement of the diameter and capacity of the pelvis.

With Antoine Dugès, professor of medicine at Montpellier and nephew of Mme. Lachapelle, she published her treatise on diseases of the uterus in 1833. This was her most important practical work on pathology. In it she discussed various diseases of the uterus as seen by the midwife. It was immediately recognized as an outstanding contribution to the subject and became a standard textbook. It was said to be as modern as possible before the advent of bacteriology. The book contained 41 plates and 116 figures, all of which she colored herself. It was translated into English in 1834 by G.O. Heming, who added an index and his own analysis. Heming's translation gives a good indication of the clarity of Mme. Boivin's prose. The book begins:

> The organs of generation of the female form a system, all the parts of which are united together, and seem to be a continuous substance; but upon examining them separately and anatomically, they are found to possess as much difference in their organisation and texture, as in their form and functions.[4]

Boivin frequently acknowledged Mme. Lachapelle's expertise and cited the writings of her peers.

The book contains the first recorded observation of cancer of the female urethra. Boivin noted that she had seen many cases of midwives diagnosed with cancer of the uterus, rectum, or stomach, and she attributed these large numbers to the fatigue to which their occupation subjected them. In this work she made numerous and heartfelt references to the pain suffered by her patients, who came from all walks of life. She railed against cruel operations performed by careless and callous surgeons. Yet none of Mme. Boivin's tribulations appeared to have affected her outwardly. She coped with the tragedies in her private life and the lack of acceptance by the French medical community by immersing herself in science.

When she retired she received a small pension, reflecting the excessively low wages she had accepted during her career. It was barely enough to live on. She succumbed to exhaustion and poor health on May 16, 1841. Even 25 years after her death, there was such a demand for her last work that it was republished in Paris in 1866.

Notes

1. Kate Hurd-Mead, *A History of Women in Medicine: From the Earliest Times to the Beginning of the Nineteenth Century* (Haddam, Conn.: Haddam Press, 1938), p. 501.

2. Charles D. Meigs, *Females and Their Diseases* (Philadelphia: Lea and Blanchard, 1848), quoted in Geoffrey Marks and William K. Beatty, *Women in White* (New York: Scribner's, 1972), p. 69.

3. Melanie Lipinska, *Histoire des Femmes Médecine Depuis Antiquité Jusqu'à Nos Jours* (Paris: G. Jacques, 1900), p. 331.

4. Marie Boivin and Antoine Dugès, *Traité Pratique des Maladies de l'Uterus et de Ses Annexes: Fondé Sur Un Grand Nombre D'Observations Clinique . . .*, trans. G.O. Heming (London: Sherwood, Gilbert and Piper, 1834), p. 1.

Bibliography

Alic, Margaret. *Hypatia's Heritage*. Boston: Beacon Press, 1986.

Cutter, Irving S. *A Short History of Midwifery*. Philadelphia: W.B. Saunders, 1964.

Delacoux, Alexis. *Biographie des Sage-Femmes Célèbres, Anciennes, Modernes et Contemporaines.* (Biography of Eminent Ancient, Modern and Contemporary Midwives.) Paris: Tringuart, 1834.

Hurd-Mead, Kate. *A History of Women in Medicine: From the Earliest Times to the Beginning of the Nineteenth Century*. Haddam, Conn.: Haddam Press, 1938.

Lipinska, Melanie. *Histoire des Femmes Médecine Depuis Antiquité Jusqu'à Nos Jours*. (History of Women Doctors from Antiquity to the Present.) Paris: G. Jacques, 1900.

MARTHA E. STONE

ROBERTA LYNN BONDAR
(1945–)
Neurologist

Birth	December 4, 1945
1968	B.Sc. (Agr.), zoology and agriculture, University of Guelph, Ontario
1971	M.Sc., experimental pathology, University of Western Ontario
1974	Ph.D., neurobiology, University of Toronto
1977	M.D., McMaster University
1981	F.R.C.P. (C), neurology, University of Western Ontario
1984–92	Canadian Astronaut, Canadian Astronaut Program, Canadian Space Agency
1985–89	Chair, Canadian Life Sciences Committee for Space Station, National Research Council of Canada
1986–93	Civil Aviation Medical Examiner, Health and Welfare, Canada

Roberta Lynn Bondar. Photo courtesy of Roberta Lynn Bondar.

1990	William R. Franks, M.D., Award, Canadian Society of Aerospace Medicine
1990–92	Prime Payload Specialist for the First International Microgravity Laboratory shuttle flight (IML-1)
1991	Sault Ste. Marie Medal of Merit
1992	Médaille de l'Excellence, L'Association des Médecins de Langue Française du Canada; NASA Space Medal; President's Award, College of Physicians and Surgeons of Ontario; Presidential Citation, American Academy of Neurology; Officer of the Order of Canada
1993	Order of Ontario
1995	D.Sc., *honoris causa*, University of Western Ontario

When people place little emphasis on science in the school curriculum or in their lives in general, they are leaving to someone else the responsibility of helping us understand, cope with, and adapt to a changing environment. Science is an ongoing learning experience, and the knowledge which it provides can ease some of our fears about the future.[1]

Dr. Roberta Bondar is a neurologist and researcher who became Canada's first woman astronaut when she flew on the space shuttle *Discovery* in January 1992. She was born and raised in Sault Ste. Marie, Ontario, where she obtained her elementary and secondary education. In 1968 she graduated with a Bachelor of Science degree in agriculture from the University of Guelph, Ontario. Subsequently she moved into research in experimental pathology at the University of Western Ontario and neurobiology at the University of Toronto, obtaining the M.Sc. and Ph.D. degrees respectively. In 1977, Dr. Bondar received her medical degree from McMaster University. After her internship in medicine at Toronto General Hospital, she completed postgraduate training at the University of Western Ontario.

Dr. Bondar was admitted as a neurology Fellow of the Royal College of Physicians and Surgeons of Canada in 1981. She specialized in neuro-ophthalmology at Tufts New England Medical Center (Boston) and at the Playfair Neuroscience Unit of Toronto Western Hospital. She was appointed assistant professor of medicine (neurology) and director of the Multiple Sclerosis Clinic for the Hamilton-Wentworth Region at McMaster University in 1982. A year later, she was selected as one of the six original Canadian astronauts. When reflecting on her life, Bondar has said:

The seeds of space exploration had been sown at an early age. Come to think of it, I spent all my free time dreaming about space, wondering how I would pull it off, never willing to give up on what I felt was my destiny. Every birthday, I asked for a plastic model-rocket kit, a chemistry set, or a doctor's bag.[2]

While training as an astronaut in 1984, Dr. Bondar was also involved in teaching, research, and administration. She taught in the Department of Nursing at the University of Ottawa and for the Department of National Defence, where she covered the biomedical aspects of space flight. She performed clinical work in the Department of Medicine (neurology) at the University of Ottawa. She served as chair of the Canadian Life Sciences Subcommittee for Space Station from 1985 to 1989 and as a member of the Ontario Premier's Council on Science and Technology from 1988 to 1989. In early 1990 she was designated a prime payload specialist for the first International Microgravity Laboratory Mission (IML-1). As a payload specialist, she performed life science and material science experiments in the Spacelab and on the mid-deck. She has said that

> Performing a science experiment, or taking a picture, is a trainable activity, but no one can prepare you for the overwhelming experience of being in visual contact with Earth from Space. Many people have asked me what it is like to float in orbit, watching the world go by, and all I can tell them is that it is an awesome adventure.[3]

Currently, Dr. Bondar is conducting research into blood flow in the brain during microgravity and various pathological states.

Dr. Bondar is very interested in physical fitness, in educating students about environmental issues, and in encouraging girls in mathematics and science. She has certification in scuba diving as well as parachuting and holds a private pilot's license.

She has received many honors, including the Officer of the Order of Canada and the Order of Ontario, as well as honorary degrees from over 20 universities, including the University of Montreal in 1994, Carleton University in 1993, and McGill University in 1992. Awards for science and technology are given in Toronto in her name, and an elementary school in British Columbia has been named after her. She has been awarded the NASA Space Medal and the Médaille de l'Excellence (Association des Médecins de Langue Française du Canada) and was named La Personalité de l'Année by *La Presse* in 1992.

Dr. Bondar currently holds the positions of distinguished professor, Centre for Advanced Technology and Education (CATE), Ryerson Polytechnic University; visiting distinguished professor, Faculty of Kinesiology, University of Western Ontario; visiting research scholar, Department of Neurology, University of New Mexico; and visiting research scientist, Universities Space Research Association, Johnson Space Center, Houston, Texas. She is held in the highest esteem for her research, leadership, and dedication by scientists, students, friends, and young people in Canada and abroad. Working in a truly avant-garde research area, she provides a brilliant role model for young women interested in nontraditional areas of science and engineering. She is recognized for her distinguished work in neurology and space science and her role in fostering an interest in science among Canada's youth.

Notes

1. Roberta Lynn Bondar, *Touching the Earth* (Toronto, Ontario: Key Porter, 1994), pp. 84–85.

2. *Ibid.*, p. 9.
3. *Ibid.*, p. 24.

Bibliography

Bondar, B.C., with R.L. Bondar. *On the Shuttle.* Toronto, Ontario: Greey de Pen-
cier, 1993.
Bondar, R.L. "Bondar Space Experiments and Teacher-Librarians: Is There a Con-
nection?" *Emergency Librarian* 183 (1991): 34–35.
———. "Cerebral Autoregulation in Spaceflight." Course #333. New York:
American Academy of Neurology, 1993.
———. "Cerebro/Cardiovascular Aspects of Spaceflight." Course #323. Wash-
ington, D.C.: American Academy of Neurology, 1994.
———. "An Overview of Space Medicine." *Space, Biomedicine and Biotechnology,
Proceedings,* 1986, pp. 16–19.
———. "Soviet Report." *Canadian Aeronautics and Space Journal* 34 (1988): 194–
95.
———. "Space Qualified Humans: The High Five." *Aviation, Space and Environ-
mental Medicine* 65, no. 2 (1994): 161–69.
———. *Touching the Earth.* Toronto, Ontario: Key Porter, 1994.
"Destiny in Space." (Film) IMAX—The National Aeronomics Space Museum.
Smithsonian Institute and Lockheed, 1994.

GLORIA I. TROYER

ELIZABETH GERTRUDE KNIGHT BRITTON
(1858–1934)
Botanist

Birth	January 9, 1858
1875	Graduated from Normal College, NY
1875–85	Teacher, Hunter College
1883	Published first scientific paper on albinism in plants
1885	Married Nathaniel Lord Britton
1886–88	Editor, *Bulletin of the Torrey Botanical Club*
1893	Became the only woman charter member of the American Botanical Society
1902	Organized the Wild Flower Preservation Society of America

1912   Honorary curator of mosses, New York Botanical Garden
Death   February 25, 1934

Elizabeth Gertrude Britton was the most important woman botanist of
the late nineteenth and early twentieth centuries. She was born in New
York in 1858, one of five daughters òf Sophia and James Knight. Her
father operated a furniture factory and sugar plantation near Matanzas,
Cuba. Elizabeth spent a large part of her childhood on this island plantation, where the profusion of tropical flora and fauna stimulated her
interest in the natural world. One of her sisters believed that her interest
in bryology, the study of mosses and liverworts, was awakened by visits
to an old well covered with ferns and moss near Matanzas.

Elizabeth attended elementary schools in Cuba until she was 11 and
then returned to New York. There she finished her education at Dr. Be-
nedict's private school and Normal College, which later became Hunter
College. Following graduation at age 17, she was appointed to the teaching staff there. She pursued her interest in botany, becoming assistant in
natural science in 1883. In 1879 she joined the Torrey Botanical Club,
named for John Torrey, a prominent New York botanist who had authored *Flora of North America* with Asa Gray, the leading botanist and
taxonomist of the time. Elizabeth published her first scientific paper on
albinism in plants in 1883. Six months later she completed a paper on
the discovery of the fruit of the moss *Eustichium norvegicum*. This was
the first of many publications on mosses. Her ability to identify ferns
and mosses attracted the attention of influential people, one of whom
was Asa Gray. She became recognized as a hard worker, meticulous
observer, and excellent teacher, with "a clear and precise way of presenting a subject" and a "patient, thorough manner, leaving no minutest
detail unexamined," according to a former student, Mary A.C. Liver-
more.[1] Other than William Starling Sullivant (1803–1873), she was the
first botanist in the United States to intensively study mosses.

In 1885 she married Nathaniel Lord Britton, an assistant geology professor at Columbia College whose growing interest in botany led to an
appointment in that field in 1886. He was to become the leading botanist
in New York after John Torrey's time. Nathaniel Britton became the author of the three-volume *Illustrated Flora of the Northern United States,
Canada and the British Possessions*, and he also became the first director of
the New York Botanical Garden. He continued his study of flowers while
his wife concentrated on bryophytes and ferns. However, they went on
a number of botanical trips together. One was to the gardens at Kew,
near London, England, where Nathaniel classified Bolivian tropical botanical collections that had been given to Columbia. They spent two and
a half months in England. Elizabeth worked on mosses at the Linnaean

Society in London, but she also spent a great deal of time at Kew Gardens, where her husband was working. Here she found an outstanding herbarium (a place where dried plants are arranged and described) as well as a botanical library and extensive gardens. She has been credited with the idea for the New York Botanical Garden, saying of Kew, "Why couldn't we have something like this in New York?"[2]

When the couple returned to New York, Elizabeth made a presentation on Kew at a meeting of the Torrey Botanical Club. In January 1889, the club issued a proposal for a botanical garden in New York. The garden was incorporated in 1891. Nathaniel Britton was appointed director and remained in that position for over 33 years. By 1895, enough money had been raised and the commissioner of public parks set aside 250 acres of Bronx Park for establishment of the garden.

Elizabeth Britton became the unofficial curator of the moss collection at Columbia and honorary curator of mosses at the New York Botanical Garden in 1912. From 1886 to 1888 she was editor of the *Bulletin of the Torrey Botanical Club*. She compiled a huge amount of taxonomic information about mosses from collecting trips with her husband and those that she took on her own. She visited the Dismal Swamp in Virginia, the Adirondacks, the mountains of North Carolina, Europe, and the West Indies. In addition, she worked full-time on the arrangement of the moss herbarium at the Botanical Garden. The Columbia collection had been transferred there, and she oversaw the acquisition of other collections, including that of the English bryologist William Mitten. The latter collection kept her busy for years. She continued writing articles for popular and scientific journals, eventually completing a total of 346 papers and reviews.

In 1889 Britton completed a series of 11 papers entitled "Contributions to American Bryology," and in 1894 she published eight articles in the *Observer* on "How to Study the Mosses." Britton also had an interest in ferns and published several articles about them. In 1893 she became the only woman charter member of the American Botanical Society.

During this time she was working on a handbook of the mosses of northeastern America. It was never finished, possibly because a doctoral student she was assisting published a similar book in 1903. This student, Joel Grout, later became the leading American bryologist, although Britton never had a high opinion of him. She thought he was not careful enough in his identifications; he believed she was too quick to criticize. Her impatience with careless work can be seen in her signed reviews. Her strong personality antagonized some, but she was devoted to helping amateurs and promoting the causes in which she believed.

Britton worked with Grout to found the Sullivant Moss Chapter, named after the leading bryologist William Sullivant. This organization eventually became the American Bryological and Lichenological Society, which still publishes the *Bryologist,* a journal that Britton helped create.

In 1902, just after the Botanical Garden officially opened, Elizabeth was instrumental in organizing a new society, the Wild Flower Preservation Society of America. She had become alarmed at the amount of vandalism in the garden. Wildflowers and other plants were disappearing. One day, visiting public school students uprooted 410 jack-in-the-pulpits as well as other flowering plants and trees. She realized the problem extended outside the Botanical Garden, especially in the vicinity of the large cities in the Northeast. In 1901, Olivia and Caroline Phelps Stokes of New York had presented the Botanical Garden a fund of $3,000. The interest was to be used "for the investigation and preservation of native plants or for bringing the need of such preservation before the public."[3] This money funded an essay contest conceived by Elizabeth on the preservation of wild plants. The winning essay was widely circulated and credited, and ultimately led to the establishment of the Wild Flower Preservation Society.

For almost 34 years Elizabeth was active in promoting the cause of wildflower preservation. She published a number of papers on the topic, including a series of 14 articles under the title "Wild Plants Needing Protection" that appeared in the *Journal of the New York Botanical Garden* between 1912 and 1929. She also gave lectures and engaged in extensive correspondence in an effort to arouse public concern. In 1925 she launched a national boycott against the use of wild American holly at Christmas. Instead, she urged the cultivation of holly from seed for commercial use. Eventually her efforts resulted in the passage of laws in a number of states for the protection of native flora and in the establishment of other preservation societies.

Although during these years she spent much of her time working for the cause of preservation, Britton did not completely abandon bryology. In 1905 she attended the Botanical Congress of Vienna and was one of three prominent bryologists appointed to a world commission dealing with the nomenclature of mosses. She also accompanied her husband on 23 botanical collecting trips to the West Indies on behalf of the Botanical Garden. Visiting such islands as Puerto Rico, St. Thomas, Cuba, Jamaica, and Trinidad, she brought back great numbers of mosses, hepatica, lichens, and ferns from tropical forests. Journeying into remote mountain and valley areas to collect these specimens appealed to her sense of adventure, which is conveyed to the reader through her articles describing the results of these trips. Many of her papers were relatively short but full of lively description. A good example is found in her report of a trip to Jamaica in which she described the Hope Botanical Gardens, "where the blossoms of the *Poinciana* sprinkled the ground with red and the humming-birds darted in and out of the arbors of *Thunbergia grandi-*

flora."[4] In 1913 Elizabeth, in collaboration with R.S. Williams and Julia T. Emerson, contributed the systematic treatment of several families of mosses to the book *Flora of North America* (Britton and Brown). Later she contributed the treatment of mosses for her husband's works *Flora of Bermuda* and *The Bahama Flora*.

The Brittons had no children, so Elizabeth was able to devote all her energies to the study of mosses, wild plant preservation, and the Botanical Garden in the Bronx. No doubt her influence would have been less if she had not married Nathaniel Britton, but she was widely recognized by botanists at the time for her knowledge and achievements in the field of bryology. She certainly was a major force in early wild plant preservation efforts, and her contributions to the New York Botanical Garden are documented in the collections of her papers held in the library there. Fifteen species of plants and the moss genus *Bryobritonia* were named for her. She died in 1934.

Notes

1. Marcia Myers Bonta, "Elizabeth Gertrude Knight Britton, Mother of American Bryology," in *Women in the Field* (College Station: Texas A&M University Press, 1991), pp. 125–26.

2. Marshall A. Howe, "Elizabeth Gertrude Britton," *Journal of the New York Botanical Garden* 35 (May 1934): 102.

3. John Hendley Barnhart, "Elizabeth Gertrude Knight Britton as a Scientist," *Journal of the New York Botanical Garden* 41 (June 1940): 139.

4. Elizabeth Britton, "A Trip to Jamaica," *Torreya* 8 (Jan. 1908): 12.

Bibliography

Barnhart, John Hendley. "Elizabeth Gertrude Knight Britton as a Scientist." *Journal of the New York Botanical Garden* 41 (June 1940): 137–42.

Bonta, Marcia Myers. "Elizabeth Gertrude Knight Britton, Mother of American Bryology." In *Women in the Field*. College Station: Texas A&M University Press, 1991.

Britton, Elizabeth. "A Trip to Jamaica." *Torreya* 8 (Jan. 1908): 9–12.

Gager, C. Stuart. "Elizabeth G. Britton and the Movement for the Preservation of Native American Wildflowers." *Journal of the New York Botanical Garden* 41 (June 1940): 142–43.

Howe, Marshall A. "Elizabeth Gertrude Britton." *Journal of the New York Botanical Garden* 35 (May 1934): 97–104.

Ogilvie, Marilyn Bailey. *Women in Science.* Cambridge, Mass.: MIT Press, 1986.

Rossiter, Margaret. *Women Scientists in America.* Baltimore: Johns Hopkins University Press, 1982.

Slack, Nancy G. "American Women Botanists." In *Uneasy Careers and Intimate Lives: Women in Science, 1789–1979*, eds. Pnina G. Abir-am and Dorinda Outram. New Brunswick, N.J.: Rutgers University Press, 1987.

Steere, William Campbell. "Elizabeth Gertrude Knight Britton." In *Notable American Women 1607–1950*, Vol. 1. Cambridge, Mass.: Belknap Press, 1971.

JOAN GARRETT PACKER

E. (ESTELLA) ELEANOR CAROTHERS
(1882–1957)
Zoologist

Birth	December 4, 1882
1911	A.B., University of Kansas
1912	A.M., University of Kansas
1913–14	Pepper Fellow, Pennsylvania State University
1914–36	Assistant Professor of Zoology, Pennsylvania State University
1916	Ph.D., Pennsylvania State University
1920–41	Independent Investigator, Marine Biological Laboratory
1921	Ellen Richards Research Prize, Naples Table Association
1927	"Starred" in fourth edition of *American Men of Science*
1936–41	Research Associate, University of Iowa
Death	1957

E. Eleanor Carothers attained a star in *American Men of Science* in 1927 for her research on the embryos of grasshoppers. The star meant that she was among the elite of American scientists, a rare tribute for a woman at that time.

Estella Eleanor Carothers was born in Newton, Kansas, in 1882 to Mary (Bates) and Z.W. Carothers. (Throughout her career she signed her name as E. Eleanor Carothers.) She attended Nickerson Normal College in Kansas, finishing her undergraduate education at the University of Kansas in 1911. She then earned a master's degree at Kansas the following year. Moving east, Carothers began work on a Ph.D. at the University of Pennsylvania, which she completed in 1916. After completing her graduate work, she became an assistant professor of zoology at the University of Pennsylvania. One career highlight was her participation in the university's scientific expeditions to the southern and southwestern states in 1915 and 1919.[1]

According to the Marine Biological Laboratory's annual report published each year in the *Biological Bulletin*, Carothers left the University of Pennsylvania around 1936. She next became a research associate for the Department of Zoology at the University of Iowa, then the State University of Iowa. While she was at the University of Iowa, Carothers received a grant from the Rockefeller Foundation Fund for research on the physiology and cytology of the normal cell. This grant helped her to conduct research on grasshopper embryos. For nearly 20 years the Marine Biological Laboratory's annual report listed Carothers among its independent investigators. She was also listed as a regular member of the Marine Biological Laboratory in Woods Hole, Massachusetts, from 1920 to 1956. Sometime in 1941, Carothers moved from Iowa to Kingman, Kansas. She spent the final three years of her life in Murdock, Kansas.

Carothers's speciality in entomology was *Orthopteran* (grasshopper) genetics and cytology. Her research specifically focused on heteromorphic homologous chromosomes, although she did contribute some insights into general research about the cytological basis of heredity. Extensive plates illustrate almost all of her work with heteromorphic chromosomes. In the bibliography of Thomas Hunt Morgan's book *The Mechanism of Mendelian Heredity* (1915), Carothers is listed among seven other women who were primary investigators in the field of genetics.[2]

Notes

1. Margaret Ogilvie, *Women in Science* (Cambridge, Mass.: MIT Press, 1986), p. 52.

2. Margaret Rossiter, "Women Scientists in America before 1920," *American Scientist* 62 (May–June 1974): 322.

Bibliography

Cattell, John McKeen, ed. *American Men of Science*, 3rd ed., s.v. "Carothers, Estella Eleanor." Garrison, N.Y.: The Science Press, 1921.

Morgan, Thomas Hunt. *The Mechanism of Mendelian Heredity*. New York: Henry Holt and Co., 1915.

Ogilvie, Margaret. *Women in Science*. Cambridge, Mass.: MIT Press, 1986.

Rossiter, Margaret. "Women Scientists in America before 1920." *American Scientist* 62 (May–June 1974): 312–23.

Who Was Who among North American Authors, 1921–1939, s.v. "Carothers, E(stella) Eleanor." Detroit: Gale Research Company, 1976.

FAYE A. CHADWELL

RACHEL CARSON

(1907–1964)

Biologist, Environmentalist

Birth	May 27, 1907
1918	First published work appeared in *St. Nicholas* magazine
1929	Graduated *magna cum laude* from Pennsylvania College for Women; went to Woods Hole on a summer study fellowship; entered Johns Hopkins University
1932	M.S., marine zoology, Johns Hopkins University
1935	Began working at the U.S. Bureau of Fisheries
1937	Published "Undersea" in *Atlantic Monthly*
1941	Published *Under the Sea Wind*
1950	George Westinghouse Science Writing Award
1951	Serialization of *The Sea around Us* in *New Yorker*; published *The Sea around Us*; Guggenheim Fellowship; National Book Award
1952	Henry G. Bryant Medal, Philadelphia Geographical Society; John Burroughs Medal; honorary doctorates from Drexel Institute of Technology, Oberlin College, and Chatham College (formerly Pennsylvania College for Women); resigned from (renamed) U.S. Fish and Wildlife Service
1955	Published *The Edge of the Sea*
1956	Achievement award, American Association of University Women
1962	Excerpts from *Silent Spring* serialized in *New Yorker*; published *Silent Spring*
1963	Received the National Audubon Society's medal and the American Geographical Society's medal; elected to American Academy of Arts and Letters; President Kennedy's Science Advisory Committee supported the findings published in *Silent Spring*
Death	April 14, 1964
1965	*A Sense of Wonder* published posthumously
1980	President's Medal of Freedom awarded posthumously

As a biologist and a writer who made significant contributions to the modern environmental movement, Rachel Carson is remembered for both her literary accomplishments and her scientific pursuits. Through her writing, which was directed to scientists and general readers alike, she endeavored to instill in her audience an appreciation of the natural world and a sense of society's responsibility for preserving it. Carson, the author of such books as *The Sea around Us, The Edge of the Sea,* and *Silent Spring,* observed, "The aim of science is to discover and illuminate truth. And that, I take it, is the aim of literature, whether biography or history or fiction, it seems to me, then, that there can be no separate literature of science."[1] Carson reflected this belief throughout her life as she combined her literary talent with her scientific knowledge to enlighten not only the scientific community but the general public. Her books about the delicate environment of the earth were best-sellers around the world.

From an early age, Carson assumed she would be a writer. As a child growing up in Springdale, Pennsylvania, in the lower Allegheny Valley, she had a dual love of books and nature. Her mother, Maria, who was a dominant influence on the young Carson, encouraged her daughter in these interests. At age 10 she had a short story published in *St. Nicholas* magazine. As a student at Pennsylvania College for Women (later Chatham College) she majored in English at first, reasoning that this was the appropriate education for a writer. After taking a required biology course, however, she became fascinated with science and switched her major to zoology. Although she believed at the time that this course of action would force her to abandon her literary aspirations, she later discovered that studying science, instead, had given her something about which to write. She graduated *magna cum laude* in 1929.

Before embarking on graduate studies, she went to the Marine Biological Laboratory in Woods Hole, Massachusetts, to spend the summer. This experience provided Carson with her first real exposure to the ocean. She was no stranger to it, however, because from an early age books had instilled in her a great passion for the sea.

Carson enrolled at Johns Hopkins University in the fall of 1929. Despite the discrimination she experienced there as a woman in what was considered a man's field, she distinguished herself and earned a master's degree in marine zoology in 1932. For the next few years she taught part-time at both Johns Hopkins and the University of Maryland. However, when her father and sister both died in the mid-1930s, Carson was forced to pursue more lucrative and steady employment in order to support her mother and two young nieces. After receiving the highest score on a civil service examination, she obtained a position as a junior aquatic biologist with the U.S. Bureau of Fisheries in Washington, D.C. She had been working there part-time, writing scripts for a series of radio broadcasts on marine life, and was one of the first two women to be hired by the department in a nonclerical capacity.

At night and on weekends she continued to write. Her first article, "Undersea," was published in the *Atlantic Monthly* in 1937. This article formed the basis of her first book, *Under the Sea Wind,* which was written "to make the sea and its life as vivid a reality for those who may read the book as it has become for me during the past decade. It was written, moreover, out of the deep conviction that the life of the sea is worth knowing."[2] Carson believed that knowledge of the creatures of the ocean could give humanity a better perspective on life. Unfortunately, this book, published in November 1941, did not sell very well because of the nation's preoccupation with World War II.

Carson contributed to the war effort by writing conservation bulletins for the government that encouraged people to eat fish in order to augment the wartime food supply. After the war she continued at the Bureau, which had been renamed the U.S. Fish and Wildlife Service. She was a contributing writer and editor of a 12-part pamphlet series entitled "Conservation in Action," which reflected a superior literary style often lacking in such bureaucratic publications.

By 1949 she had been promoted to biologist and chief editor. Although she excelled at her work, she still longed to devote more time to her personal writing but could not afford to go without her small but steady income. She managed to complete her second book, *The Sea around Us,* by writing at night and during brief leaves from her job. This book was intended to popularize the subject of oceanography. Carson, who believed that all living creatures (including human beings) have a strong link to the ocean, wrote, "for the sea lies all about us . . . in its mysterious past it encompasses all the dim origins of life and receives in the end, after, it may be, many transmutations, the dead husks of that same life. For all at last return to the sea—to Oceanus, the ocean river, like the everflowing stream of time, the beginning and the end."[3]

Published in the summer of 1951, this book was a great success, in part because it had been serialized in the *New Yorker* prior to publication. It hit the *New York Times* best-seller list and stayed there for 86 weeks. *Under the Sea Wind* was then re-released and also became a best-seller.

For Carson, an extremely private person, her sudden exposure in the public eye was a bit overwhelming. She received many accolades, including the John Burroughs Medal and the National Book Award. *The Sea around Us* was also named outstanding book of the year by a *New York Times* poll. In addition to numerous television appearances, Carson even had exposure to the limelight of Hollywood. A film version of her book won an Academy Award for best full-length documentary. However, Carson was not satisfied with this project because of the many scientific errors that she could not prevent RKO from including.

The success of *The Sea around Us* finally enabled Carson to pursue her writing endeavors full-time. She resigned from her position at the U.S. Fish and Wildlife Service, purchased some land in West Southport, Maine, and had a small cottage constructed there. She now devoted herself to writing *The Edge of the Sea.* For this book, which focused on the sea creatures of the shore, she was able to conduct a good deal of her research in the tide pools near her cottage. While engaged in these scientific pursuits practically in her own backyard, she also had more time to reflect on the larger meaning of her work. She wrote, "Contemplating the teeming life of the shore, we have an uneasy sense of the communication of some universal truth that lies just beyond our grasp. . . . The meaning haunts and ever eludes us, and in its very pursuit we approach the ultimate mystery of Life itself."[4]

In all of Carson's writings about the sea, humans were either on the periphery or played no part at all. This reflected her belief that humanity is not the center of the natural world. In the following passage, she explained that by trying to attain this central position, human beings have alienated themselves from nature: "The ocean has nothing to do with humanity. It is supremely unaware of man, and when we carry too many trappings of human existence with us to the threshold of the sea world, our ears are dulled and we do not hear the accents of sublimity in which it speaks."[5]

As Carson became more and more aware that the living things she loved were being threatened by society's dominance of nature, a new project formulated in her mind that would have a profound effect on issues of global importance. *Silent Spring*, published in 1962, was the end result of her efforts. This influential work addressed the problem of pesticides, particularly DDT, and their effects on the delicate balance of nature. In *Silent Spring* she raised the consciousness of her readers by conveying hauntingly desolate images such as the following: "Over increasingly large areas of the United States, spring now comes unheralded by the return of the birds, and the early mornings are strangely silent where once they were filled with the beauty of bird song."[6] Although her findings met with intense criticism from the chemical industry, the book became another best-seller. Perhaps more important, it motivated President Kennedy to form a government committee to study the effects of DDT and other pesticides. In this monumental work, Carson encouraged her readers to change their ways in order to save the planet, despite the difficulties this might entail:

> We stand now where two roads diverge. But unlike the roads in Robert Frost's familiar poem, they are not equally fair. The road we have long been traveling is deceptively easy, a smooth superhighway on which we progress with great speed, but at its end lies

disaster. The other fork of the road—the one "less traveled by"—
offers our last, our only chance to reach a destination that assures
the preservation of our earth.[7]

With words like these, the environmental movement, still in its for-
mative stages, found a staunch supporter and spokesperson in Rachel
Carson. Unfortunately, Carson's final years were marked by a series of
ailments; she died of cancer and heart failure in 1964 at age 57. Her last
book was published posthumously in 1965. *A Sense of Wonder* was actually a reprint in book form of an article, "Help Your Child to Wonder,"
she had written for *Women's Home Companion* in 1956. It addressed the
importance of instilling a love of nature in children. This work was inspired by her great-nephew, Roger, whom she adopted after the death
of her niece. In 1969, in memory of her efforts to preserve the environment, the Coastal Maine Wildlife Refuge, located not far from her cottage, was renamed the Rachel Carson Natural Wildlife Refuge. This
tranquil spot exemplifies Carson's belief as expressed in the following
words:

> The pleasures, the values of contact with the natural world are not
> reserved for the scientists. They are available to anyone who will
> place himself under the influence of a lonely mountain top—or the
> sea—or the stillness of a forest, or who will stop to think about so
> small a thing as the mystery of a growing seed.[8]

Throughout her life, Rachel Carson strived to fulfill what she considered the purpose of both science and literature: "to discover and illuminate truth." She accomplished this as an observant scientist who
uncovered some of the secrets of nature, and as an eloquent writer who
shared her observations with the entire world.

Notes

1. Rachel Carson, National Book Award acceptance speech, excerpts in Paul
Brooks, *The House of Life* (Boston: Houghton Mifflin, 1972), pp. 127–29.
2. Rachel Carson, *Under the Sea Wind* (New York: Simon and Schuster, 1941),
Foreword.
3. Rachel Carson, *The Sea around Us* (New York: Oxford University Press,
1951), p. 19b.
4. Rachel Carson, *The Edge of the Sea* (Boston: Houghton Mifflin, 1955), p. 250.
5. Rachel Carson, "Our Ever-Changing Shore," *Holiday* (July 1958): 70–71ff.
6. Rachel Carson, *Silent Spring* (Boston: Houghton Mifflin, 1962), p. 103.
7. *Ibid.*, p. 277.
8. Rachel Carson, speech to Theta Sigma Pi, excerpts in Brooks, *The House of
Life*, pp. 324–26.

Bibliography

Brooks, Paul. *The House of Life: Rachel Carson at Work*. Boston: Houghton Mifflin, 1972.

Carson, Rachel. *The Edge of the Sea*. Boston: Houghton Mifflin, 1955.

———. "Help Your Child to Wonder." *Women's Home Companion* (July 1956): 24–27ff.

———. "Our Ever-Changing Shore." *Holiday* (July 1958): 70–71ff.

———. "Rachel Carson Answers Her Critics." *Audubon* (Sept. 7, 1963): 262–65ff.

———. *The Sea around Us*. New York: Oxford University Press, 1951.

———. *A Sense of Wonder*. New York: Harper and Row, 1965.

———. *Silent Spring*. Boston: Houghton Mifflin, 1962.

———. *Under the Sea Wind*. New York: Simon and Schuster, 1941.

———. "Undersea." *Atlantic* (Sept. 1937): 322–25.

Graham, Frank, Jr. *Since Silent Spring*. Boston: Houghton Mifflin, 1970.

Hynes, H. Patricia. *The Recurring Silent Spring*. New York: Pergamon Press, 1989.

McCay, Mary A. *Rachel Carson*. New York: Twayne Publishers, 1993.

ARLENE RODDA

MARY AGNES MEARA CHASE

(1869–1963)

Agrostologist

Birth	April 20, 1869
1888	Married William Ingraham Chase, January 21
1889	Widowed, January 2
1901	Meat Inspector, Chicago stockyards
1903	Agrostological Artist, U.S. Department of Agriculture (USDA), Washington, DC
1905	Began collaboration with Albert S. Hitchcock
1918–19	Arrested for suffragette activities
1922	Published *First Book of Grasses*
1923	Assistant Botanist, USDA
1925	Associate Botanist, USDA
1936	Principal Scientist in charge of systematic agrostology and Senior Botanist, USDA

1939	Officially retired; Research Associate, National Herbarium, Smithsonian Institution
1956	Certificate of Merit, Botanical Society of America
1958	Honorary D.Sc., University of Illinois; named eighth Honorary Fellow of the Smithsonian Institution; awarded medal for service to the botany of Brazil
1961	Fellow, Linnaean Society
Death	September 24, 1963

Agrostology, the field of botany concerned with grasses, owes a tremendous debt to Agnes Chase, who collected and catalogued over 10,000 specimens never before described and who encouraged younger colleagues in their studies. Called "the dean of American agrostologists," Chase achieved this with a grammar school education, a single-minded devotion to her subject, and sheer hard work.[1]

Mary Agnes Meara was born in Iroquois County, Illinois, in 1869, one of six children. Her father died when she was 2, and the family moved to Chicago.[2] In a rare interview, Agnes related how as a young girl she made tiny bouquets of grass flowers and gave them to her grandmother. Her grandmother told her grass didn't have flowers, but young Agnes was sure her grandmother was wrong.[3]

Agnes attended grammar school but then had to work to help support the family. "Girls didn't get to go to college when I was young," she said in an interview. "I just had to pick up my education as I went along."[4] She proofed and set type for a paper called the *School Herald*. At age 19 she married the editor, William Chase. He died of tuberculosis within a year, leaving her a mountain of unpaid debts that she paid off by living on a diet of oatmeal and beans.[5]

Agnes developed a close relationship with her young nephew, Virginius Chase, and together they developed a keen interest in botany. As a hobby she began to study the plants of northern Illinois and Indiana. Virginius later became a botanist in his own right.[6]

While collecting in an Illinois swamp, Agnes met the Reverend Ellsworth Hill, a retired minister and bryologist (one who studies mosses). Hill discovered Agnes had talent as a botanical artist and enlisted her to illustrate his papers. He taught her about botany and the use of a microscope. He also introduced her to Charles Frederick Millspaugh, the curator of botany at Chicago's Field Museum of Natural History. Agnes illustrated two museum publications for Millspaugh as well.[7]

Encouraged by Hill, Agnes landed a job in which she could use her skill with the microscope. She became a meat inspector in the Chicago stockyards for the U.S. Department of Agriculture (USDA). After a few years she became a botanical illustrator at the USDA in Washington, D.C.

She studied in the grass herbarium and worked on her first scientific paper after hours.[8]

Grasses intrigued Agnes. "If it were not for grasses," she said, "the world would never have been civilized."[9] Primitive man had hunted animals that grazed on grasses. When man learned to cultivate, it was grasses he grew for bread and livestock fodder. Wheat, sorghum, rice, millet, rye, oats, sugar cane, bamboo, corn, hay—all these plants of economic value are grasses, utilized for food, shelter, oil, alcohol, and sugar, and used to feed animals that provide meat, milk, wool, leather, and horsepower. As they grow, grasses also control erosion and build up the land.[10] And grasses took a firm hold in Agnes's life.

She began working with Albert Spear Hitchcock, the USDA's principal scientist in charge of systematic agrostology, first as an illustrator and then as his scientific assistant. She spent her spare time, and often her own money, traveling throughout the United States collecting and taking note of which species grew where. This information aided Hitchcock in preparing his *Manual of Grasses of the United States.* Their professional friendship lasted until his death nearly 30 years later, when she succeeded him as senior botanist at the Smithsonian Institution's grass herbarium. Like Hitchcock, Agnes collected botanical specimens in the field as well as studied them in the laboratory. As she put it, "The best place to study grasses is in the field."[11]

Agnes ventured to Mexico, Puerto Rico, and Europe to collect specimens and study in herbariums. She wrote professional articles about her expeditions, but she also wanted to teach the nonprofessional how to identify grasses. Thus she wrote the *First Book of Grasses.* She hoped it would make American grasses better known and their worth and beauty more fully appreciated.[12]

Occasionally her interests did range beyond grasses. In 1918 she was arrested for picketing the White House in the cause of women's suffrage. In the following year she was among those arrested for maintaining a continuous fire fed by copies of President Woodrow Wilson's speeches mentioning the words "freedom" or "liberty." The goal of the protesters had been to keep the fire burning until women had the right to vote.[13] Agnes was a pacifist, a prohibitionist, a socialist, and a contributor to a number of organizations, including the Fellowship of Reconciliation, the National Association for the Advancement of Colored People, the National Woman's Party, and the Women's International League for Peace and Freedom.[14]

Yet grasses always remained the major focus of her life. She made two trips to Brazil, one in 1924 and one in 1929 at the age of 60. Her aim was to collect grasses in areas largely ignored by botanists. She traveled by boat, train, car, hand-pushed trolley, dugout canoe, horseback, muleback, and on foot. She climbed mountains and navigated marshes, often camping out, besieged by biting insects and heavy rains, the "hardest physical feat of my life." Her experiences were "glorious, fatiguing, joyous, harrowing."[15] On one occasion she was detained as a suspected lunatic, having been observed crawling on her hands and knees pulling up clumps of grass.[16]

In 1940, a year after she had officially retired, Agnes went to Venezuela at the invitation of the Venezuelan Ministry of Agriculture to collect and study grasses and to help develop a range management program.[17] During her trips to South America, she met students of botany and urged them to study in the United States. She even boarded some in her Washington home.[18]

Agnes did not believe in retirement. She continued to work almost every day, without pay.[19] She revised Hitchcock's *Manual of Grasses* and her *First Book of Grasses*, continued writing, and eventually completed 70 publications. She also catalogued her specimens. It was reported that her first question when meeting someone was, "And what grasses do you work on?"[20]

Agnes's contributions to botany were recognized by the scientific community. In 1956 the Botanical Society of America presented her a certificate of merit. Two years later the University of Illinois awarded her an honorary Doctor of Science degree and the Smithsonian Institution named her its eighth honorary fellow. For her contributions to the study of botany in Brazil, she was presented a medal by the nation of Brazil in that same year. The Linnaean Society named her a fellow in 1961.[21]

Agnes Chase finally quit working in 1963, at age 94. She died five months later on her first day in a nursing home.[22] A woman barely 5 feet tall, Agnes had worked her way to become a giant in botany.

Notes

1. Marcia Bonta, *Women in the Field: America's Pioneering Women Naturalists* (College Station: Texas A&M University Press, 1991), p. 132.

2. Michael T. Stieber, "Chase, Mary Agnes," in *Notable American Women, The Modern Period* (Cambridge, Mass.: Harvard University Press, 1980), p. 146.

3. Bess Furman, "Grass Is Her Liferoot," *New York Times* 107 (June 12, 1985): 37.

4. *Ibid.*

5. Stieber, *Notable American Women*, pp. 146–47.

6. *Ibid.*, p. 147.

7. *Ibid.*

8. Pamela M. Henson, "Case, (Mary) Agnes Merrill," *Dictionary of American Biography*, Supplement 7 (New York: Charles Scribner's Sons, 1981), p. 119.

9. Liz Hillenbrand, "87-Year-Old Grass Expert Still Happy with Subject," *Washington Post and Times Herald* 79 (Apr. 30, 1956): 7.

10. Agnes Chase with A.S. Hitchcock et al., *Smithsonian Scientific Series*, Vol.

11: *Old and New Plant Lore* (Washington, D.C.: Smithsonian Institution, 1934), pp. 201–49.

11. Hillenbrand, "87-Year-Old," p. 7.
12. Bonta, *Women in the Field*, p. 135.
13. Furman, "Grass Is Her Liferoot," p. 37.
14. Stieber, *Notable American Women*, p. 148.
15. Bonta, *Women in the Field*, pp. 137–41.
16. *Magnificent Foragers* (Washington, D.C.: Smithsonian Institution, 1978), p. 25.
17. Stieber, *Notable American Women*, p. 148.
18. Bonta, *Women in the Field*, p. 142.
19. Furman, "Grass Is Her Liferoot," p. 37.
20. Henson, *Dictionary of American Biography*, p. 120.
21. Stieber, *Notable American Women*, p. 148.
22. *Magnificent Foragers*, p. 25.

Bibliography

Bonta, Marcia. *Women in the Field: America's Pioneering Women Naturalists*. College Station: Texas A&M University Press, 1991.

Fosberg, F.R., and J.R. Swallen. "Agnes Chase." *Taxon* 8 (June 1959): 145–51.

Furman, Bess. "Grass Is Her Liferoot." *New York Times* 107 (June 12, 1958): 37.

Henson, Pamela M. "Chase, (Mary) Agnes Merrill." In *Dictionary of American Biography*, Supplement 7. New York: Charles Scribner's Sons, 1981.

Hillenbrand, Liz. "87-Year-Old Grass Expert Still Happy with Subject." *Washington Post and Times Herald* 79 (Apr. 30, 1956): 7.

The Magnificent Foragers. Washington, D.C.: Smithsonian Institution, 1978.

"Mrs. Agnes Chase, Botanist, Is Dead." *New York Times* 113 (Sept. 26, 1963): 35.

Ogilvie, Marilyn Bailey. *Women in Science: Antiquity through the Nineteenth Century*. Cambridge, Mass.: MIT Press, 1986.

Siegel, Patricia Joan, and Kay Thomas Finley. *Women in the Scientific Search: An American Bio-Bibliography, 1724–1979*. Metuchen, N.J.: Scarecrow Press, 1985.

Stieber, Michael T. "Chase, Mary Agnes." In *Notable American Women: The Modern Period*. Cambridge, Mass.: Harvard University Press, 1980.

TERESA R. FAUST

LIN CH'IAO-CHIH

(1901–1983)

Physician

Birth	December 23, 1901
1921	Entered Peking Union Medical College (PUMC) School of Medicine as its first female enrollee
1929	Graduated from PUMC; the first woman retained on the staff of PUMC and its hospital
1932–33	Studied at Manchester University, England
1939–40	Studied at London University, England
1942	Joined staff of the Zhonghe Hospital, where she founded the Department of Obstetrics and Gynecology
1948	PUMC and its hospital reopened after World War II with Dr. Lin as head of the Department of Obstetrics and Gynecology
1949	Declined Chairman Mao's invitation to attend a state dinner
1953	Vice President, Democratic Women's Association, Beijing's Chapter
1954	Deputy, first National People's Congress (NPC); Member, Chinese National Committee of Protection for Children
1955	Fellow, Chinese Academy of Biological Sciences
1957	Vice Chairman, Chinese Medical Association; Member, Birth Control Guidance Committee
1958	Deputy, second NPC
1958–66	"Barefoot Doctors Movement" was begun and then was reinvigorated during the Cultural Revolution under Mao
1959	Served as Peking's delegate to the Chinese People's Political Consultative Conference (CPPCC)
1960	Vice Chairman, Peking's Municipal Committee, CPPCC
1963	Vice President, Democratic Women's Association, Beijing's Chapter
1964	Deputy and Member, Standing Committee, third NPC; Vice Chairman, Chinese Academy of Medical Sciences
1972	Head of Ob/Gyn Dept., Shoutou (Capital) Hospital of Beijing; served on a special Chinese delegation to Canada and France

1975	Member, Standing Committee, fourth NPC
1978	Member, Standing Committee, fifth NPC; Vice Chairman, National Women's Association
1980	Vice President and Founding Member, Family Planning Association of China
Death	April 23, 1983

Medical doctor Lin Ch'iao-chih was one of the most beloved public figures in the People's Republic of China. (Her name is also spelled variously Lim Kh-at'i in the Amoy dialect, Lin Chiao-chi, and Lin Qiaozhi.) She was active as an obstetrics and gynecology physician, teacher, researcher, and community service leader until her death at age 82 in 1983.[1] "Mother Lin" was esteemed by her colleagues and loved by her patients. In over 55 years of practice she delivered thousands of babies, many of whom as adults often wrote her or visited the hospital where she worked.[2] In explaining why she chose medicine as a career, Dr. Lin wrote in 1972 that "our motherland was . . . tormented by natural disasters and diseases. Out of sympathy and pity for my fellow countrymen and cherishing the ideology of 'kindness' and 'happiness in serving and helping people' [I resolved] to become a doctor with 'conscience.' "[3] Dr. Lin shared a heightened sense of social consciousness with many other Chinese women scientists. It has been observed that "nearly all come from scholarly families and are devoted to their chosen fields . . . they exhibit a deeply felt desire to serve their country."[4]

Although details of her early childhood years are not revealed to us at this time, one catalyst for the young Lim Kh-at'i from the port city of Amoy in Fukien Province may have been the typhoon of 1917 that destroyed the 75-year-old Amoy Hospital and many other buildings when the young woman was just 16 years old.[5] The trade city of Amoy is much like Venice, Italy, in that several islands in the bay are connected by bridge or ferry. The semitropical climate is temperate; warm, humid days give way to cool evenings and mornings. The devastation visited upon the beautiful city of Amoy obviously affected Lim Kh-at'i.

Four years later, at age 20, Lim passed her entrance exams and entered the Medical School of the Peking Union Medical College (PUMC). In fact, 1921 was the first year PUMC admitted women.[6] The Rockefeller Foundation supported the PUMC with grants given through the China Medical Board, Inc. (CMB). "The PUMC was created in 1915 by CMB to stimulate formation of a medical elite of western medicine in China, who would themselves be teachers of medicine."[7] As one of the PUMC's most distinguished graduates in 1929, Lin Qiaozhi (Lim Kh-at'i) joined the Department of Obstetrics and Gynecology as its first woman Chinese doctor. She continued her studies in 1932–1933 at Manchester University and London University. She also studied at Chicago University in 1939–

1940, but she spent most of her professional life at PUMC and became head of Ob/Gyn there in 1948. Dr. Lin had a lifelong association with the PUMC, which was renamed the Anti-Imperialist Hospital from 1966 until early 1972, when its name was changed to the Shoutou (Capital) Hospital of Beijing.[8]

Alumni of the PUMC, including Lin Ch'iao-chih and her fellow 1929 graduate Chung Huei-Ian, were active researchers and scholars. Lin Ch'iao-chih became vice president of the Chinese Medical Association in 1957. In 1964 she was elected vice president of the Chinese Academy of Medical Sciences (CAMS), the premier organization for medical research in China. The CAMS is directly under the jurisdiction of China's Ministry of Public Health. Dr. Lin's scholarship included pure research on tumors in women, family planning, and public health policy.[9]

Although she was certainly distinguished in her professional life, Dr. Lin's achievements become even more remarkable when her student days and professional career are viewed within the context of the changing political scene in China. There were subtle and overt instances of gender discrimination, such as the two-year delay after PUMC opened in 1919 before enrolling women. Dr. Lin said that "women were especially oppressed and discriminated against, but I cherished this ideology [of "kindliness and love for all"] and buried myself in medical studies day and night for years."[10] She overcame such gender obstacles primarily because there were other "more important" threats to personal and professional stability—both political and military. The Chinese Communist Party gained a strong foothold over the country with the formation of the National Government in Canton on July 1, 1925. During the following summer, General Chiang Kai-shek undertook his historic Northwest Expedition. Finally, after years of war and political instability, the city of Peking fell to the Nationalists. It was renamed Peiping in June 1928.

With the escalation of Sino-Japanese hostilities during the 1930s as a precursor to World War II, Peiping was taken over by Japanese troops on August 4, 1937. The seizure was followed by two more years of increasingly oppressive occupation by the Japanese, during which "friends and colleagues suddenly disappeared" and were rumored to be "held without charges or trials" in Japanese jails. Despite this turmoil, the PUMC remained actively engaged in research and medical care. Japanese soldiers finally entered the PUMC Hospitals and occupied the school complex on Pearl Harbor Day. This spelled the beginning of the end of the PUMC for the duration of the war. Eventually the staff was disbanded and all medical activities suspended. The PUMC shut down on January 19, 1942, for all but military use. Private practice became the only option for many staff members. In 1942, Dr. Lin joined the staff of Zhonghe Hospital, an institution previously affiliated with PUMC, where she founded the Department of Obstetrics and Gynecology.[11]

No sooner had World War II ended and the PUMC reopened in 1948, with Dr. Lin as head of obstetrics and gynecology, than the next Chinese political shift occurred. By early February 1949, Peiping had been taken over by the Chinese Communist army. The PUMC continued to operate for two more years until it was nationalized on January 20, 1951, when the Chinese government seized control of it. By now the Rockefeller Foundation and CMB's control and financial support of the PUMC was at an end.[12]

As the Union Medical College's hospital continued to operate under successive political regimes, Lin Ch'iao-chih began expanding her public role. Although her participation during the early 1950s may have been politically expedient initially, she met the challenge with enthusiasm. "As an old intellectual, coming from the old society, I too gained a new lease on life," she said when the party liberated China in 1949.[13] Later, the Cultural Revolution of the 1960s shone a spotlight on "the relationship of science and technology to the distribution of power." This resulted in years of "disruption within the science system [1966–1969] followed by three years [1969–1971] of reconstruction along Maoist lines."[14] In order to achieve a leveling of status differences between intellectuals and peasants, the cultural elite was sent to rural areas to help in agriculture, manual labor, education, and health issues while being indoctrinated to the new order. These "assignments" to rural May 7th Cadre [student] Schools could be anywhere from six months to one year in duration and were repeated at annual intervals. The hardship of extended placement in the countryside was "directed far more against the science administrators who made science policy than against the scientists themselves."[15] Lin did spend some time in re-education work camps "in response to Chairman Mao's call" when she was in her sixties, but she did not mention feeling any hardship on her part.[16]

Because the Rockefeller Foundation and the PUMC had supported mass education programs such as public literacy and public health between 1929 and 1949, it is not surprising that Dr. Lin became a more public advocate of such efforts in the 1950s, 1960s, and 1970s. The Barefoot Doctors movement, an effort in the late 1950s to train rural citizens in first aid and routine medical procedures, was a response to an "inadequate supply of doctors to meet the needs of all the people." What began in 1958 with the training of 2,300 lay people grew to over 29,000 Barefoot Doctors by 1968.[17]

Until 1949, Lin had primarily lived and worked in a large urban setting. She felt she had "become more and more out of touch with the broad masses of the people . . . still serving the minority of the city."[18] She believed that gaining the perspective of helping those for whom professional medical care was scarce broadened her understanding and appreciation for patients' needs beyond medicine. Once, while assisting a young midwife in the delivery of a woman who lived in a rural area, Dr. Lin became "a little impatient" because the mother wasn't quite ready to deliver. But, said Dr. Lin, the Barefoot Doctor midwife "treated the patient like one of the family . . . seeing the water vat empty, she went to fetch water . . . her warmth and manner moved me."[19]

Articles describing Lin's experiences served as guidelines for the rest of China's modern medical profession. Lin Ch'iao-chih never married, and the lack of children and a family of her own was her one regret.[20] However, she "felt career and family were not compatible."[21] Seventy years ago, she believed she didn't have the choice that women enjoy today—having both a family and a career.

Dr. Lin became extensively involved in public service through her participation in the committee governance system of the People's Republic of China. Highlights include her two decades of work for the National People's Congress as well as the Chinese People's Political Consultative Conferences. Although these organizations focused primarily on social and public policy issues, Dr. Lin was also active in family planning and women's and family health issues through her role in groups such as the Birth Control Guidance Committee in 1957 and the National Women's Association from 1953 to 1978. Lin visited Canada and France in 1972 as part of a special Chinese delegation on a cultural exchange. She became vice president and a founding member of the Family Planning Association of China in May 1980. When she died in 1983, she left her body to science.[22]

Lin Ch'iao-chih's medical career and community service contributions were so varied and numerous that she once "set aside" an invitation from Chairman Mao to attend a state ceremony in 1949—celebrating the formation of the People's Republic of China—because she was "too busy" to think about it. Most people would have hesitated to decline an invitation from a head of state, especially in a time of political upheaval when small slights to party officials could mean dire consequences. Yet Lin survived this minor faux pas unscathed and became known as the woman who had said "no" to Mao.[23] The Chairman must have understood that Lin's was a singular devotion in the service of her fellow citizens. Indeed, her life embodied compassion, dedication, constancy, and conscience—the goals she had set for herself as a young medical student.

Notes

1. Yiming Wu, ed., *People's Republic of China Year-Book 1984* (Hong Kong, China: Evergreen Publishing, 1984), p. 840.

2. Qichang Su, "Lin Chiao-chih (1901–1983)," *Biographical Literature* 42, no. 6 (1984): 146–47 (translations by Qichang Su, November 28, 1994).

3. Lin Ch'iao-chih, "Why the Party Keeps Me Young," in *New Women in China* (Peking, China: Foreign Language Press, 1972), p. 22.

4. Susan V. Meschel, "Teacher Keng's Heritage: A Survey of Chinese Women Scientists," *Journal of Chemical Education* 69, no. 9 (Sept. 1992): 729–30.

5. K. Chimin Wong and Wu Lien-Teh, *History of Chinese Medicine* (Shanghai, China: National Quarantine Service, 1936), p. 647.

6. Mary E. Ferguson, *China Medical Board and Peking Union Medical College: A Chronicle of Fruitful Collaboration, 1914–1951* (New York: Rockefeller Foundation, 1970), p. 38.

7. Thomas Rosenbaum, *The Archives of the China Medical Board and the PUMC at the Rockefeller Archive Center: Some Sources on the Transfer of Western Science, Medicine, and Technology to China during the Republican Period* (New York: Rockefeller Archive Center, 1989), p. 65.

8. Ferguson, *China Medical Board*, pp. 236, 245; Wu, *People's Republic of China Year-Book 1984*, p. 840; Victor Sidel and Ruther Sidel, *Serve the People: Observations on Medicine in the People's Republic of China* (New York: Josiah Macy, Jr., Foundation, 1973), p. 65.

9. *People's List of Names in Communist China* (Taipei, Taiwan: National Cheng-chih University, 1983), p. 341 (translations by Qichang Su, November 28, 1994); "Lin Ch'iao-chih," in *Who's Who in Communist China* (Hong Kong, China: Union Research Institute, 1966), p. 380; Sidel and Sidel, *Serve the People*, p. 183; "Lin Chiao Chih (1901–1983)," in *PRC Biographical Dictionary* (Peiching, Taiwan: Chinese Economic Publishers, 1989), p. 330 (translations by Shang-Fen Ren, November 18, 1994).

10. Lin Ch'iao-chih, "Why the Party Keeps Me Young," p. 22.

11. Ferguson, *China Medical Board*, pp. 152, 176, 177; Wu, *People's Republic of China Year-Book 1984*, p. 840.

12. Rosenbaum, *The Archives of the China Medical Board*, p. 17; Ferguson, *China Medical Board*, p. 206.

13. Lin Ch'iao-chih, "Why the Party Keeps Me Young," p. 24.

14. Richard P. Suttmeier, *Research and Revolution: Science Policy and Societal Change in China* (Lexington, Mass.: DC Heath and Company, 1974), p. 101.

15. *Ibid.*, p. 107.

16. Lin Ch'iao-chih, "Why the Party Keeps Me Young," p. 25.

17. Sidel and Sidel, *Serve the People*, p. 78.

18. Lin Ch'iao-chih, "Why the Party Keeps Me Young," p. 24.

19. *Ibid.*, p. 25.

20. Mary B. Bullock, "A Brief Sketch of the Role of PUMC Graduates in the People's Republic of China," in *Medicine and Society in China* (New York: Josiah Macy, Jr., Foundation, 1974), p. 101; Jingrong Huang and Grace Foote Johns, electronic mail interview, Columbia University, October 11, 1994.

21. "Lin Chiao Chih (1901–1983)," in *PRC Biographical Dictionary* (Peiching, Taiwan: Chinese Economic Publishers, 1989), p. 146 (translations by Shang-Fen Ren, November 18, 1994).

22. Institute of China Studies, "A Survey of the National People's Congress," in *Inside China Mainland* (Taipei, Taiwan: Institute of China Studies, 1983), pp. 10–11; "Lin Ch'iao-chih," in *Who's Who in Communist China*, p. 380; *People's List of Names in Communist China*, p. 341; "Population and Birth Control," in *China*

Official Annual Report, 1981 (Kowloon, Hong Kong: Kingsway International Publications, 1981), p. 731.

23. "Lin Chiao Chih (1901–1983)," in *PRC Biographical Dictionary*, pp. 146–47.

Bibliography

Bullock, Mary B. "A Brief Sketch of the Role of PUMC Graduates in the People's Republic of China." In *Medicine and Society in China*. New York: Josiah Macy, Jr., Foundation, 1974.

Ferguson, Mary E. *China Medical Board and Peking Union Medical College: A Chronicle of Fruitful Collaboration, 1914–1951.* New York: Rockefeller Foundation, 1970.

Institute of China Studies. "A Survey of the National People's Congress," in *Inside China Mainland*, pp. 10–11. Taipei, Taiwan: Institute of China Studies, 1983.

Kleinman, Arthur. *Medicine in Chinese Cultures: Comparative Studies of Health Care in Chinese and Other Societies*, Publication #75-653. Washington, D.C.: National Institutes of Health, U.S. Department of Health, Education, and Welfare, 1975.

Lin Ch'iao-chih. "Why the Party Keeps Me Young," in *New Women in China*, pp. 21–30. Peking, China: Foreign Language Press, 1972.

"Lin Ch'iao-chih." In *Who's Who in Communist China*. Hong Kong, China: Union Research Institute, 1966.

MacFarquhar, Roderick, and John K. Fairbanks. *The Cambridge History of China.* Vol. 15, *The People's Republic,* Part 2; "Revolutions within the Chinese Revolution, 1966–1982." New York: Cambridge University Press, 1991.

Meschel, Susan V. "Teacher Keng's Heritage: A Survey of Chinese Women Scientists." *Journal of Chemical Education* 69, no. 9 (Sept. 1992): 723–30.

"Population and Birth Control." In *China Official Annual Report, 1981*. Kowloon, Hong Kong: Kingsway International Publications, 1981.

Rosenbaum, Thomas. *The Archives of the China Medical Board and the PUMC at the Rockefeller Archive Center: Some Sources on the Transfer of Western Science, Medicine, and Technology to China during the Republican Period.* New York: Rockefeller Archive Center, 1989.

Sidel, Victor, and Ruther Sidel. *Serve the People: Observations on Medicine in the People's Republic of China.* New York: Josiah Macy, Jr., Foundation, 1973.

Spence, Jonathan D. *The Search for Modern China.* New York: W.W. Norton and Co., 1990.

Suttmeier, Richard P. *Research and Revolution: Science Policy and Societal Change in China.* Lexington, Mass.: D.C. Heath and Co., 1974.

Wong, K. Chimin, and Wu Lien-Teh. *History of Chinese Medicine.* Shanghai, China: National Quarantine Service, 1936.

GRACE FOOTE JOHNS

CORNELIA M. CLAPP

(1849–1934)

Zoologist

Birth	March 17, 1849
1868–71	Student, Mount Holyoke Seminary
1872–1916	Teacher/Professor, Mount Holyoke Seminary/College
1874	Attended Anderson School on Penikese Island
1888–1934	Spent summers at Woods Hole Marine Biological Laboratory
1889	Ph.D., Syracuse University
1896	Ph.D., University of Chicago
1916	Retired to Florida
Death	December 31, 1934

Cornelia Clapp was one of the pioneers in American biology. As one of the first women to earn a doctorate in science, she proved that women were capable of advanced work in the sciences. Her research at Woods Hole was instrumental in the development of the fields of marine biology and embryology, and her teaching inspired several generations of women biologists.

Cornelia Maria Clapp was born in Montague, Massachusetts, in 1849, the daughter of Richard C. and Eunice Amelia (Slate) Clapp. She was the eldest of three brothers and three sisters. Her early education was in the public and private schools of Montague.

In 1868 she entered Mount Holyoke Seminary, one of the first academies for young women, from which she graduated in 1871. For one year after graduation she taught Latin at a boys' school, Potter Hall in Andalusia, Pennsylvania. Then she was invited back to teach at Mount Holyoke. She recalled that "to my consternation, I didn't know what I was to teach!"[1] She taught mathematics for the first year but began the following year to teach natural history, which would soon be known as zoology. Additionally, she taught gymnastics from 1876 to 1891.

During the school year 1872–1873, Clapp and her colleague and former teacher, **Lydia Shattuck**, spent their spare time searching for the microscopic organism known as the amoeba, which they knew only from descriptions in books. Having taken water from a nearby pond, they

Cornelia M. Clapp. Photo courtesy of the Mount Holyoke
College Library Archives.

studied it drop by drop, taking turns at the microscope. They found the
amoeba and many other plants and animals as well.

In 1874, Clapp was selected to go with Lydia Shattuck to Penikese
Island in Massachusetts. The Penikese Island school, also called the Anderson School, had been founded by the distinguished naturalist Louis
Agassiz during the previous year and was the first seaside school of
natural history, later marine biology. Agassiz's philosophy in opening
the school was that scientists should study nature, not books. The Anderson School was also one of the first places where women scientists
could obtain advanced education in science. Said Clapp, "I had an opening of doors at Penikese. I looked and saw a thousand new doors. Every-
body was talking. Discussions in every corner. I felt my mind going in
every direction."[2] After that summer she discarded the textbook she had
been using in her teaching and began to base her teaching on the animals
themselves.

During that year she introduced the study of embryology to the college. Clapp rented a hen, made a nest for her, and placed an egg under
her every day for 21 days. At the end of that time, she had samples of
all stages of growth of the chick from one day's growth to a chick that

peeped and walked. She arranged them in dishes on a table and charged admission to see them, thereby earning back the money she had spent to rent the hen.

In the summer of 1878 she joined a group of zoology professors for a walking tour of the South led by Dr. David Starr Jordan, a fellow Penikesan and already a well-known biologist. Clapp recalled, "At that time, I had short hair and wore short dresses. I could walk in and out of the water any time I wanted to. We went through the rapids in one stream. Jordan and some others held the rope and I went hand over hand through the rapids."[3] In the summer of 1879 she joined the same group for a walking tour of Europe, including Switzerland and Italy. In 1882–1883 she worked for a period with Professor William T. Sedgwick of MIT on the embryology of the chick; in 1883–1884, with Professor E.B. Wilson of Williams College on the earthworm.

Cornelia Clapp dedicated herself to learning throughout her life. In 1889 she earned one of the first Ph.D. degrees awarded to a woman, from Syracuse University. In 1888 she spent her first summer at the Woods Hole Marine Biology Laboratory in Massachusetts, where she spent every subsequent summer until her death in 1934. She started there as an investigator, later became a lecturer, and still later became the only woman on the board of trustees. (Clapp Road in Woods Hole is named after her.) She was the first person at Woods Hole to be given her own problem to study: the toad fish, which became known to zoologists as "Dr. Clapp's fish." From 1893 to 1896 she did further study at the University of Chicago, where at age 47 she obtained her second Ph.D. In 1901 she worked in Italy at the famed Naples Station in biology. She was a member of the American Association for the Advancement of Science, the Society of American Zoologists, and the Association of American Anatomists. In 1906 she was one of only six women named in the first edition of *American Men of Science*. Of herself, Clapp said, "I was all bent on one thing, then another . . . first an entomologist, then a conchologist and then a fish woman."[4]

As a teacher, Clapp was an inspiration to her students. One student remembered being told on her acceptance to Mount Holyoke in 1891, "You ought to study under Dr. Clapp. She keeps live frogs in tanks." When the student got there, she said, "I came; I saw; she conquered. Her bounding vitality and thirst for knowledge were contagious. I felt then and have felt ever since that I was never fully alive until I knew her."[5] Another wrote of her in 1899, "Whatever she goes into, she goes into it head over heels."[6] A third student remembered, "One day as we were seated in the classroom, she left the room, leaving the door a little ajar. In a few minutes a slight noise attracted our attention and in came a thin line of fiddler crabs, sidling along. From behind the door she was sending them in, one at a time. There was not one of us who was not eager to study crabs after such a gay introduction."[7]

Clapp retired from Mount Holyoke in 1916, spending the rest of her life wintering in Mount Dora, Florida, and summering at Woods Hole. In 1923, Mount Holyoke's new building, Clapp Laboratories, was dedicated to her. At this she said, "I am pleased, of course, at the honor paid me in giving the name to the building, but I almost wish the name was only for the rooms especially for Zoology. It seems almost too much Clapp for the whole house."[8] At the time of her death on December 31, 1934, she was the last surviving member of the Penikese School. Of her life Clapp said, "I have always had an idea that if you want to do a thing, there is no particular reason why you shouldn't do it."[9]

Notes

1. Cornelia Clapp, interview (1921). In Cornelia Clapp Papers, Mount Holyoke College Library Archives.

2. *Ibid.*

3. *Ibid.*

4. *Ibid.*

5. Louise Baird Wallace, "Cornelia Clapp," *Mount Holyoke Alumnae Quarterly* 19 (May 1935): 4.

6. Sarah French, letter dated April 9, 1889. In Cornelia Clapp Papers, Mount Holyoke College Library Archives.

7. Mrs. J.E. Leslie, "She Was My Teacher," *Detroit News* (January 8, 1935). (From clipping in Cornelia Clapp files at Mount Holyoke College.)

8. Clapp, on the naming of Clapp Laboratories, February 17, 1923. In Cornelia Clapp Papers, Mount Holyoke College Library Archives.

9. Clapp, interview.

Bibliography

Carr, Emma Perry. "One Hundred Years of Science at Mount Holyoke College." *Mount Holyoke Alumnae Quarterly* 20 (1937): 135–38.

Clapp, Cornelia. "Some Recollections of the First Summer at Woods Hole." *Collecting Net* 2, no. 4 (1927): 2–10.

Cole, Arthur C. *A Hundred Years of Mount Holyoke College: The Evolution of an Educational Ideal.* New Haven, Conn.: Yale University Press, 1940.

Rossiter, Margaret W. *Women Scientists in America: Struggles and Strategies to 1940.* Baltimore and London: Johns Hopkins University Press, 1982.

Wallace, Louise Baird. "Cornelia Clapp." *Mount Holyoke Alumnae Quarterly* 19 (1935): 1–9.

CAROLE B. SHMURAK

EDITH JANE CLAYPOLE

(1870–1915)

Physiologist, Pathologist

Birth	January 1, 1870
1892	Ph.B., Buchtel College, Akron, OH
1893	M.S., Cornell University
1894–99	Instructor in Physiology and Histology, Wellesley College
1896	M.D., University of California, Los Angeles
1901–1902	Instructor, Throop Polytechnic Institute (now California Institute of Technology, Pasadena)
1902–11	Pathologist, Los Angeles, CA
1912–15	Research Associate, University of California, Berkeley
Death	1915

Edith Jane Claypole did pioneering research in the fields of blood and tissue histology. Her work on infectious diseases was especially valuable before the discovery of antibiotics. Although the various specialities of medicine offered more employment to women than did the other sciences, Edith had to struggle before finally finding a position that allowed her to engage in research. Her tenacity and drive, together with her great personal charm, helped her overcome the obstacles talented women were facing at the time; unfortunately, her tragic death cut short a promising career.

Edith and her identical twin sister, Agnes Mary Claypole, were born in 1870 in Bristol, England. (For family background and education, see entry for **Agnes Mary Claypole Moody**.) She and her sister moved to America in 1879 and joined their father, Edward Waller Claypole, who had come to teach natural sciences at the college level. Edith and her sister prepared for college under their parents' guidance and instruction. They studied at Buchtel in Akron, Ohio, where their father had been teaching since 1883, and graduated with bachelor's degrees in biology in 1892.

The influence and example of their parents and their own talents motivated the sisters to continue their studies at an institution that admitted women to graduate school. During the two decades before the turn of the century, opportunities for women seeking advanced degrees had increased; the Claypole sisters enrolled at Cornell University in Ithaca, New York. Both also attended the Marine Biology Laboratory at Woods Hole, Massachusetts, a seaside summer school for teachers and researchers in the biological sciences. At Cornell, Edith Claypole became interested in researching blood cells. In 1893 she wrote her master's thesis, "Blood of Necturus and Cryptobrachus," about the blood cells of the mudpuppy and a species of salamander.

At that time, women who wanted a career in a scientific field found their best chances for employment at the independent women's colleges. Edith Claypole followed a typical pattern when she started teaching physiology and histology at Wellesley College, Massachusetts, from 1894 to 1899. For two of those years, she also was acting head of the zoology department. Her research interests led her toward the study of medicine, and in the fall of 1899 she went back to Cornell University for a medical degree. While she was a student in the medical department, she also taught physiology as an assistant. After two years she left Cornell and moved to Pasadena, California, to be with her mother, who was ailing.

Edith's father died in August 1901, and her mother survived him by only a few weeks. Edith then taught until 1902 at the Throop Polytechnic Institute at Pasadena, where her sister Agnes had taken over her father's position. Later in 1902 she resumed her interrupted studies when she moved to Los Angeles and enrolled in the medical school of the University of California, Southern Branch, with a specialty in pathology. For two years she worked part-time after her classes as a pathologist for a group of physicians and surgeons associated with a hospital in Los Angeles. After she received her doctorate in medicine in 1904, she continued working in the same medical office. For eight years she performed the everyday chores of laboratory testing in cramped facilities. Through her work, however, she gained an exceptional expertise in the diagnosis and knowledge of bacterial cultures and vaccines.

Edith wanted to apply these skills to research projects. In 1912 she volunteered her services to the Department of Pathology at the University of California, Berkeley, and to its director, Professor Frederick Parker Gay. Despite her superior qualifications, evidently she could not find a job at a university and decided that she could pursue her research interests only by doing volunteer work. Professor Gay let her use his department's facilities and eventually offered her the position of research associate in pathology.

Edith Claypole began investigating methods for differentiating certain infectious diseases caused by a group of bacteria from tuberculosis, an illness they resembled. She designed a simple skin reaction test that allowed a distinction between these infections and tuberculosis in human cases. At the time of her death, she was working on a method of treatment. Her articles based on this work on bacterial lung diseases appeared in two medical journals.

After the outbreak of World War I, Professor Gay started working on a new typhoid vaccine for the English and French troops in Europe. Because Edith Claypole had already done so much research on infectious diseases, including typhoid fever, she supervised the actual lab work involved in developing the vaccine. Together with Professor Gay, she reported their findings and progress in four journal articles published between 1913 and 1914. Although she herself had been vaccinated against typhoid, the prolonged and unusually concentrated exposure took its toll. The onset of the disease was very gradual, and the infection was diagnosed too late. Edith Jane Claypole died of typhoid fever contracted in the course of her research when she was only 45 years old.

Friends and relatives endowed the Edith J. Claypole Memorial Research Fund in Pathology in her memory. The income from the fund was to be used for annual awards to encourage the investigation of pathological problems, preferably by women, for the diagnosis and therapy of infectious diseases.

Bibliography

Bailey, Martha J. *American Women in Science: A Biographical Dictionary*. Santa Barbara, Calif.: ABC-CLIO, 1994.

Herzenberg, Caroline L. *Women Scientists from Antiquity to the Present: An Index*. West Cornwall, Conn.: Locust Hill Press, 1986.

Moody, Agnes Claypole, and Marian E. Hubbard. *In Memoriam: Edith Jane Claypole*. Berkeley, Calif.: [s.n.], 1915.

National Cyclopedia of American Biography. New York: James T. White, 1906.

["Obituary."] *Science* 41, no. 1058 (Apr. 1915): 527, 754.

Ogilvie, Marilyn Bailey. *Women in Science: Antiquity through the Nineteenth Century*. Cambridge, Mass.: MIT Press, 1986.

Rossiter, Margaret W. *Women Scientists in America: Struggles and Strategies to 1940*. Baltimore: Johns Hopkins University Press, 1982.

Siegel, Patricia Joan, and Kay Thomas Finley. *Women in the Scientific Search: An American Bio-Bibliography, 1724–1979*. Metuchen, N.J.: Scarecrow Press, 1985.

Williamson, Mrs. Burton M. "Some American Women in Science." *The Chautauquan* 28 (Nov.–Mar. 1898–1899): 361–68.

Woman's Who's Who of America, ed. John William Leonard. New York: American Commonwealth, 1914. (Reprinted: Detroit: Gale Research, 1976.)

IRMGARD H. WOLFE

JEWEL PLUMMER COBB

(1924–   )

Cell Biologist

Birth	January 17, 1924
1944	B.S., biology, Talladega College
1947	M.S., cell physiology, New York University
1950	Ph.D., cell physiology, New York University
1950–52	Postdoctoral Fellow, National Cancer Institute; joined the Harlem Hospital Cancer Research Foundation
1952–54	Appointed to the Anatomy Dept. faculty, University of Illinois Medical School; established the first tissue culture research laboratory and course in cell biology
1954	Married Roy Raul Cobb; returned to New York and the Harlem Hospital Cancer Research Foundation
1955	Designed and established the Tissue Culture Research Laboratory
1960	Published a five-year cytological study on cultures; left NYU to assume a full-time teaching position at Sarah Lawrence College
1962	Presented paper at the eighth International Cancer Congress in Moscow
1969	Dean of the College and Professor of Zoology, Connecticut College; established Postgraduate Premedical and Predental Program for Minority Students
1974	Member, National Science Board
1976–81	Dean and Professor of Biological Science, Douglass College, Rutgers University
1981–90	President, California State University, Fullerton

Jewel Plummer was the only child of an upwardly mobile, middle-class Chicago family. Her father, Frank Plummer, was a physician who graduated from Rush College in 1924, a year before Cobb was born. Her mother, Carriebel Plummer, studied interpretive dance and physical education at Sargeants, a physical education college affiliated with Harvard University. In 1944, when Jewel received her bachelor's degree from Talladega College, her mother received her bachelor's from Roosevelt College. Throughout her career, Jewel modeled both parents' struggle for knowledge and excellence. And although both her parents supported her, she shared a special interest in science with her father.

Despite the positive support system her parents offered, Jewel, being an African American, was often subjected to racism over which her parents had little control. As a fifth grader she spent a year in a "portable classroom heated by a pot-bellied stove." She considered as overt racism her transfer from "Sexton Elementary School in a predominantly white neighborhood to overcrowded, old and dilapidated Betsy Ross Elementary School because the city had shifted school districts." She contended that it happened because the school districts wanted to be "certain that fewer black families were eligible to attend Sexton." She recalls having to attend Englewood High School under a "double shift scheme, starting at 11:45 A.M. and ending at 5:15 P.M., often walking home in the dark in the winter months." Her experiences with discrimination continued when she enrolled at the University of Michigan, "a disaster for black students . . . in terms of dormitory living." Jewel and other African Americans found themselves segregated into a "League House" (an "official" residence for blacks). In fact, black students were not allowed to live in dormitories. Moreover, the housing situation was exacerbated by other social restrictions: blacks were also barred from the "popular grills" and eateries. Jewel and her parents sought to rectify this "disaster" by transferring her to Talladega College in Alabama, where she graduated in 1944.[1]

Thus, although politics and racism to some extent influenced where Jewel lived and how she was educated, she managed to achieve against the odds. As a research scientist, educator, and administrator, Jewel Plummer Cobb's life is an extraordinary one in which the overarching influence of racism is mediated by singular acts of concern and mentoring combined with her own innate intelligence and ability to focus on her goals. Crucial to her success, of course, was the fine parental support she received; but her intelligence also attracted the interest of educators who wanted to serve as mentors to her. A pattern established early in her childhood, this continued throughout her professional career. The roles of these enablers varied. Some introduced her to the joys of science and instilled the importance of excellence; others guided her toward meritorious scientific research. Ms. Hyman, the first of these educators, was "a gentle yet strong woman" who taught biology in a "large high school in Chicago at Englewood High." Her influence set Jewel on the scientific path she would follow until 1976 when she stopped doing research. It was in Hyman's class that Jewel was given a microscope and discovered "an entirely new world beyond my normal viewing capacity."[2]

During this same period, science played a crucial part in validating her potential. She and other students were the subjects in a research project entitled "A Study of 100 Gifted Negro Children." The study was conducted by Martin Jenkins, who later became president of Morgan State College in Baltimore, Maryland. The research format, according to Cobb, "involved selecting 100 gifted Negro children in grammar school" and giving them a battery of standard intelligence tests for several days. The tests were followed by an in-depth study of the social, intellectual, personal, and family environment of each child tested. Cobb was one of those children. Certainly this "scientific" validation of her intelligence at such an early and impressionable period reinforced her parents' expectations as well as her own. Throughout her high school career, Cobb was a member of the honor society.[3]

Cobb balanced the good "pre-college education she received" against "the covert racial segregation that was apparent in her schools" and concluded that her pre-college education was good and that it was sufficient preparation for the demanding studies that lay before her.[4] However, pre-college training alone was not sufficient to buffer Cobb when the social difficulties of racism became apparent at the University of Michigan. But even this disastrous experience resolved itself in a positive manner, because her transfer to Talladega was the first step toward an illustrious career in science. Cobb acknowledged that Talladega College had a "strong science program," but as in many black colleges, there were certain deficits. For example, the college had laboratories but no laboratory assistants. These deficits challenged Cobb and her classmates. She and other students had to prepare their own "media and plates."[5]

Whatever deficiencies may have existed at Talladega, Cobb's professional journey was again facilitated by a helpful mentor there. Having spent two years taking courses such as comparative anatomy and bacteriology, she captured the attention of her bacteriology professor.[6] This professor advised her to pursue graduate studies at New York University. Cobb did so, but not without difficulty. Rejected for a teaching fellowship, she responded with the confidence and poise she had developed over the years. When "she appeared at the institution, armed with her excellent credentials," including a bachelor's degree in biology from Talladega, "she was offered a fellowship that she held for the next five years."[7]

After six years at New York University, Cobb earned a master's degree in cell physiology. To satisfy the requirements for the degree, she had to develop an original research project and write a thesis about it. Her work "dealt with a series of organic molecules, aromatic amidines, and their effect on the respiration of yeast cells." This research "was an original laboratory project using the intricate Warburg respirometer apparatus." Cobb also used "the Warburg apparatus to measure the biochemical reactions as a part of the methodology" for her doctoral dissertation, which she also completed at New York University.[8]

In developing her doctoral research work, "which dealt with the way melanin pigment granules could be formed *in vitro*," Cobb worked closely with her advisor, who was concerned that she present a "perfect" dissertation. Their collaboration was successful. Cobb was awarded her Ph.D. in biology from New York University in 1950. So successful was this relationship that Cobb received her first postdoctoral fellowship in the same year. "On July 1," she wrote, "I began my post-doctoral fellowship awarded to accomplish my research, by the National Cancer Institute of the National Institutes of Health." Cobb prefers theoretical research rather than pathological and considers herself primarily "a cell rather than a molecular biologist." Cobb conducted her research at the Harlem Hospital Cancer Research Foundation headed by Dr. Louis T. Wright, who was a "pioneer in treating advanced cancer patients with the newly synthesized folic acid antagonist Aminopterin and Amethopterin or Methotrexate." This first foray into independent research began the career of Jewel Plummer Cobb as a researcher.[9]

In 1952, Cobb was appointed to the anatomy department faculty of the University of Illinois Medical School as an instructor. There she established the first tissue culture research laboratory and course in the field of cell biology. During this period the National Cancer Institute awarded Cobb her first research grant as an independent investigator.

It was my first independent research activity and it was exciting to design new experiments growing human tumor tissue in flasks or test tubes outside of the body (in vitro). I used surgical specimens or autopsy specimens ... for this research. As part of the cancer research team at Harlem Hospital, I undertook tissue culture (in vitro) studies of the tumors of cancer patients.[10]

For nearly 24 years Cobb would build laboratories and conduct research wherever she found herself as professor. She authored and co-authored over 40 articles. In 1960, when she assumed a faculty position, Cobb continued her own laboratory research, which was "tied closely to helping students do research with mouse melanomas under the Undergraduate Research Participation (URP) Program of the National Science Foundation." She is proud of the fact that two students who worked with her in this program now have Ph.D.s and "are in the forefront of research in their respective fields of biology." In 1969, as dean of the college and professor of zoology at Connecticut College, she built a new laboratory and with her lab assistants "developed melanotic and amelanotic strains of the S91 mouse melanoma in cell culture." But what Cobb describes as her most "interesting work" took place at New York University, where she worked with Dorothy G. Walker, her "research assistant and scientific colleague." Cobb conducted an "in-depth analysis of Thio-Tepa, a promising anti-cancer drug," and later presented a paper on her findings at the eighth International Cancer Congress in 1962. Cobb continued her research through 1976, when she became dean of Douglass College at Rutgers University and a full professor of biology.[11]

By this time Cobb had enjoyed an illustrious career. She had experienced the difficulties of managing a career and marriage as well as motherhood. In 1954 she married Roy Raul Cobb. (Although that marriage did not last—they divorced in 1967—they had one son, Roy Jonathan Cobb.) After her marriage, Cobb returned to New York and the Harlem Hospital Cancer Research foundation, combining marriage and motherhood. The woman who enjoyed the triumph of presenting a paper to colleagues in Moscow was also a mother who was concerned that she had to leave her 5-year-old son behind in the United States even though he was with his father. The impact of mentors and family on Cobb's life seems to have influenced her decision to shift to administration and policy enactment. She has expressed concern about the lack of women and minorities in science and has acted to address this problem by serving on the National Science Board, the policy body for the National Science Foundation. She describes her six-year tenure there as the only "minority woman" as both exciting and frustrating.

Cobb points out that during the 1970s the minority participation in the hard sciences was increasing, but that since 1978 the number has declined from 2.1 percent to only 1 percent. She sees these "dismal results" as "part of a very complicated mix of social injustice, racism, discouragement, and in some instances a poor K–12 course of study," along with other unfortunate school practices such as social promotion, lowered teacher expectations, and early tracking as "slow learners." Cobb's reading of a complex phenomenon is informed by both research and personal experience. She clearly recognizes her own success as a combination of strong parental support and educational foundation.[12]

Jewel Plummer Cobb's is a life of intelligence and perseverance, a life in science balanced by both research and service, which came to its peak when she assumed the presidency of California State University. Although she was not doing research herself, she continued a tradition at California State of enabling in science when she "established the first privately funded gerontology center in Orange County, [and] lobbied the state legislature for the construction of a new engineering and computer science building and a new science building" while working to ensure opportunities for underrepresented groups.[13]

Notes

1. Jewel Plummer Cobb, "A Life in Science: Research and Service," *SAGE: A Scholarly Journal on Black Women* 6 (1989):39.
   2. *Ibid.*

3. Dona L. Irvin, "Jewel Plummer Cobb," in *Epic Lives: One Hundred Black Women Who Made a Difference,* ed. Jessie Carney Smith (Detroit: Visible Ink Press, 1993), p. 92.

4. *Ibid.*

5. Cobb, "A Life in Science," pp. 40–41.

6. *Ibid.*, p. 41.

7. Irvin, "Jewel Plummer Cobb," p. 93.

8. Cobb, "A Life in Science," p. 40.

9. *Ibid.*, p. 41.

10. *Ibid.*

11. *Ibid.*, p. 42.

12. *Ibid.*, p. 43.

13. Irvin, "Jewel Plummer Cobb," p. 94.

Bibliography

Cobb, Jewel Plummer. "A Life in Science: Research and Service." *SAGE: A Scholarly Journal on Black Women* 6 (1989): 39–43.

Irvin, Dona L. "Jewel Plummer Cobb." In *Epic Lives: One Hundred Black Women Who Made a Difference,* ed. Jessie Carney Smith. Detroit: Visible Ink Press, 1993.

F. ELAINE DE LANCEY

RITA ROSSI COLWELL
(1934–)
Microbiologist

Birth	November 23, 1934
1961	Ph.D., genetics, University of Washington
1961–64	Research Assistant Professor, University of Washington
1972–	Professor of Microbiology, University of Maryland, College Park
1975	Phi Sigma Service Award, American Chemical Society
1975–	Consultant, Environmental Protection Agency
1981	Annual Achievement Award, Sigma Xi
1983–87	Vice President for Academic Affairs, University of Maryland System

1983–90	Member, National Science Board
1984	Research Award, Sigma Xi; Certificate of Recognition, NASA
1985	Fisher Award, American Association of Microbiologists
1987	Honorary D.Sc., Hood College; Honorary D.Sc., Heriot-Watt University, Edinburgh, Scotland
1987–91	Director, Center for Marine Biotechnology, Maryland
1988	Alice Evans Award, American Society of Microbiology, Committee of the Status of Women; Phi Kappa Phi Scholar of the Year
1989–	Chair, Board of Governors, American Academy of Microbiologists
1990–	Vice Chair, Polar Research Board; President, Sigma Xi
1990	Civic Award, Government of Maryland; Gold Medal, International Biotechnology Institute
1991	Purkinje Gold Medal for Achievement in Sciences, Czechoslovakian Academy of Sciences
1993	Honorary D.Sc., Purdue University

Dr. Rita Colwell is an internationally recognized pioneer in marine biotechnology as well as in the use of computers to identify bacteria. Through her research, the promise of harvesting ocean resources for the good of both society and the environment comes closer to fruition.

Rita Barbara Rossi was born in 1934 in Beverly, Massachusetts, the daughter of Louis and Louise (DiPalma) Rossi. Her family was strongly oriented toward education and very supportive of her research interests.[1] Her interest originally was chemistry, despite the fact that she was told in high school that it was not a field for women. She enrolled at Purdue University as a chemistry major but, finding the classes large and poorly taught, switched to bacteriology. She received her undergraduate degree with distinction in bacteriology in 1956. She served as a research assistant in the genetic lab while pursuing her degree. She was accepted into medical school, but she chose to continue her education at Purdue and earned her master's degree in genetics in 1958. From Purdue, Colwell moved to the University of Washington with her husband, Jack Colwell, to pursue her doctoral degree in genetics, which was awarded to her in 1961. She worked as a research assistant at the University of Washington and later, from 1959 to 1960, as a predoctoral associate. From 1961 to 1964, Colwell was an associate professor at the university.

After several years at the University of Washington, Colwell accepted a teaching and research position at Georgetown University as assistant professor of biology (1961–1964) and later as associate professor of biology (1966–1972). She joined the staff of the University of Maryland, College Park, in 1972 as professor of microbiology, a position in which she continues to serve.

Colwell has continued her career at the University of Maryland. She became interested in the field of marine biotechnology, a very new field at the time, while doing her graduate degree. Her thesis advisor and mentor was Dr. John Liston, whose work concentrated on marine technology and world food problems. Colwell's primary research interests are marine biotechnology; marine and estuarine microbial ecology; survival of pathogens in the aquatic environment; ecology of *Vibrio chlorae* and related organisms; microbial systematics; marine microbiology; antibiotic resistance; antibiotic resistance indexing of *E. coli* to identify sources of contamination in water; and certain environmental aspects of *Vibrio chlorae* in the transmission of cholera. Colwell most enjoys the research work of microbiology, looking at bacteria under a microscope and "decoding" them. Because marine biotechnology was such a new field when Colwell entered it, she feels she had an advantage in being able to pursue her own interests in her own way. Colwell is also a pioneering scientist in the use of computers to identify bacteria.[2]

Colwell's achievements in her field are evident in the positions she has attained. Between 1977 and 1983 she was the director of the University of Maryland Sea Grants Program, and from 1987 to 1991 she was the director of the Center for Marine Biotechnology in Maryland. Since 1987, Colwell has served as director of the Maryland Biotechnology Institute. She has also been a member of a Coastal Resources Advisory Board, advising the Department of Natural Resources of Maryland. In these positions she has played an important leadership role in encouraging future investment and study in the field of marine biotechnology, a field she helped pioneer, on a global basis. She writes:

> The potential of marine biotechnology is clear and investment in this field is being made by a number of countries. Unfortunately, the major oceanographic institutions in the U.S. have not moved dramatically in the field, probably because marine biology at oceanographic institutions and biological stations tends to be classical and very traditional, the inertia against change being difficult to overcome.[3]

Through her writings, it is evident that Colwell has a deep concern for the environment and a strong commitment to using ocean resources in a positive way. She sees the future of marine biotechnology in new drugs made from marine sources, new methods of cost-effective fish culture, seaweed genetics, and improved biotechnological waste recycling. Colwell is a strong advocate of the use of genetic engineering in the marine sciences, particularly in the area of marine pharmaceuticals. "The greatest opportunity of all is represented by the application of genetic engineering to the marine sciences to pursue the untapped gene pool representing transport systems for minerals, metal concentration, novel photosynthetic systems, and marine pheromones by marine animals."[4] Colwell also holds two patents for marine biotechnology.

In addition to her research, Colwell has worked on numerous committees and is an active member of many organizations. Most recently she has served as president of Sigma Xi; as vice president of the International Union of Microbiological Sciences (1986–1990); and as chair of the Board of Governors of the American Academy of Microbiologists. She has also been president of the American Society for Microbiology (1984–1985).

Colwell has been active at the University of Maryland, serving as vice president for academic affairs from 1983 to 1987 and since 1964 sitting on a committee as external examiner of various universities abroad. She has also served as a consultant to the Bureau of Higher Education. In addition, she has been a member of the editorial staff of several prominent journals in her field, including the *Journal for Aquatic Living Resources, Microbiological Ecology, Johns Hopkins University Oceanographic Series,* and the *Journal of the Washington Academy of Sciences.*

Her extensive and productive career has earned her national and international recognition in the form of fellowships, honorary degrees, and prestigious awards. She received the Gold Medal of the International Biotechnology Institute in 1990 and the Purkinje Gold Medal Achievement Award from the Czechoslovakian Academy of Sciences in 1991. She received the Fisher Award from the American Association of Microbiologists in 1985; the Sigma Xi Annual Achievement Award in 1981; and, for her environmental work, the Alpha Chi Sigma Award for Ecology. Colwell is most proud of her honorary degree from Purdue University, the degree of doctor of science, which was awarded to her in 1993.[5] Her dedication as an educator and role model is evident: she received the Outstanding Woman on Campus Award from the University of Maryland in 1979 and the Civic Award from the government of the State of Maryland in 1990 for her contributions to education and the environment of the state.

Dr. Colwell currently resides in Maryland. She is married and has two daughters, in whose success she takes a great deal of pride.

Notes

1. Rita R. Colwell and Stefanie Buck, telephone conversation, October 14, 1994.
2. *Ibid.*
3. Rita Colwell, "Marine Biotechnology in the Marine Sciences," *Sea Technology* (Jan. 1988): 29.

4. Rita Colwell, "Biotechnology in the Marine Sciences," *Science* 22 (Oct. 7, 1983): 23.

5. Colwell and Buck, telephone conversation.

Bibliography

American Men and Women of Science, 18th ed., pp. 353–54. New York: Bowker, 1989.

Colwell, Rita. "Biodiversity—An International Challenge." *FASEB Journal* 7 (Sept. 1, 1993): 1107.

———. "Marine Biotechnology—The Future Is Here; Time for Action Is Now." *Sea Technology* 34 (1993): 15ff.

———. *Marine Biotechnology and the Developing Countries.* Vienna, Austria: United Nations Industrial Development Organization, 1986.

Colwell, Rita, and D. Palmer. "Back to the Future with UNESCO." *Science* 265 (Aug. 19, 1994): 1047.

Who's Who in Technology, 6th ed., p. 270. Woodbridge, Conn.: Research Publications, 1986.

Who's Who of American Women, 17th ed. Chicago: Marquis Who's Who, 1992.

STEFANIE BUCK

ELIZABETH CAROLINE CROSBY
(1888–1983)
Anatomist

Birth	October 25, 1888
1910	B.S., mathematics, Adrian College, Adrian, MI
1915	M.S., anatomy, University of Chicago; Ph.D., neuroanatomy, *magna cum laude,* University of Chicago
1918–20	Superintendent of Schools, Petersburg, MI
1920–83	Faculty appointments in Anatomy, University of Michigan Medical School: Instructor, 1920–26; Assistant Professor, 1926–29; Associate Professor, 1929–36; Professor of Anatomy and first woman professor in the Medical School, 1936–58; Professor Emerita, 1958–1983
1946	First woman awarded the Henry Russell Lectureship
1950	Achievement Award, American Association of University Women

Elizabeth Caroline Crosby. Photo courtesy of the Elizabeth Crosby Collection, Bentley Historical Library, University of Michigan.

1972	Henry Gray Award in Neuroanatomy, American Association of Anatomists
1979	Awarded the National Medal of Science for 1979 in January 1980 by President Carter
1981	National Medical Women's Association Award
Death	July 28, 1983

Elizabeth Caroline Crosby worked at what she loved until the end of her life in 1983 at age 94. During the dedication of the University of Michigan neurosurgery research laboratories in her name in 1982, Dr. Julian R. Hoff, professor and head of the Section of Neurosurgery, noted, "She is still one of the most productive scientists at the University."[1] She was not only a world traveler but also a National Medal of Science winner and a renowned teacher and authority in comparative neuroanatomy.

Elizabeth Caroline was born a preemie in 1888 to Lewis Frederick and

Frances Kreps Crosby in Petersburg, Michigan. She seemed to have an ordinary childhood—reading, taking piano lessons, playing basketball, working puzzles, and liking geometry. Her father gave her a high school graduation gift of four years of college. She finished her undergraduate degree in mathematics from Adrian College in only three years because she wanted to continue schooling following the encouragement of Professor Elmer Jones. He had studied with Professor C. Judson Herrick of the University of Chicago, and Elizabeth so impressed Herrick that he let her take a basic medical school anatomy course and his neuroanatomy course at the same time. She passed them both and then did her thesis work under him on the forebrain of the alligator. She became his most famous student. Herrick worried about Elizabeth's health and took away her key to the laboratory, but, undaunted, she merely carried the microscope, books, and slides back to her living quarters.[2] When Crosby completed her Ph.D. degree in 1915, she returned to Petersburg to help care for her mother and act as teacher, principal, and superintendent in the local high school. After her mother died in 1918, she sought and obtained a junior instructorship in anatomy in 1920 at the University of Michigan.[3]

Elizabeth remarked once in an interview that she had a long instructorship in anatomy and was surprised each time she was promoted. In response to questions about discrimination, she said that she had never felt it; yet she noted that she was paid less and did as much teaching as others but "less work" (i.e., research).[4] Her extensive bibliography of articles and books, co-authored primarily with her graduate students and colleagues and written from 1918 until 1982, seems to suggest otherwise. Her writing was succinct and scientific, but she could also interject her feelings—as in the following statement: "The thoughtless statement that research on the nervous system began in the late sixties (or thereabout) is on a par with the statement that American history began with the termination of World War II."[5]

Dr. Crosby is said to have taught neuroanatomy to 8,500 medical students between 1920 and 1958. Her students often presented her with the ritual gift of a box of roses to "the angel of the medical school."[6] Her heavy schedule usually listed full mornings Monday through Saturday for MicroAnat and 16 hours of appointments with graduate students in the afternoons and evenings. But her compensation did not reflect her work load. Over her 38 years as a regular faculty member, she trained 39 students for the Ph.D. and at least 30 more postgraduate students in her laboratory. Fifteen of her former students became department heads at the University of Michigan. After her retirement, she worked as a consultant paid by various departmental accounts.

During her career at Michigan she had at least two offers to move. But her department chair and others apparently prevailed each time, because she always stayed in Michigan or came back. (In 1937 she offered to resign over the promotion of another faculty member that appeared to signal her demotion. Later she praised the department chair for raising the man's salary and promoting him.) In 1938 she had an offer from Marischal College at the University of Aberdeen, Scotland; she spent 1939–1940 on unpaid leave from Michigan guiding the staff at Marischal as they set up teaching and research in neuroanatomy. There was a great deal of letter writing while she was on leave in Scotland that promoted her hopeful return. She asked several times in letters whether the staff at Michigan really wanted her to come back.

Crosby commented in a 1980 oral interview that she usually stayed in Ann Arbor instead of attending many meetings, because she just wanted to work.[7] But Professor Muriel Ross, a former student, stated that Professor G. Carl Huber, Crosby's first mentor at the University of Michigan, had let her know he did not think it "quite proper" for a woman to speak before a group of people.[8] Long after his death in 1934, Crosby gave her first public lecture as the recipient of the Henry Russell Lectureship in 1946 and then did more lecturing all over the world. Whenever she spoke as a lecturer or as an instructor, she never used notes.

Her many letters reveal a very thoughtful person. In one letter, dated April 30, 1969, to the chair of the Petersburg Centennial, she thanked him for inviting her to speak but apologized that she was not very good when she had to speak about something besides anatomy. A letter from 1961, which she obviously typed and made written corrections on, described for a doctor certain nerve pathways that might explain his patient's symptoms. She thanked the doctor for sending the information, because she was interested. On the basis of materials collected for Crosby's 90th birthday celebration, Professor Ross has noted that she always took a very personal interest—this seems to be evident in all of Crosby's correspondence with colleagues, former students, and other professionals.

Despite her great commitment to her profession, Dr. Crosby did have outside interests. These included literature, detective stories, history, and poetry. She retained a childhood love of puzzles. Her travel included excursions, and she enjoyed attending plays.[9]

But science was her life. One paper on the mammalian midbrain, coauthored with Professor Russell T. Woodburne, included many comparisons of various organisms. "The two parts of the Edinger-Westphal nucleus form a continuous band in all forms except the rodents and man. . . . The rostral part is bilateral in the opossum, the shrew, the cat and the dog, the monkey and man, and is an unpaired median mass in the armadillo, various rodents, the pig and the mink."[10] In the early years, Crosby's clinical and experimental approach to neuroanatomy was controversial. In 1949 she started to help lead medical students on clinical rounds. In 1962 she co-authored a textbook on neuroanatomy that was directed to medical students and graduate students in anatomy.

A more clinical paper describing a case of a former woman astronaut candidate illustrates Crosby's later consultative work. There is a detailed description of the case, clear views of the brain vasculature, and a discussion of possible tests to discover whether astronaut candidates might have potentially disastrous brain anomalies. The book that Crosby worked on during her later years, *Comparative Correlative Neuroanatomy of the Vertebrate Telencephalon*, was directed to biologists, behavioral scientists, and medical scientists.

Crosby was a member of numerous editorial boards, including the *Journal of Comparative Neurology*. As a member of the Women's Research Club at the University of Michigan, she gave three research presentations between 1945 and 1962 and was one of eight distinguished members honored at the 75th anniversary in 1978. She was also the recipient of numerous honorary degrees from such institutions as Marquette University (1957); Denison University (1959); Smith College (1967); Woman's Medical College (1967); the University of Michigan (1970); and Wayne State University (1958).

When she had to retire in 1958, her sixty ninth year, Dr. Crosby was not happy. Guests invited to her retirement party were asked not to discuss it in their remarks. A memorial volume of the *Journal of Comparative Neurology*, known as *The Crosby Volume*, paid for by her former students and research associates, included 13 research papers by students whom she had trained. As a professor emerita, she held concurrent consulting positions in neurosurgery at the University of Michigan and at the University of Alabama, where her former student, Tryphena Humphrey, was a professor of anatomy. Amazingly, Crosby commuted to be at each location for half of each month.

Dr. Crosby was a small person who appeared frail. She had some leg and hip problems as she aged. However, she outlived some of her earlier students and was productive very late in her life. Although she did not marry, in 1940 she adopted an 11-year-old girl, Kathleen, from Scotland. Another young girl, Susan McCotter, came to live with them in 1944. Crosby cared for and educated both girls. At the time of her death, Crosby lived with her daughter and had five grandchildren and one great-granddaughter. In keeping with her modest life style, she wanted memorials to be made to neurosurgery at the University of Michigan.

Elizabeth Caroline Crosby really had three careers: high school teacher, University of Michigan faculty member, and clinical consultant.[11] She excelled in all three. "Many brilliant investigators are regarded as authorities on their particular segment of the brain or spinal cord, but no one in the world rivals Doctor Crosby's knowledge of the entire nervous system of animals throughout the vertebrate phylum."[12]

Notes

1. William B. Treml, "Dr. Crosby Was 'Simply a Genius,'" *Ann Arbor News* (July 29, 1983): A2.
2. See C. Judson Herrick, "Elizabeth C. Crosby," *Journal of Comparative Neurology* 112 (1959): 13–17.
3. See Russell T. Woodburne, "Elizabeth C. Crosby; A Biographical Sketch," *Journal of Comparative Neurology* 112 (1959): 19–29.
4. Lee Katterman, "Interview with Elizabeth Caroline Crosby," January 23, 1980. Tape in Box 1, Michigan Historical Collections, Bentley Historical Library, University of Michigan.
5. Elizabeth C. Crosby and H.N. Schnitzlein, *Comparative Correlative Neuroanatomy of the Vertebrate Telecephalon* (New York: MacMillan, 1982), p. 795.
6. "Roses for Anatomy," *Newsweek* 46, no. 2 (1955): 54.
7. Katterman, "Interview."
8. Leslie Lin, "Elizabeth Crosby: Laying the Foundations of Neuroscience," *Research News* 34 (1983): 10.
9. See E. Carl Sensenig and Tryphena Humphrey, "Elizabeth C. Crosby." *Alabama Journal of Medical Sciences* 6 (1969): 357–63.
10. Elizabeth C. Crosby and Russell T. Woodburne, "The Mammalian Midbrain and Isthus Region, Part 1, The Nuclear Pattern," *Journal of Comparative Neurology* 78 (1943): 507.
11. Michigan Historical Collections, Bentley Historical Library, University of Michigan, Boxes 1–5. Original papers, letters, citations, pictures, employment records, curriculum vitae, and more. All references to and the original materials were found here; the author is most grateful for their crucial support.
12. Woodburne, "Elizabeth C. Crosby," p. 28.

Bibliography

Burns, Scott M. "Elizabeth C. Crosby: A Biography." *Alabama Journal of Medical Sciences* 22 (1985): 317–23.
Herrick, C. Judson. "Elizabeth C. Crosby." *Journal of Comparative Neurology* 112 (1959): 13–17.
Lin, Leslie. "Elizabeth Crosby: Laying the Foundations of Neuroscience." *Research News* 34 (1983): 3–21.
Rosenthal, Marilynn M., and Ruth M. Brend. *Ninety Years with the Women's Research Club, University of Michigan: A Retrospective.* Ann Arbor: Women's Research Club of the University of Michigan, 1992.
"Roses for Anatomy." *Newsweek* 46, no. 2 (1955): 54.
Rossiter, Margaret W. *Women Scientists in America: Struggles and Strategies to 1940.* Baltimore: Johns Hopkins University Press, 1982.
Sensenig, E. Carl, and Tryphena Humphrey. "Elizabeth C. Crosby." *Alabama Journal of Medical Sciences* 6 (1969): 357–63.
Treml, William B. "Dr. Crosby Was 'Simply a Genius.'" *Ann Arbor News* (July 29, 1983): A2.

Woodburne, Russell T. "Elizabeth C. Crosby: A Biographical Sketch." *Journal of Comparative Neurology* 112 (1959): 19–29.

JUDITH E. HEADY

GLADYS ROWENA HENRY DICK
(1881–1963)
Physician, Biomedical Researcher

Birth	December 18, 1881
1900	B.S., University of Nebraska
1907	M.D., Johns Hopkins University School of Medicine
1911	Moved to Chicago; began study of scarlet fever
1914	Married George Dick on January 28; became a member of the John R. McCormick Memorial Institute for Infectious Diseases in Chicago
1918	Founded the Cradle Society in Evanston, IL—the first professional child adoption agency in the United States
1923	Proved (with her husband) that hemolytic streptococci cause scarlet fever
1925	Nominated (with her husband) for the Nobel Prize in medicine for the Dick test for susceptibility to scarlet fever
1926	Received (with her husband) the Mickle Prize, University of Toronto
1933	Received (with her husband) the Cameron Prize, University of Edinburgh
1953	Retired to Palo Alto, CA
Death	August 21, 1963

The die for Gladys Dick's life was cast when her mother took her, at age 5, to see a neighbor's sick child. While young Gladys watched, the child had a fit, a sight that fascinated her and awakened a desire to become a physician.[1] She eventually realized her ambition as a leading researcher (with her husband) of scarlet fever, succeeding at a time when the deck was stacked against women.

Gladys Dick, M.D., was born Gladys Rowena Henry in 1881 in Pawnee

City, Nebraska. She was the youngest of three children and the second girl in the family. Her mother was Azelia Henrietta (Edson) Henry. Her father, William Chester Henry, had served as a cavalry officer during the Civil War. At the time of Gladys's birth he was a banker and grain dealer who also raised carriage horses.[2]

After Gladys was born, the family moved to Lincoln, Nebraska. Here she attended the public schools, eventually earning her baccalaureate degree from the University of Nebraska in 1900. As an adult, Gladys still held her ambition to become a physician but faced opposition from her mother as well as society at large. Very few medical schools at the turn of the century admitted women as students. She spent two years teaching high school biology in Carney, Nebraska, and taking graduate courses at the University of Nebraska.[3]

Finally, in 1903, Gladys gained admission to Johns Hopkins University School of Medicine in Baltimore, Maryland. Once there, however, she faced another obstacle. Johns Hopkins admitted women as students, but it did not provide housing for them. Gladys provided her own housing by organizing with other women students to buy a house. In 1907 she completed her M.D. degree and remained at Johns Hopkins for her internship. This internship, combined with a year of postgraduate training in Berlin, introduced her to biomedical research, the field in which she eventually made her greatest contribution.[4]

In 1911 Gladys moved to Chicago. There, while in charge of the Children's Memorial Hospital laboratory, she began the study of scarlet fever that ultimately became her major contribution to medicine. This work exposed her to the disease so much that she eventually contracted it herself.[5] It was also during this time that she met another doctor, George Frederick Dick, who shared her interests. They married in January 1914, and her life and work became intertwined with his.

In the early part of the twentieth century, scarlet fever was greatly feared. It was a common childhood disease that killed up to 30 percent of the children under age 5 who caught it, and it could also be fatal to older children and adults. Its complications, for those who survived, included deafness, pneumonia, and heart or kidney disease. Others had worked in the past to understand, treat, or prevent scarlet fever, but without much success. It was not known what bacterium caused the disease, although a type of streptococcus was suspected to be responsible.

The Dicks took on the task of determining with certainty the cause of scarlet fever and then working to find a vaccine and a cure. Later, in talking about their research, they said:

> We decided that scarlet fever offered the most hopeful field. The conquest of diphtheria through antitoxin had already been achieved. The vaccine for typhoid was being worked out. But little

had been accomplished with scarlet fever although a great amount of work had been done by many investigators.[6]

In 1914, both working at the John R. McCormick Memorial Institute for Infectious Diseases in Chicago, Gladys and George Dick set out to identify exactly which organism was responsible for scarlet fever according to Koch's postulates. Koch's postulates say that to identify the organism causing a disease, the researcher must (1) show that the organism is always present in patients with the disease; (2) isolate the organism and use it to produce the disease in another, previously healthy patient; and (3) then re-isolate the organism from the newly infected patient.[7]

The Dicks were faced with several problems. First, although hemolytic streptococci—the organisms they suspected—were nearly always found in patients with scarlet fever, it was not certain whether they were the cause of the disease or merely opportunistic invaders. Second, there were different strains of hemolytic streptococci, not all of which might be responsible for the disease. Third, no one had been able to produce a case of scarlet fever in animals, so human volunteers might have to be used. Fourth, even in humans, exposure to scarlet fever produced the disease less than half the time.[8] Finally, even if hemolytic streptococci turned out to be the culprit in the disease, no one knew what the bacterium's mechanism of action might be.

The Dicks began working with animals, trying to produce the disease with cultures taken from scarlet fever patients. When they had no more success than their predecessors did, they turned to experimenting on human volunteers, many of whom were their friends. Now the Dicks felt they were getting close. They produced scarlet fever in a human volunteer with one of their cultures, and the only question was whether the culture had indeed been pure. They planned a series of controlled experiments that would remove this doubt. Unfortunately, World War I and an influenza epidemic intervened. George went overseas to serve in a medical unit, and Gladys carried on alone until she became ill with influenza. When she was well enough to go back to the laboratory, she found that the precious cultures that had been so promising had dried up during her absence.[9] The work would have to begin again.

After the war the Dicks went back to work with a fresh supply of volunteers, many of whom had lost family members to the disease. They also included themselves in the trials. "In all these experiments," they said, "we first inoculated ourselves with whatever variety of material was to be used in inoculating the volunteers. Both of us had had scarlet fever and the preliminary inoculations of ourselves were done to determine whether or not the material to be used had any pathogenic action other than the production of scarlet fever."[10]

Finally, in 1923, they had a stroke of luck in the form of a nurse who contracted scarlet fever from one of her patients. The Dicks took cultures from a lesion on her finger, isolated the hemolytic streptococcus, and used the bacterium to produce scarlet fever in two volunteers.[11]

Having proved that hemolytic streptococci were responsible for the disease, the Dicks went on to investigate the mechanism by which they did so. They found that the bacteria produced a toxin that, by itself, could produce the symptoms of scarlet fever. This toxin was more useful to study than the bacteria themselves because it eliminated the problem of dealing with different strains of hemolytic streptococci. To determine if a particular strain was harmful, it was only necessary to test it for production of the toxin.[12]

In 1924 the Dicks used the toxin to develop a skin test for susceptibility to scarlet fever. This test, similar to the Schick test for diphtheria, came to be known as the Dick test and is generally thought to be the Dicks' most important contribution. Gladys and George went on, though, to develop ways to use the toxin as a scarlet fever vaccine and to produce an antitoxin to be used in the treatment of those who already had the disease.

Others were working on the problem of scarlet fever when the Dicks were, and others also manufactured the scarlet fever toxin. The Dicks, concerned about the quality of some of these preparations, filed for patents on their own methods; the patents were to be controlled by a nonprofit Scarlet Fever Committee, which the Dicks formed.[13] This action sparked a series of letters and editorials for and against the Dicks in U.S. and British medical journals, with some writers accusing the researchers of commercialism.

The Dicks' research made them reluctant celebrities, as articles about them appeared in the popular press. They were nominated for the Nobel Prize in medicine in 1925. In 1926 they were awarded the University of Toronto's Mickle Prize, and in 1933 they received the Cameron Prize from the University of Edinburgh for valuable additions to the field of practical therapeutics.

Gladys Dick had other interests beyond scarlet fever. She helped found the first professional adoption society in the United States, the Cradle Society of Evanston, Illinois. She worked with the society from 1918 until her retirement in 1953, adopting two children of her own at age 49. Her later research was on polio, another disease that threatened primarily children but whose effects might last a lifetime.

Gladys Dick and her husband retired to Palo Alto, California, in 1953. She died of cerebral arteriosclerosis on August 21, 1963. Although the height of her fame had been four decades in the past, her obituary nevertheless appeared in the *Chicago Tribune*. During her life she overcame both social and scientific obstacles to achieve her dream of helping children like the one she had seen when she herself was only 5 years old.

Notes

1. Ernest Gruening, "The Conquest of Scarlet Fever," *Harper's Monthly Magazine* 150 (Dec. 1924): 107.
2. Lewis P. Rubin, "Gladys Rowena Henry Dick," in *Notable American Women: The Modern Period*, eds. Barbara Sicherman and Carol Hurd Green (Cambridge, Mass.: Harvard University Press, 1980), p. 191.
3. *Ibid.*
4. *Ibid.*
5. Gruening, "The Conquest of Scarlet Fever," p. 108.
6. Ernest Gruening, "Another Germ Bites the Dust!" *Collier's* 74 (Oct. 4, 1924): 26.
7. Ives Hendrick, "Conquering Scarlet Fever," *World's Work* 50 (May 1925): 33.
8. George F. and Gladys H. Dick, "Scarlet Fever," *American Journal of Public Health* 14 (1924): 1022–23.
9. Gruening, "The Conquest of Scarlet Fever," pp. 108–10.
10. *Ibid.*, p. 110.
11. George F. and Gladys H. Dick, "Experimental Scarlet Fever," *Journal of the American Medical Association* 81 (Oct. 6, 1923): 1166–67.
12. Dick and Dick, "Scarlet Fever," pp. 1023–24.
13. George F. and Gladys H. Dick, "The Patents in Scarlet Fever Toxin and Antitoxin," *Journal of the American Medical Association* 88 (Apr. 23, 1927): 1341–42.

Bibliography

Baker, S. Josephine, M.D. "Conquering the Contagious Diseases." *Ladies' Home Journal* 43 (Feb. 1926): 41, 204.
Dick, George F., M.D., and Gladys H. Dick, M.D. "The Patents in Scarlet Fever Toxin and Antitoxin." *Journal of the American Medical Association* 88 (Apr. 23, 1927): 1341–42.
———. "Scarlet Fever." *American Journal of Public Health* 14 (1924): 1022–28.
———. "A Scarlet Fever Antitoxin." *Journal of the American Medical Association* 82 (Apr. 19, 1924): 1246–47.
———. "A Skin Test for Susceptibility to Scarlet Fever." *Journal of the American Medical Association* 82 (Jan. 26, 1924): 265–66.
"Ethics and Patents." *American Journal of Public Health* 16 (1926): 919–20.
Gruening, Ernest. "Another Germ Bites the Dust!" *Collier's* 74 (Oct. 4, 1924): 26, 30.
———. "The Conquest of Scarlet Fever." *Harper's Monthly Magazine* 150 (Dec. 1924): 107–14.
Henrick, Ives. "Conquering Scarlet Fever." *World's Work* 50 (May 1925): 30–34.
Obituary. *Chicago Tribune* 116 (August 23, 1963): 1A, p.2.
Obituary. *Journal of the American Medical Association* 186 (Dec. 28, 1963): 120.
Rossiter, Margaret W. *Women Scientists in America*. Baltimore: Johns Hopkins University Press, 1982.
Rubin, Lewis P. "Gladys Rowena Henry Dick." In *Notable American Women: The*

Modern Period, eds. Barbara Sicherman and Carol Hurd Green. Cambridge, Mass.: Harvard University Press, 1980.

Stanley, Autumn. *Mothers and Daughters of Invention: Notes for a Revised History of Technology.* Metuchen, N.J.: Scarecrow Press, 1993.

MARILYN R.P. MORGAN

ALICE EASTWOOD

(1859–1953)

Botanist

Birth	January 19, 1859
1879–90	Teacher, East Denver High School
1892–1949	Curator of Botany, California Academy of Sciences
1950	Honorary President, VIIth International Botanical Congress
Death	October 29, 1953

A grove of giant redwoods is named after Alice Eastwood, as well as a rare California shrub, *Eastwoodia elegans.* One writer recently gave her the title "Grand Old Botanist of the Academy."[1] A lifetime of hard work, dedication, and love for learning and life characterize this American pioneer botanist.

Alice's childhood was a difficult one, as evidenced by her solemn promise to her dying mother, delivered at age 6, to take care of her younger sister and brother. However, immediately after the loss of their mother, Alice's father left all three children with relatives. Alice stayed with her uncle, Dr. William Eastwood, who was an experimental horticulturist. In addition to teaching her the Latin plant names and giving her books on plants, he shared with Alice his passion for botany. Throughout her entire adult life, Alice shared her consuming interest with many others internationally. When speaking of her vision for the herbarium, she said, "Looking forward to the future greatness of San Francisco, I wanted this new herbarium to be a great one, founded on a broad basis, a herbarium containing not only plants of North America, but of the whole world."[2]

Alice was elected valedictorian of her high school class after years of overcoming difficult situations. The Eastwood children were shifted several times between living with their father when he sent for them and other locations of his choice. Alice's early botanical interest was nurtured by a French priest-gardener in the convent where she and her sister lived between stays with their father. This was also where she learned music from a French Canadian nun, whose love for and knowledge of music permanently influenced the young girl.

When Alice was 14 she was summoned to her father's home in Denver, Colorado, where she worked to help with family expenses. With the convictions of hard work and dedication that came to epitomize her life, Alice was a nursemaid for a 2-year-old and an infant in the Scherrer family. Association with the Scherrers provided her with access to their comprehensive multilingual library as well as camping trips in the Colorado mountains.

Anna Palmer, a teacher and musician, helped Alice with her studies and enabled her to enter high school. However, she had to go to work again when her father's store lost money. Once again Alice Eastwood's hard-work ethic prevailed as she managed the furnace fires in the school building, rising at 4 o'clock every morning. Every afternoon from 2 to 6 o'clock, and all day Saturday, Alice worked as a cutter's assistant in the ready-made dress department of a downtown store. Managing to pay all her own expenses, she re-entered East Denver High School as a senior and was an early graduate as well as class valedictorian in 1879.

During her first summer after graduation, Alice learned to ride horseback so she could pursue her collections of botanical specimens in the most inaccessible places. She even designed her own outfits: buttoned denim skirts with bustles, which were heavy cotton nightgowns. Alice traveled light, expending her energies on gathering specimens and carrying them with her in heavy wooden plant presses.

Alice soon began her teaching career, scraping together enough money to spend her summers collecting and exploring. The 1880s were frontier times. Alone on foot or horseback, Alice would travel in the High Rockies traversing great distances beyond the stage stops. "Never in all my experiences have I had the slightest discourtesy and I have never had any fear. I believe that fear brings danger."[3] She never carried a gun because she felt it was a sign of fear, although the cowboys, miners, and ranchers she met while collecting probably did.

By age 31, thanks to investments in real estate, Alice quit teaching and dedicated herself full-time to studying botany. She had already served as a guide for Alfred Russel Wallace, the famous British naturalist; she had established her reputation as a botanist in Denver; and she had organized the only herbarium in the state of Colorado.

Free to visit the California Academy of Sciences in San Francisco where T.S. and Kate Brandegee (botanists with the Academy and authors of their own journal) did research, Alice was asked to contribute articles.

In the October 1890 issue of *Zoe*, the Brandegees' journal, appeared one of Alice Eastwood's essays, "Mariposa Lilies of Colorado." The following excerpt from that essay illustrates her passionate writing and love for Colorado and botany:

> So distinct, so individual are those blossoms that they seem to have souls. They speak a wonderfully enticing language to draw the wandering insects to their honeyed depths. . . . The bands of color on both divisions of the perianth are bewildering, impossible to describe; but more than aught else, they cause each flower to say proudly, with uplifted head, "I am myself; there is no other like me."[4]

Alice wrote this essay in San Francisco while on temporary leave to help organize the Academy herbarium. In 1892 she returned to Colorado, where, with dear friend Al Wetherill (Mesa Verde was on the Wetherill property), she took trips into the desert country. Traveling with her companion, Alice was the first botanist to penetrate one remote area of the Great American Desert and collect new specimens, some of which she named. The hardships that she encountered without complaint over her years of collecting were difficult for many of her male companions, but this pioneer woman viewed them as adventures and recounted her trips with fervor and enjoyment.

In the summer of 1892, Kate Brandegee offered Alice Eastwood her entire salary of $75 a month for acceptance of the position of joint curator of the California Academy. However, one of Alice's companions was a journalist from the East who was living in Colorado owing to poor health. Theirs was one of the few romances in Alice's life, and it almost kept her from going to San Francisco. Once again, though, the death of someone dear to her acted as a major turning point in her life. She accepted the Academy position after the death of the young man; and one year later, after the Brandegees moved, Alice became the Academy's curator of botany, a position she held for the next 55 years.

Alice said, "My desire is to help, not to shine," and this she did throughout her life; but in 1906 she did both during the earthquake in San Francisco.[5] Because of her habit of keeping irreplaceable species types in their own special unit, Alice saved over 1,000 of these most-valued plant specimens as she entered the shattered Academy building. Climbing the broken marble staircase as fires roared in the buildings nearby, she and a friend grasped the iron railing and used the rungs as a ladder to reach the priceless pieces. They lowered them by improvised rope, and Alice stayed with them until they were removed to her home. However, when fire threatened there also, she carried the collection by hand to a safer place. In addition to the collection, books from other departments and unbroken records dating back to the first meeting of the Academy in 1853 were saved. Alice's determination and presence of mind rendered a tremendous service, but at personal cost. "Not a book was I able to save, nor a single thing of my own, except my favorite lens, without which I should feel helpless."[6]

Alice Eastwood retired from the California Academy of Sciences in 1949 at age 90. She remained at the Academy for 57 years, and to this day her name is known and spoken with reverence by those in close association with the Academy and knowledgeable in botany. She was active until her death in San Francisco in 1953 at age 94.

Notes

1. See Bonta, Marcia Myers, *Women in the Field: America's Pioneering Women Naturalists* (College Station: Texas A&M University Press, 1991).

2. Carol Green Wilson, "The Eastwood Era at the California Academy of Science," *Leaflets of Western Botany* (Aug. 28, 1953): 63.

3. Bonta, *Women in the Field: America's Pioneering Women Naturalists*, p. 94.

4. Carol Green Wilson, *Alice Eastwood's Wonderland: The Adventures of a Botanist* (San Francisco: California Academy of Sciences, 1955), p. 38.

5. Susanna Bryant Dakin, *The Perennial Adventure: A Tribute to Alice Eastwood, 1859–1953* (San Francisco: California Academy of Sciences, 1954), p. 2.

6. Leroy Abrams, "Alice Eastwood—Western Botanist," *Pacific Discovery* 2 (1949): 15.

Bibliography

Abrams, Leroy. "Alice Eastwood—Western Botanist." *Pacific Discovery* 2 (1949): 14–17.

"Alice Eastwood, Noted Botanist, 94." *New York Times* 103 (Oct. 31, 1953): 17.

Bonta, Marcia Myers. *Women in the Field: America's Pioneering Women Naturalists.* College Station: Texas A&M University Press, 1991.

Dakin, Susanna Bryant. *The Perennial Adventure: A Tribute to Alice Eastwood, 1859–1953.* San Francisco: California Academy of Sciences, 1954.

Hollingsworth, Buckner. *Her Garden Was Her Delight.* New York: Macmillan, 1962.

Howell, John T. "Alice Eastwood, 1859–1953." *Taxon* 3 (1954): 98–100.

———. "I Remember, When I Think . . ." *Leaflets of Western Botany* 7 (1954): 153–64.

———. "Memorials & Correspondence. Alice Eastwood, 1859–1953." *Sierra Club Bulletin* 39 (1954): 78–80.

Notable American Women, 1607–1950: A Biographical Dictionary. Cambridge, Mass.: Harvard University Press, 1974.

Reitter, Victor, Jr. "Horticulture and the California Academy of Science." *Leaflets of Western Botany* (Aug. 28, 1953): 79–84.

Sexton, Veronica J. "Books and Botany." *Leaflets of Western Botany* (Aug. 28, 1953): 85–88.

Valjean, Nelson. "Alice Eastwood, Hardy Perennial." *Nature Magazine* 42 (1949): 361–62.

Wilson, Carol Green. *Alice Eastwood's Wonderland: The Adventures of a Botanist.* San Francisco: California Academy of Sciences, 1955.

———. "The Eastwood Era at the California Academy of Science." *Leaflets of Western Botany* (Aug. 28, 1953): 58–64.

———. "A Partial Gazetteer and Chronology of Alice Eastwood's Botanical Explorations." *Leaflets of Western Botany* (Aug. 28, 1953): 65–68.

                                                                    CONNIE H. NOBLES

SOPHIA HENNION ECKERSON

(?–1954)

Microchemist, Botanist, Plant Physiologist

Birth	Date unknown
1905	A.B., Smith College
1905–1906	Fellow, Smith College
1907	A.M., Smith College
1908–1909	Assistant First Fellow and Demonstrator, Smith College
1911	Ph.D., University of Chicago
1911–15	Assistant Plant Physiologist, University of Chicago
1916–20	Instructor, University of Chicago
1919–22	Bureau of Plant Industry, USDA, Washington, DC
1921–22	Cereals Division, USDA, Washington, DC
1921–23	University of Wisconsin
1923–40	Plant Microchemist, Boyce Thompson Institute, NY
1935–36	Chair, Physiological Section, Botanical Society of America
1940	Retired
Death	July 19, 1954

At a time when the accomplishments of very few women in science were nationally acknowledged, plant physiologist Sophia Eckerson was recognized as one of the 1,000 best scientists in America when her name was starred in the sixth edition of *American Men of Science.*

Sophia Hennion Eckerson was born in Tappan, New Jersey, to Albert

Bogert and Ann Hennion Eckerson. The date of her birth is unknown. After helping to put her younger brothers through school, Eckerson began her career as a microchemist at Smith College. She received her baccalaureate degree in 1905 and her master's degree in 1907 from Smith.

At Smith, Eckerson studied with William Francis Ganong, who was both her teacher and mentor and probably the greatest influence in her academic life.[1] In fact, her earliest works are listed in the second edition of Ganong's *The Teaching Botanist*. From 1906 to 1908, Eckerson was a demonstrator at the college who became well known as an enthusiastic and inspirational instructor. Both of her degrees were specialized in physiology and microchemistry, but her publications show a wide area of interest covering plant germination, mineral nutrition, reduction of nitrates by plants, cell walls, endophytic fungi, and starch grains.[2]

While she was at Smith, Eckerson wrote *Outlines of Plant Microchemistry*, a work she declined (out of modesty and a desire for perfection) to have published, but which was heavily used in mimeographed form by students as a textbook. She wrote this work to teach the metabolic processes in plant physiology. Many of her special methods for "following the metabolic processes in plants by detection of the products through crystallization or by color reaction" found their way into other works of botany.[3]

Eckerson went on to the University of Chicago for her Ph.D. degree, which was awarded to her in 1911. She was apparently considered one of the outstanding graduate students there. After graduation, she accepted a position in the plant physiology department as an assistant plant physiologist at that institution, which was not known to have a liberal attitude toward women scientists.[4] Eckerson remained at the University of Chicago for five years, from 1911 to 1915. Because of her outstanding work in plant physiology and microchemistry, she was invited to the State College of Washington for a term in 1914 as a plant microchemist. In 1916 she was employed as a microchemist at the University of Chicago, a position she held until 1920.

In 1919 she joined the staff of the USDA Bureau of Plant Industry, an agency that employed many noted women scientists—particularly in the area of plant physiology. She remained in Washington, D.C., until 1922. She worked at the Bureau until 1921, then joined the USDA Cereals Division from 1921 to 1922. Eckerson was also employed as a physiologist/microchemist at the University of Wisconsin between 1921 and 1923.

Eckerson was a charter member of the Boyce Thompson Institute in New York when it was founded in 1924, and she remained on staff there until 1940. "Her selection and continued employment during the Depression years confirm her reputation as an outstanding scientist."[5] Her work was recognized with a star next to her name in the sixth edition of *American Men of Science*.

Eckerson was an active member of the scientific community through organizations such as Phi Beta Kappa and Sigma Xi. She also served as chair of the Physiological Section of the Botanical Society of America from 1935 to 1936, a unique honor because the position had rarely been held by a woman.

Sophia Eckerson was described by her colleague, Norma Pfeiffer, as "a quiet, reserved friend" and "a versatile person with a wide interest in letters and art as well as science." Pfeiffer noted further that "throughout her career, [Sophia] gave generously and enthusiastically of her time and experience to many in organizing and pursuing botanical problems as well as in the careful presentation of the finished work."[6] She died on July 19, 1954, after a week-long illness in Pleasant Valley, Connecticut.

Notes

1. Patricia Joan Siegel and Kay Thomas Finley, *Women in the Scientific Search: An American Bio-Bibliography, 1724–1979* (Metuchen, N.J.: Scarecrow Press, 1985), p. 74.

2. Norma E. Pfeiffer, "Sophia H. Eckerson, Plant Microchemist," *Science* 120 (July–Dec. 1954): 821.

3. *Ibid.*, p. 820.

4. *Ibid.*

5. Martha J. Bailey, *American Women in Science: A Biographical Dictionary* (Santa Barbara, Calif.: ABC-CLIO, 1994), pp. 96–97.

6. Pfeiffer, "Sophia H. Eckerson," pp. 820–21.

Bibliography

Bailey, Martha J. *American Women in Science: A Biographical Dictionary.* Santa Barbara, Calif.: ABC-CLIO, 1994.

Herzenberg, Caroline L. *Women Scientists from Antiquity to the Present.* West Cornwall, Conn.: Locust Hill Press, 1986.

Howes, Durward, ed. *American Women: The Standard Biographical Dictionary of Notable Women.* Teaneck, N.J.: Zephyrus Press, 1974.

Pfeiffer, Norma E. "Sophia H. Eckerson, Plant Microchemist." *Science* 120 (July–Dec. 1954): 820–21.

Rossiter, Margaret W. *Women Scientists in America.* Baltimore: Johns Hopkins University Press, 1982.

Siegel, Patricia Joan, and Kay Thomas Finley. *Women in the Scientific Search: An American Bio-Bibliography, 1724–1979.* Metuchen, N.J.: Scarecrow Press, 1985.

STEFANIE BUCK

Rosa Smith Eigenmann

(1858–1947)

Ichthyologist

Birth	October 7, 1858
1880	Summer study with Dr. D.S. Jordan
1887	Married Dr. Carl H. Eigenmann
1887–88	Student of Dr. W.G. Farlow at Harvard
1888–91	Independent research, San Diego, CA
1891–93	Research with husband at Indiana University
Death	January 12, 1947

Rosa Eigenmann's scientific research career was cut short by depression and the demands of children with birth defects, yet she is recognized as the first American woman to attain distinction in the field of ichthyology.

On October 7, 1858, in Monmouth, Illinois, Charles Kendall and Lucretia (Gray) Smith became the proud parents of a daughter, Rosa. Charles Kendall Smith was a printer and the clerk of the San Diego school board.

Rosa Smith attended business college in San Francisco. In 1880 she entered Indiana University, where she remained until 1882. During the summer of 1880, she studied fishes on the Pacific coast as a special student under the esteemed scientist Dr. David Starr Jordan. During the following summer, Rosa Smith and 33 other students traveled with Jordan to Europe. Her studies then brought her to Harvard University. At Harvard from 1887 to 1888, she was a special student in cryptogamic botany with Dr. W.G. Farlow.

On August 20, 1887, Rosa Smith married Dr. Carl H. Eigenmann, dean of the Graduate School of Indiana University and professor of zoology in San Diego, California. They eventually had five children: Lucretia Margaretha, Charlotte, Theodore, Adele, and Thora.

Rosa's career began as a reporter for the *San Diego Union,* a newspaper owned by her brother and brother-in-law. After marrying Professor Eigenmann, she devoted her time to conducting research with him. At Harvard's Museum of Comparative Zoology in 1887, they studied South American fishes. While at Harvard, they also co-authored a paper on South American nematognathi. They jointly wrote many papers on ich-

thyology, but this was their largest. In 1888 the Eigenmanns returned to
San Diego. There, in their own small biological facility, they conducted
independent research. Three years later they moved to Bloomington, Indiana, where Carl was appointed professor of zoology. For several years
Rosa and Carl Eigenmann continued to carry out research together. Rosa
discontinued research in 1893 in order to care for her children, but she
continued to assist her husband in editing his scholarly research. Many
years later, in 1926, she returned to San Diego. While dividing her time
between helping her husband with his research and the demands of family members, she somehow managed to find time for recreation. Rosa
enjoyed the delicate hobby of lace pillow making and the bold adventure
of mountain climbing.

Her scientific contributions to the field of ichthyology began in 1880
with her first publication. A paper she delivered to the members of the
San Diego Society of Natural History at a meeting in 1880 was well
received. From 1888 to 1893 she conducted joint research with her husband. They co-authored 15 papers during this period, examined classification, and explored questions of embryology and evolution. Their
most significant works were on the fishes of western North America and
South America. As an independent author, Rosa Smith Eigenmann published 20 scientific papers. Her major personal interest was the fishes of
the San Diego region.

During her scientific career she was a member of Sigma Xi (Indiana
Chapter); the first woman member of the San Diego (California) Society
of Natural Science; and a lifetime member of the California Academy of
Sciences (San Francisco). Writing in 1895, she noted that

> science is exacting, requiring the devotion of months and even of
> years to the completion of a series of observations which frequently
> must be carried on with little or no interruptions. Women are sel-
> dom in a position to do work in this way; therefore we oftener find
> that women popularize the results of students of science rather than
> add to a positive knowledge by studies and researches of their
> own.[1]

Certainly Rosa herself was not in a position to carry out uninterrupted
research. Nevertheless, she also believed that

> women eminent in science have received more praise than is their
> due. Comparatively speaking, so few women have entered this field
> of knowledge that when one does accomplish somewhat she is as
> loudly lauded as the precocious child; but in science, as everywhere
> else in the domain of thought, woman should be judged by the

same standard as her brother. Her work must not be simply well done for a woman.[2]

Rosa Smith Eigenmann died on January 12, 1947, in San Diego.

Notes

1. Rosa Smith Eigenmann, "Women in Science," *Proceedings of the National Science Club* 1 (Jan. 1895): 16.
2. *Ibid.*, p. 13.

Bibliography

Biographical Dictionary of American Science. Westport, Conn.: Greenwood Press, 1979.
Eigenmann, Rosa Smith. "Women in Science." *Proceedings of the National Science Club* 1 (Jan. 1895): 13–17.
Martin, Hemme N. "San Diego Woman One of the Earliest Discoverers of Pt. Loma Blindfish." *San Diego Union* (June 14, 1935).
Notable American Women, 1607–1950. Cambridge and London: Belknap Press, 1971.
"Noted Woman Expert on Fish Dies Here at 88." *San Diego Journal* (Jan. 13, 1947): 14.
Raridan, W.J. "First Leaders in Natural History Society Recall Unit's Early Days." *San Diego Union* (Jan. 16, 1945): 4.
Woman's Who's Who of America: A Biographical Dictionary of Contemporary Women of the United States and Canada, 1914–1915. New York: American Commonwealth Co., 1914.

MARGARET A. IRWIN

CORNELIA BONTÉ SHELDON AMOS ELGOOD
(1874–1960)
Physician

Birth	1874
1900	M.D., London University

1901	First woman appointed in the service of the Egyptian government
1907	Married Major Percy Elgood
1918	Officer, Order of the British Empire
1923	Decoration of the Nile
1939	Commander, Order of the British Empire
Death	November 21, 1960

Miss Sheldon Amos received her M.D. degree from London University in 1900. Her father was a judge in the Egyptian judicial system; her brother, Sir Maurice Amos, was a judicial advisor to the same system of justice. Sheldon spoke Arabic. In 1901 she accepted an appointment with the International Quarantine Board of Egypt, the first woman doctor ever to be accepted into service by the Egyptian government. She worked in Suez for two years at the Quarantine Hospitals at El-Tor. Pilgrims returning from Mecca were required to stay for a period here, and Sheldon observed many infectious diseases, most notably dysentery. She wrote several papers on this disorder.

Sheldon was transferred to Alexandria in 1902. Her replacement in Suez was another woman English doctor. In the new city, Sheldon established a private practice and opened an outpatient clinic for women and children. The clinic was located in a government hospital, and she did her work there gratis.

In 1906 Sheldon was transferred again, this time to Cairo, with directions from the Ministry of Education to begin educating Egyptian girls. What began as administration of three schools with 600 female pupils blossomed into 106 schools with 20,000 students by 1923. This system eventually was implemented throughout the entire country. Sheldon headed an expanding staff to administer these schools. The facilities were modern, and Sheldon guarded against poor sanitation and overcrowding. A year after establishing the school system, in 1907, she married Major Percy Elgood.

In addition to directing the schools, Sheldon served as woman medical member on a commission established by the Countess of Cromer. Its mission was to build the first free children's dispensaries in Egypt. These dispensaries were located in Cairo and Alexandria, but they proved so popular that the colonial government initiated them country-wide. Sheldon also served on a board called by Field Marshal Viscount Kitchener, a good family friend, to found a special school for maternity training. The project was so successful that provincial officials replicated it, training many Egyptians to be midwives. Sheldon also helped sponsor six Egyptian women to study medicine in England. In 1919 she testified before the Balfour Commission on Public Health in Egypt, and for many years she served on the board of the Victoria Hospital.

During World War I, Sheldon earned the Order of the British Empire (officer) and the French Medaélle de la Recomissance Française for her services to Allied troops in Cairo and throughout Egypt. She also received a silver medal from the Union des Femmes de France. Having boundless energy, Sheldon was active with the Cairo Red Cross Committee. She organized the Cairo Voluntary Aid Detachment, serving as commandant as well as medical officer.

In 1923, at the end of 22 years of service, Sheldon received the Decoration of the Nile (third class). It was the first time a woman had received any honor for public service by an official Egyptian agency. After her retirement, she and her husband lived in Heliopolis, where she continued to live after his death in 1941. She was promoted to commander, Order of the British Empire, in 1939. In 1956 she was forced from Egypt during the Suez Crisis. The Suez Canal was successfully nationalized by the Egyptian government, and Great Britain lost much of its influence in the Middle East. Elgood fled to Cyprus, then to London, where she lived until her death at age 86.

Bibliography

Levin, Beatrice. *Women and Medicine: Pioneers Meeting the Challenge*. Lincoln, Neb.: Media Publishing, 1988.
"Mrs. Elgood." *Medical Woman's Journal* 30 (Oct. 1923): 309–10.
"Obituary. Bonté S. Elgood, C.B.E., M.B." *British Medical Journal* (Dec. 17, 1960): 1813.

KELLY HENSLEY

KATHERINE ESAU
(1898–)
Botanist

Birth	April 3, 1898
1918	Emigrated from Russia to Germany
1922	Diploma, Landwirtschaftliche Hochschule, Berlin; emigrated from Germany to the United States

Katherine Esau. Photo courtesy of the Davidson Library, Department of Special Collections, University of California. Reprinted by permission of Jennifer Thorsch.

1923–27	Employed by Spreckels Sugar Co., Oxnard, CA
1931	Ph.D., University of California, Berkeley
1931–63	Faculty Member, University of California, Davis
1949	Elected to American Academy of Arts and Sciences
1963–	Faculty Member, University of California, Santa Barbara
1965–	Professor Emerita, University of California, Santa Barbara
1971–	Foreign Member, Swedish Royal Academy of Sciences

Katherine Esau's work in botany earned her membership in both the National Academy of Sciences and the American Academy of Arts and Sciences. Her book, *Plant Anatomy*, is a classic in its field.

Katherine was born in Ekaterinoslav, now called Dnepropetrovsk, Ukraine, in 1898. Her father, John Esau, was mayor of the city, where he established the streetcar system and waterworks. Both her father and her mother, the former Margarethe Toews, came from Mennonite families who had emigrated to Ukraine (which was then a part of Russia) from Prussia at the beginning of the nineteenth century. Having learned to read and write at home as a small child, Katherine was sent for four years to a Mennonite school. After that, she studied at a gymnasium (a European secondary school that prepares pupils for university-level studies) in Ekaterinoslav, from which she graduated in 1916. She then entered the Golitsin Women's Agricultural College in Moscow. Ironically, for a woman who was to become known as the "Grande Dame of Botany," Katherine Esau chose to study agriculture because she had the mistaken idea that botany was only concerned with the naming of plants.

The Russian Revolution in 1917 prevented her from returning to Moscow from Ekaterinoslav after her first year at the college. Shortly thereafter, the German army occupied Ukraine in the course of World War I. The revolutionary government had already deposed John Esau from the mayoralty. When the war ended in 1918, the Esaus were among the many refugees who fled from the Bolsheviks. The family settled in Berlin, where Katherine became a student at the Landwirtschaftliche Hochschule, the Berlin agricultural college. She received her diploma in 1922. The famous geneticist Erwin Baur, with whom she had studied plant breeding, naively suggested that Katherine return to Russia because she could make a great contribution to Russian agriculture. Having experienced the Revolution, the Esau family understood that this was impossible and emigrated to Reedley, California, in October 1922.

Katherine's father had chosen Reedley for its heavily Mennonite population and considered buying a farm where his daughter could apply her agricultural knowledge. Meanwhile, Katherine sought employment with a seed company and worked as a housekeeper in nearby Fresno. After a year as manager of a seed farm in Oxnard, California, during which she learned Spanish to communicate with the Mexican workers, she was hired by the Spreckels Sugar Company to improve the company's patented P19, curly-top disease-resistant sugar beets.

The chair of the Botany and Truck Crops Divisions at the University of California at Davis came to see the Spreckels work in 1927. Dr. W.W. Robbins, the botany chair, offered to appoint Katherine as an assistant at the University of California, Davis. In 1927 she entered the university as a graduate student, but because Davis lacked a graduate school she was technically registered at Berkeley. In 1931 she received her Ph.D. degree and was elected to Phi Beta Kappa. Dr. Esau had never heard of the honor society and stunned her advisor by asking him if she should join it.

Katherine Esau became an instructor in botany and a junior botanist in the Experiment Station of the College of Agriculture at the University of California, Davis, as soon as she graduated. Among the subjects she taught were plant anatomy, systematic botany, and morphology of crop plants. Until she achieved full professorship in 1949, Dr. Esau remained in each rank for the maximum six years. It is a matter of speculation whether this resulted from her chair's disapproval of early promotions or from discrimination against women.

Katherine Esau had become interested in research on phloem, the plant tissue that is primarily responsible for support, storage, and translocation. Her angle of research concerned the control of phloem-limited viruses. In pursuit of these viruses, Dr. Esau became involved in electron microscopy. Space and equipment for photomicrography were severely limited at Davis; and without air conditioning, the small darkroom was unusable much of the time. Esau bought her own equipment and set up her own darkroom at home. All her published microphotographs in the 1940s and 1950s were made in this way, including the illustrations for the first edition of her book, *Plant Anatomy*.

In 1963, Katherine Esau moved to the University of California, Santa Barbara, to join her co-researcher on phloem, Dr. Vernon I. Cheadle, who had accepted the position of chancellor there. It was from Santa Barbara that Esau officially retired in 1965, although as professor emerita she continued to do research and to come into her office regularly until early 1994. When Dr. Ray F. Evert was writing the introduction for the first presentation of the Katherine Esau award, which was established for student papers in 1985, she told him that she considered her years at Santa Barbara to be her most productive.

When Esau began to rewrite the latest edition of *Plant Anatomy* in 1985, she did so on a newly purchased personal computer, which she learned to operate at age 86. Her awards and honors include being elected to deliver the Faculty Research Lecture at Davis in 1946; election to the National Academy of Sciences; presidency of the Botanical Society of America; the Merit Award Certificate for the 50th Anniversary of the Botanical Society of America; the 11th International Botanical Congress Medal in 1969; and foreign membership in the Swedish Royal Academy of Sciences from 1971. She has been awarded honorary doctorates by Mills College and the University of California.

Bibliography

Abir-am, Pnina G., ed. *Uneasy Careers and Intimate Lives: Women in Science 1789–1979*. New Brunswick, N.J.: Rutgers University Press, 1987.
American Men and Women of Science, 18th ed. New Providence, N.J.: Bowker, 1992.
Bailey, Martha J. *American Women in Science.* Denver: ABC-CLIO, 1994.
Esau, Katherine. *Plant Anatomy.* New York: Wiley, 1965.
Evert, Ray F. "Katherine Esau." *Plant Science Bulletin* 5 (1985).

Howes, Durward, ed. *American Women: The Standard Biographical Dictionary of Notable Women*. Teaneck, N.J.: Zephyrus Press, 1974.

International Who's Who. London: Europa Publications, 1935–.

Rossiter, Margaret W. *Women Scientists in America*. Baltimore: Johns Hopkins University Press, 1982.

Visher, Stephen S. *Scientists Starred 1903–1943 in "American Men of Science."* Baltimore: Johns Hopkins University Press, 1947.

Who's Who in America. Chicago: Marquis Who's Who, 1899–.

LEE MCDAVID

ALICE CATHERINE EVANS
(1881–1975)
Bacteriologist

Birth	January 29, 1881
1909	First woman to be awarded a scholarship in bacteriology from the College of Agriculture, University of Wisconsin, Madison
1910	Employed by the Dairy Division of the Bureau of Animal Industry of the U.S. Department of Agriculture
1913	First woman to hold a permanent position in the Dairy Division in Washington, DC
1917	Discovered *Micrococcus melitensis*
1918	Bacteriologist, Hygienic Laboratory, U.S. Public Health Service (later National Institutes of Health, or NIH)
1922	Became infected with brucellosis
1928	Elected as the first woman president of the Society of American Bacteriologists (now American Society for Microbiology)
1945	Honorary President, Inter-American Committee on Brucellosis; retired from NIH
Death	September 5, 1975

Alice Catherine Evans was a prominent and prolific bacteriologist in the early twentieth century who published nearly 100 scientific papers during her 35 years as a government-employed researcher. Her most sig-

Alice Catherine Evans. Photo courtesy of the Library of Congress.

nificant contribution was her discovery in 1917 that drinking raw cow's milk and handling infected animals could cause undulant fever, a debilitating and sometimes fatal human disease, which was later named brucellosis.

At the 1917 annual meeting of the Society of American Bacteriologists in Washington, D.C., Evans reported to a skeptical audience that the bacterium *Micrococcus melitensis*, known since the late 1800s to cause what was then called Mediterranean or Malta fever in people who drank raw goat's milk, was closely related to *Bacillus abortus*, which caused contagious abortion in cattle, known as Bang's disease. Discussing her research results, she commented insightfully:

Considering the close relationship between the two organisms, and the reported frequency of virulent strains of *Bact. abortus* in cows' milk, it would seem remarkable that we do not have a disease resembling Malta fever in this country. Are we sure that cases of glandular disease, or cases of abortion, or possibly diseases of the

respiratory tract may not sometimes occur among human subjects
in this country as a result of drinking raw cows' milk?[1]

Evans spent most of her career researching the genus *Brucella*, as it
became known, and the disease it caused. She was an ardent, although
not immediately compelling, advocate of pasteurizing milk. In the 1930s,
after years of hostile resistance, the U.S. dairy industry finally began to
pasteurize all milk and cases of brucellosis subsequently decreased.
Evans explained in the unpublished memoirs she wrote at 82 years of
age: "Some of the [dairymen] found it easy to believe that the bacteriologist who warned against the use of raw milk was collaborating with
the manufacturers of pasteurizing equipment."[2] Similarly, she wrote in
a letter to Dr. O.N. Allen, a microbiologist at the College of Agriculture,
University of Wisconsin, Madison, on June 19, 1950, "as bacteriologists
of 25 or 30 years ago remember, there was a storm of protest against my
publication of papers between 'B. abortus' and 'M. melitensis.' "[3]

Alice Catherine Evans was born in 1881 in rural Neath, Pennsylvania,
to parents of Welsh descent. She had one older brother. Her father was
a surveyor, teacher, and farmer. She attended secondary school at the
Susquehanna Institute in Towanda, Pennsylvania. In her memoirs she
wrote that her "dreams of going to college were shattered by lack of
means. Because teaching was almost the only profession open to a
woman, I had no thought of doing anything else."[4]

She taught grade school for four years and then in 1905 began a two
year, tuition-free nature study course for rural schoolteachers at the College of Agriculture at Cornell University in Ithaca, New York. That
experience changed her life.

When the course in nature study was completed, I was no longer
interested in obtaining the certificate for which I was eligible. My
interest in science had been whetted by the basic courses I had
taken, and I wanted to continue to study science—any branch of
biological science would satisfy me.[5]

Evans continued her education at Cornell and in 1909 graduated with
a bachelor of science degree in agriculture, specializing in bacteriology.
Following graduation, the College of Agriculture at the University of
Wisconsin, Madison, awarded her a scholarship to pursue her master's
degree in bacteriology. She was the first woman to be awarded this scholarship. At Madison, Dr. E.G. Hastings was her advisor. She also studied
with Dr. Elmer V. McCollum, who taught the chemistry of nutrition and
later discovered vitamin A. In 1910, Evans earned a master of science
degree in bacteriology. Dr. McCollum urged her to pursue a Ph.D. in

chemistry at Wisconsin on a university fellowship, which she declined. She explained her decision in her memoirs:

> College had been a financial strain, and a physical strain too, for each year I had earned money to pay a part of my expenses. . . . There were other reasons why I did not want to spend two more years . . . studying to obtain a higher academic degree. In those days the Ph.D. degree was not a *sine qua non* for advancement. Further, my college education was over-balanced with science, of necessity, not by preference. I wanted to read history and other literature, to acquire greater appreciation of classic music, and to browse in various fields. It seemed more important to me to broaden my education than to acquire another degree in science.[6]

On July 1, 1910, Evans began working for the Dairy Division of the Bureau of Animal Industry of the U.S. Department of Agriculture at the state agricultural experiment station in Madison, Wisconsin. She studied the bacteriology of milk and cheese. Three years later she was transferred to the Washington, D.C., office, where she was the first woman to hold a permanent position in the Dairy Division. She studied the bacterial contamination of milk products and the bacterial flora of milk.

In 1917 she discovered that the "causal organism of Bang's disease [contagious abortion] in cattle [is] closely related to that derived from goats' milk, which causes human disease commonly known then as Malta or Undulant fever."[7] Her research results were published in the *Journal of Infectious Diseases* in July 1918. She wrote in her memoirs that "the reaction to my paper was almost universal skepticism, usually expressed by the remark that if these organisms were closely related, some other bacteriologist would have noted it."[8]

As Evans's research interests expanded from *Brucella* to include the diseases that it causes, she transferred in 1918 to the Hygienic Laboratory of the U.S. Public Health Service (later National Institutes of Health), where she also studied epidemic meningitis and influenza. But after six years of researching the bacterium that causes brucellosis, she herself became infected with the disease in 1922. During the next 23 years she had chronic brucellosis and was hospitalized often, sometimes for months at a time. By the late 1920s, brucellosis was recognized worldwide as a threatening and increasingly prevalent disease.

The Society of American Bacteriologists (now American Society for Microbiology) elected Evans its first woman president in 1928. In 1930 she was one of two delegates from the United States to the First International Congress in Microbiology at the Pasteur Institute in Paris. (Dr. Robert E. Buchanan of Iowa State University was the other U.S. delegate. Evans and Dr. Lydia Rabinowich of Russia were the only two women delegates at the Congress.[9]) In 1934 the Woman's Medical College of Pennsylvania awarded Evans an honorary M.D. degree, and in 1936 she was awarded an honorary Sc.D. degree from Wilson College. That year she was also one of two delegates from the United States to the Second International Congress in Microbiology held in London.

In 1939, Evans began research on immunity to streptococcal infection, which she continued until she retired in 1945 from her position as senior bacteriologist at the National Institutes of Health. She was elected honorary president of the Inter-American Committee on Brucellosis, a position she held from 1945 to 1957. In 1947 she presented evidence "indicating that the actual number of cases of brucellosis occurring in the United States was at least ten times the number of reported cases [because] in acute cases diagnosis is difficult and is sometimes incorrect; mild cases of brief duration are commonly mistaken for influenza."[10] A year later, the University of Wisconsin, Madison, awarded Evans an honorary Sc.D. degree.

Evans wrote her unpublished memoirs in 1963 after Dr. Wyndham D. Miles, historian at the National Institutes of Health, asked if he could interview her. Recalling the incident, she said she responded: "I would not want to make an extemporaneous record, but that I would be willing to write some memoirs if he should like to have me do so."[11] Her memoirs and papers reveal a colorful, intellectually curious person who was highly dedicated to science and to humanity. She described herself in 1961 as a "spry octogenarian" in a letter to Dr. McCollum, her former professor at Wisconsin who was her contemporary.[12] "Spry" characterizes Evans perfectly. Many members of the scientific community considered her the pre-eminent expert on brucellosis and an exceptionally astute researcher. As testimony to this, physicians and researchers from all over the world continued to seek her advice by mail well after she retired in 1945. Even at 80 years of age, she continued to publish an occasional scientific paper on brucellosis.

In 1969 she moved to Goodwin House, a retirement home in Alexandria, Virginia. On September 5, 1975, Alice Catherine Evans died at age 94 after suffering a stroke.[13]

Notes

1. A.C. Evans, "Memoirs," 1963, p. 22, in Alice Evans Papers, 1908–1965, #2552, Division of Rare Manuscript Collections, Carl A. Kroch Library, Cornell University.

2. *Ibid.*, p. 5.

3. Alice Evans to Dr. O.N. Allen, June 19, 1950, in Evans Papers, Carl A. Kroch Library, Cornell University.

4. "Memoirs," p. 2.

5. *Ibid.*, p. 5.

6. *Ibid.*, p. 10.
7. *Ibid.*, p. 35.
8. *Ibid.*, p. 22.
9. Evans Papers.
10. "Memoirs," p. 70.
11. "Memoirs," author's note.
12. Alice Evans to Dr. E. McCollum, April 4, 1961, Evans Papers.
13. The advice of Dr. Margaret Rossiter, Cornell University, was most helpful in completing this entry.

Bibliography

Burns, V. *Gentle Hunter: A Biography of Alice Evans.* Laingsburg, Mich.: Enterprise Press, 1993.
Current Biography. New York: H.W. Wilson, 1943.
DeKruif, P. "Before You Drink a Glass of Milk: The Story of a Woman's Discovery of a New Disease." *Ladies Home Journal* 46 (Sept. 1929): 8–9, 162, 165–166, 168–169.
MacKaye, M. "Undulant Fever: Are You Unaccountably Tired and Depressed? The Answer May Be in an Innocent Looking Bottle of Unpasteurized Milk." *Ladies Home Journal* 61 (Dec. 1944): 23, 69–70.
Notable American Women: The Modern Period. Cambridge, Mass.: Belknap Press of Harvard University Press, 1980.
O'Hearn, E.M. *Profiles of Pioneer Women Scientists.* Washington, D.C.: Acropolis Books, 1985.

SHARON SUE KLEINMAN

MARILYN GIST FARQUHAR

(1928–   )

Cell Biologist

Birth	July 11, 1928
1949	A.B., zoology, University of California, Berkeley
1952	M.A., pathology (experimental), University of California, Berkeley
1954–55	Research Associate, Dept. of Anatomy, University of Minnesota Medical School
1955	Ph.D., pathology (experimental), University of California, Berkeley

Marilyn Gist Farquhar. Photo courtesy of Marilyn Gist Farquhar.

1956–58	Assistant Research Pathologist, University of California, San Francisco
1958–62	Research Associate, Dept. of Cell Biology, The Rockefeller University
1962–64	Associate Research Pathologist, University of California, San Francisco
1964–68	Associate Professor of Pathology in Residence, University of California, San Francisco
1968–70	Professor of Pathology in Residence, University of California, San Francisco
1970–73	Professor of Cell Biology, The Rockefeller University
1973–87	Professor of Cell Biology and Pathology, Yale University School of Medicine
1984	Elected to the National Academy of Sciences

1987	E.B. Wilson Medal, American Society for Cell Biology
1987–89	Sterling Professor of Cell Biology and Pathology, Yale University School of Medicine
1988	Homer Smith Award, American Society of Nephrology; NIH Merit Award
1990–	Professor of Pathology and Coordinator, Division of Cellular and Molecular Medicine, University of California, San Diego
1991	Elected to the American Academy of Arts and Sciences

I was born in 1928 and grew up in Tulare, California, in the heart of California's great Central Valley, which boasts of the richest farmland in the world. I am quite proud of my heritage as a third-generation Californian from a pioneer family that I can trace to my great-grandmother Zumwalt, who settled in California in the 1880s. My father, Brooks Dewitt Gist, was descended from Christopher Gist, a scout for George Washington in the French and Indian and Revolutionary Wars. Like his father and grandfather before him, my father was raised on a farm. He later left agricultural life to become an insurance agent, but he lived on a farm and farmed as a hobby throughout his life. Although he never went to college, my father was a writer who produced novels detailing the settling of the Central Valley. He loved to raise and ride horses, often embarking on long excursions on horseback into the wilderness areas of California's Sierra Nevada mountains to write and observe. I often went with him, and he imprinted me indelibly with a love for nature, especially the flowers, animals, and sweeping views found in the "high country"—those special places found above timberline in mountains of the West that constituted his cathedral. Appropriately, he died at age 87 riding his horse over a high mountain pass called "Farewell Gap." My mother, Alta Green Gist, who was of Irish descent, was also from a pioneer family that relocated to California from Kansas. She was highly artistic, making her living managing a floral shop in Tulare and taking up oil painting after retirement. Our home is filled with her paintings. She imprinted me with a love of flowers and introduced me early to great music and art by taking me to concerts and museums whenever possible. It was my mother who insisted from our early childhood that my older sister, Janette, and I should go to college. My mother's closest friend, Frances Zumwalt, was a pediatrician. Knowing this woman when I was a child undoubtedly influenced me to pursue a career in medicine and biology. There were very few women physicians in those days to serve as role models.

I completed all my schooling in Tulare, attending a one-room, country school for grades one through four, then Wilson Elementary School,

Cherry Avenue Junior High School, and Tulare Union High School. All were within three blocks of my home. One of the achievements of which I am most proud is that I was chosen one of ten people to be included in the Tulare High School Hall of Fame. My cousin, Elmo Zumwalt, who was head of the U.S. Navy during the Vietnam era, attended the same high school and was also enshrined in the same Hall of Fame. My high school years occurred during World War II, which were difficult years for the country. Many of our older family members went off to war. Nevertheless, we were reasonably untouched in our daily student lives, which were filled with the usual activities: athletics, marching band (where I played clarinet), concert orchestra, yearbook, debating society, honor society, and so on. I participated in all these activities but made sure, at the prompting of my mother and my sister (who preceded me at Berkeley), that I satisfied all the scholastic and course requirements necessary to attend the University of California. I also worked summers and maintained a part-time job during the school year to save money for college. It was at Tulare Union High School that my interest in biology was further sparked by an inspiring biology teacher, and the pursuit of academic excellence was reinforced by a dedicated English teacher, Lois Thompson.

My education and professional development were marked by good luck at key junctures. For example, I was lucky to live in a state with an excellent public university, the University of California, because my parents could not afford to send me to a private university. I attended the Berkeley campus of the University of California and graduated in 1949. I majored in zoology as a premedical student with the intention of becoming a medical doctor—a dream that I only partially realized. During college, I found time to participate in extramural activities such as symphony orchestra as well as intramural athletics and to hold a part-time job to help with my educational expenses. In the year I graduated from Berkeley, I was one of three women admitted to medical school at the University of California, San Francisco. I attended for about two years, and during that period I was most fascinated by learning about the nature of diseases. I then shifted to a Ph.D. program in experimental pathology and received my Ph.D. in 1955. I changed career directions because in the meantime I had married another medical student, John Farquhar, whose name I took professionally. Between us we decided that my getting a Ph.D. and doing research would be preferable to continuing in medicine, because it would allow more flexibility and would be more compatible with raising a family. During graduate school I was lucky to start working with a professor who had the only electron microscope in the entire medical center. This allowed me to get in on the very beginning of the applications of the electron microscope in the new field of cell biology. My thesis dissertation was devoted to a study with the electron microscope of the secretory process in the anterior pituitary gland. For my postdoctoral work I followed my husband to the University of Minnesota, where I was involved in kidney research. Later we went to The Rockefeller University where I joined the Palade and Porter laboratory, which had the most active and productive group working in cell biology in the country. At the time, it was the most desirable place to seek training in the field of cell biology. Here again I was lucky that the institutions my husband chose for his career development were also ideal for mine. During those years I had two sons, Bruce and Douglas, neither of whom manifests any interest in science.

My graduate and postgraduate periods were during the early days of electron microscopy. This was a very exciting period because the electron microscope, with a resolving power 1,000 times that of the finest light microscope, provided the opportunity to see and discover things that had never been seen before. Everything one looked at was new. Cell organs that were barely visible with the light microscope could now be visualized in exquisite detail. I was lucky to make a number of lasting discoveries in basic biomedical research—for example, on the mechanisms of kidney disease, the organization of junctions that attach cells to one another, and the mechanisms of secretion (mechanisms by which cells produce and release their products).

After my postdoctoral work I returned in 1962 to the University of California, San Francisco, as a faculty member and rose through the ranks over the next eight years to become a full professor of pathology in the year before I turned 40. Somewhere along the way my first marriage ended in divorce. I relocated to the Rockefeller University in 1969, where I became a professor of cell biology and the only woman professor at this institution. On June 7, 1970, I married George Palade, whose wife had died a few years earlier. In 1973 we moved to the Yale University School of Medicine to become professors of cell biology. We did so to face the challenge of beginning a new department of cell biology with the opportunity to recruit six new faculty members. We spent 17 years at Yale and built one of the premier departments of this type in the country. We also trained 17 classes of first-year medical students and graduate students in cell biology. This was the first extensive teaching I had ever done, and I discovered that it was very rewarding. In 1974 my husband, George Palade, received the Nobel Prize in Physiology and Medicine for his discovery of ribosomes (the cell organ that synthesizes proteins) and for the insights he had provided into the mechanisms of secretion. He is generally considered to be one of the founding fathers of cell biology. His receipt of the Nobel Prize changed both of our lives forever.

In 1990 we were actively recruited to leave Yale and move to the University of California's San Diego campus in La Jolla. I resigned my position as Sterling Professor of Cell Biology and Pathology at Yale and returned to the West to face another challenge—starting a new division of cellular and molecular medicine at the medical school in San Diego. George Palade and I were recruited as a team to further build cell and molecular biology in the medical school, which is a project still in progress. Throughout these moves I have maintained the same research interests (i.e., in defining the mechanisms of secretion, now called membrane trafficking, and in defining the cellular and molecular mechanisms of kidney disease). In 1987 I received the Wilson Medal of the American Society of Cell Biology for my work on secretion and membrane trafficking, and in the same year I received the Homer Smith Award of the American Society of Nephrology for studies on the cellular and molecular mechanisms of renal disease. I received the most coveted milestone of my career in being elected to the National Academy of Sciences in 1988.

As I look at our daily lives, I realize that both my husband and I fit the description of "workaholics." The demands of science are so great in teaching, research, and service that there is little choice. Nevertheless it is a rewarding and satisfying career. However, in addition to working, we make sure that we find time to indulge in recreation. I have remained a passionate lover of both nature and music. My husband and I vacation every summer in Aspen, Colorado, where we find the perfect vacation mixture—hiking in the mountains above timberline and attending the Aspen Music Festival. During the rest of the year we take long walks on the beach beneath the sandstone cliffs of Torrey Pines Park in La Jolla and Del Mar and regularly attend the San Diego and San Francisco Opera.

As I reflect on my career, I am proud of the research contributions I have made in my field. I am just as proud of the people I have trained—from the graduate students, medical students, and postdoctoral fellows who have worked in my laboratory in four different universities to the medical students and graduate students I have introduced to cell biology at Yale and UCSD. Perhaps the fact that I am a woman who has achieved some success has put me in a position to provide special encouragement as a role model to young women beginning their careers. There are still many more men in science at the senior level than women.

One of the advantages of getting older, achieving recognition, and becoming established in one's field is that one can pick and choose how to spend one's time. Throughout my career I have chosen to divide my time between scientific investigation, teaching, and service to the scientific community. The overriding driving force, however, that has motivated my career in science has been the joy of discovery. My greatest satisfaction has come from those moments when I obtained a novel insight into the workings of the cell. I have been fortunate to live and work in a time when the opportunities for discovery and the growth in knowledge of biological sciences has been greater than at any other time in history. I have also been fortunate to be able to make a personal contribution to this growth of knowledge. I like to think that it is the same personal characteristics that led my ancestors to explore life in new places that has led me to explore the workings of the cell—which is, after all, the basic unit of life.

Bibliography

American Men and Women of Science, 1995–96, 19th ed. New Providence, N.J.: Bowker, 1994.
Who's Who in America, 1995, 49th ed. New Providence, N.J.: Marquis Who's Who, 1994.

MARILYN GIST FARQUHAR

MARGARET CLAY FERGUSON
(1863–1951)
Botanist

Birth	August 20, 1863
1885	Graduated from Genesee Wesleyan Seminary, Lima, NY
1891	Bachelor's degree, Wellesley College
1893–1932	Instructor, Wellesley College
1901	Ph.D., science, Cornell University
1929	President, Botanical Society of America
1932	Professor Emerita, Wellesley College
Death	August 28, 1951

Margaret Clay Ferguson's original research on the study of native pine trees and her other contributions to the field of botany ensure her a place not only in American history but also in worldwide scientific history. This unique woman started her career early and was a trailblazer as one of the few women botanists of the late 1800s.

Ferguson was born in 1863 in Orleans, New York, one of six children. Her parents, Robert Bell and Hannah Mariah (Warner) Ferguson, were a farming couple in the township of Phelps, an area in the Finger Lakes region known for crops of cabbage, wheat, and potatoes. Her parents' farming activities provided an ideal backdrop for Ferguson's long, productive career as a botanist.

With Ferguson's parents' farm providing opportunities for exploration, experimentation, and the study and exhibition of plants, at age 14 she began teaching botany in the public schools in the Phelps area, becoming an assistant principal in 1887.[1] During her early teaching years she continued her education, graduating from Genesee Wesleyan Seminary in Lima, New York, in 1885. She never married, rigorously embracing her career instead. Her education continued at Wellesley College from 1888 to 1891, where her disciplinary concentrations were botany and chemistry. She then spent two years as head of the science department at a seminary in Ohio. Ferguson's most meritorious achievements began when she accepted an instructor's position at Wellesley. Her extraordinary performance as a teacher and researcher led to promotions to higher administrative positions: assistant principal, head of the science department at Harcourt Place Seminary, and head of the botany department at Wellesley after obtaining her doctorate in science from Cornell in 1901.

As a teacher, Dr. Ferguson earned her place in the history of women in science and teaching. Not merely relying on her love of botany, she went the distance in educating herself to provide her students with quality information and instruction in botany and genetics. In the *Wellesley Magazine*, Sophie Hart described Ferguson's teaching ethic as follows:

In teaching botany to undergraduates, Dr. Ferguson's aim has been to make it a study of life problems and life adaptations so as to engender a sympathy and understanding of all that lives—the whole unfolding evolutionary plan. It is this larger conception held ever in view, above the details of the work, that is the lasting deposit left on the minds of her students—the significance and wonder and majesty of this world of law and living things. The training of the imagination through science, the cultural ends which science serves, are a very real part of Miss Ferguson's emphasis as a teacher.[2]

Along with teaching and administration, Dr. Ferguson mentored 42 graduate students pursuing master's and Ph.D. degrees. Many of those students obtained appointments in universities throughout the world.

If her teaching was noteworthy, Dr. Ferguson's scholarship and research were equally impressive. Her cytological and genetic studies of Petunia led to ground-breaking findings for higher plant genetics. And Ferguson's doctoral thesis research on the native pine *Pinus strobus*, coupled with her teaching and training of many botany students, resulted in her election to the presidency of the Botanical Society of America in 1929, the only woman so honored.[3] She published a total of 27 papers. Two of her most well known works are *Life History of the Pinus* and *Germination of the Spores of the Basidio-myceetes,* both of which have international reputations.

At the close of her illustrious career, Wellesley College paid tribute to Dr. Ferguson by naming newly constructed greenhouses in her honor. Records show that she convinced the Wellesley College administration of the most productive design for the greenhouse laboratories that were built in 1922 and 1923. Harriet Creighton eloquently explained why the greenhouses were named for Ferguson:

> They typify every phase of her life. She was a planner not only of greenhouses but of the department laboratories and library and courses of instruction. She was a teacher, and the greenhouses exemplify her understanding of the best ways to present a life science with living plants. She was recognized as one of the outstanding women in biological science.[4]

Even though Dr. Ferguson officially retired from Wellesley in 1932, her research activities continued through 1938. Throughout her entire life, Ferguson demonstrated talents in research and teaching. During her last years she spent time in Florida with her family. In 1951, after having moved to San Diego, California, she died of a heart attack.

Notes

1. See *Notable American Women: The Modern Period. A Biographical Dictionary* (Cambridge, Mass.: Harvard University Press, Belknap Press, 1980).

2. Sophie C. Hart, "Margaret Clay Ferguson," *Wellesley Magazine: An Alumnae Publication* (June 1932): 408–9.

3. Harriet B. Creighton, "The Margaret C. Ferguson Greenhouses," *Wellesley Magazine: An Alumnae Publication* (Feb. 1947): 173.

4. *Ibid.*

Bibliography

Cattell, J. McKeen, ed. *American Men of Science: A Biographical Dictionary,* 9th ed. Lancaster, Pa: Science Press, 1955.

Creighton, Harriet B. "The Margaret C. Ferguson Greenhouses." *The Wellesley Magazine: An Alumnae Publication* (Feb. 1947): 172–73.

———. "Obituary of Margaret Clay Ferguson." *Wellesley Alumnae Magazine* (Jan. 1952): 106.

Ferguson, Margaret Clay. "Contribution to the Knowledge of the Life History of *Pinus* with Special Reference to Sporogenesis, the Development of the Gametophytes and Fertilization." *Proceedings of the Washington Academy of Sciences* 6 (1904): 1–102.

————. "On the Development of the Pollen Tube and the Division of the Generative Nucleus in Certain Species of Pines." Ph.D. dissertation, Cornell University, 1901.

Ferguson, Martin Luther. *The Ferguson Family in Scotland and America.* Canandaigua, N.Y.: 1905.

Hart, Sophie C. "Margaret Clay Ferguson." *The Wellesley Magazine: An Alumnae Publication* (June 1932): 408–10.

Notable American Women: The Modern Period. A Biographical Dictionary. Cambridge, Mass.: Belknap Press of Harvard University Press, 1980.

Ogilvie, Marilyn Bailey. *Women in Science: Antiquity through the Nineteenth Century.* Cambridge, Mass.: MIT Press, 1986.

Rossiter, Margaret. "Women Scientists in America before 1920." *American Scientist* 62 (May–June 1974): 312–23.

THURA R. MACK

KATHERINE FOOT
(ca. 1852–?)
Cytologist

Birth	ca. 1852
1892	Studied at the Marine Biological Laboratory, Woods Hole, MA
1892–1921	Regular Member, Marine Biological Laboratory
1906	"Starred" in first edition of *American Men of Science*
1921–44	Life Member, Marine Biological Laboratory
Death	Unknown

There is a dearth of biographical data for cytologist Katherine Foot, even though she was considered one of America's top scientists as early as 1906. Foot was born in 1852 in Geneva, New York. *American Men of Science* lists her education only as "private schools," although she did receive some instruction at the Marine Biological Laboratory in 1892 at age 40.[1] Between 1892 and 1921 she was a regular member of the Marine Biological Laboratory, according to its annual report published each year in the *Biological Bulletin.* Her addresses during this period changed from

Denver, Colorado, to Evanston, Illinois, and then to New York City.[2] Beginning in 1921, the annual report of the Marine Biological Laboratory lists Foot as a life member. Her name remains on the roster until 1944; the address listed is Morgan, Harjes Cie, Paris, France. However, her entries in *American Men of Science* for the period 1927–1938 give her address as London, England. Her final address is listed as Camden, South Carolina.[3] Biographers have yet to establish an exact death date for Foot.

Katherine Foot's work focused on microscopical observations of the developing eggs of *Allobophora fetida*. Foot performed much of her research on the maturation and fertilization of *Allobophora fetida*'s egg with Ella Church Strobell, who died in 1920.

Foot and Strobell's research is unique for several reasons. They were among the first to photograph their research samples. Most of these photomicrographs are included in the reprinted collection of their work, entitled *Cytological Studies*. Foot and Strobell also were among the first to develop a way of making extremely thin sections of materials at low temperatures for use under the microscope.[4] Finally, although Foot and Strobell worked at the Marine Biological Laboratory, it seems that neither woman was ever affiliated with a university. Possibly they funded their own research. Strobell may even have left a legacy enabling Foot to continue some of her research for a 1920 paper.[5]

Foot's research was highly regarded, at least during her lifetime. In the bibliography of Thomas Hunt Morgan's *The Mechanism of Mendelian Heredity* (1915), he identified Foot as one of seven women who were primary investigators in the field of genetics.[6] H.J. Mozans, author of *Woman in Science*, noted Foot's work in cellular morphology. James McKeen Cattell placed a star by her name in his reference source, *American Men of Science*. In his preface, Cattell explained that the stars indicate the 1,000 scientists in the United States whose work he considered most important.[7]

Notes

1. G. Kass-Simon, "Katherine Foot and Ella Strobell: 'Worth a Thousand Words,' " in G. Kass-Simon and Patricia Farnes, eds., *Women of Science: Righting the Record* (Bloomington: Indiana University Press, 1990), p. 228.

2. *Ibid.*

3. John McKeen Cattell, ed., *American Men of Science*, 7th ed., s.v. "Foot, Katherine" (Lancaster, Penn.: The Science Press, 1944).

4. Kass-Simon, "Katherine Foot," p. 228.

5. *Ibid.*, p. 229.

6. Margaret Rossiter, "Women Scientists in America before 1920," *American Scientist* 62 (May–June 1974): 322.

7. Cattell, ed., *American Men of Science*, 7th ed., Preface.

Bibliography

Cattell, John McKeen, ed. *American Men of Science*, 7th ed., s.v. "Foot, Katherine."
    Lancaster, Penn.: The Science Press, 1944. (See also the first six editions.)
Kass-Simon, G. "Katherine Foot and Ella Strobell: 'Worth a Thousand Words.' "
    In *Women of Science: Righting the Record*, eds. G. Kass-Simon and Patricia
    Farnes. Bloomington: Indiana University Press, 1990.
Morgan, Thomas Hunt. *The Mechanism of Mendelian Heredity*. New York: Henry
    Holt and Co., 1915.
Mozans, H.J. *Woman in Science*. Cambridge, Mass.: MIT Press, 1974.
Ogilvie, Margaret. *Women in Science*. Cambridge, Mass.: MIT Press, 1986.
Rossiter, Margaret. "Women Scientists in America before 1920." *American Sci-
    entist* 62 (May–June 1974): 312–23.

FAYE A. CHADWELL

ROSALIND ELSIE FRANKLIN
(1920–1958)
Biologist

Birth	July 25, 1920
1945	Ph.D., Cambridge University
1947	Began work at Laboratoire Central des Services Chimiques de l'État, Paris
1951	Began work on DNA at King's College
1953	Moved to Birkbeck College; publication of Watson/Crick structure of DNA appeared in *Nature*
Death	April 16, 1958

Rosalind Elsie Franklin lost the race to discover the structure of DNA (deoxyribonucleic acid) before she fully realized that she was in a competition. DNA is the substance in cells that records the genetic structure of all living things and enables them to pass on their traits to their offspring. On February 23, 1953, Franklin recorded in her laboratory notebooks that DNA had a helical structure of two chains. She had previously deduced that the phosphates in the structure must be on the outside of the chains, leaving the base-pairs on the inside. By the time the manuscript recording this was typed for submission on March 17, 1953, the race was over.[1] On March 6, 1953, James Watson and Francis Crick had submitted their description of the structure of DNA to *Nature*. Their paper, however, contained two more vital points about the structure of DNA: the way in which the bases paired off, and the fact that one chain of the helix runs up whereas the other runs down. On March 18, 1953, Franklin received a call from the editors of *Nature* regarding her interest in submitting a paper to accompany the Watson-Crick effort.[2]

As a child, Rosalind Franklin demonstrated an analytical mind and intense interest in the physical world through her preference for carpentry and building sets over dolls and games of make-believe. She attended St. Paul's Girls' School in London and spent part of one semester in Paris improving her French. At age 15 she had already decided to become a scientist, owing in part to the physics and chemistry classes offered in the rigorous St. Paul's curriculum.

Her father touched off an intense family argument when he refused to pay for her education at Cambridge, because he disapproved of women receiving a university education. Alice Franklin, Rosalind's favorite aunt, and Muriel, her mother, both informed Ellis Franklin that they would pay for Rosalind's education out of their own family money. Rosalind never entirely forgave her father, although he finally did grudgingly agree to pay for her university education.

In 1938, Franklin entered Newnham College in Cambridge University, where she became friends with Adrienne Weill, a distinguished physicist who had worked with Marie Curie. Franklin's Ph.D. research on coals and carbons during World War II helped establish the science of high-strength carbon fibers. The developing field of X-ray crystallography then attracted Franklin's attention as a way of revealing the positions of atoms in matter. In this technique, X-rays are aimed at crystalline solids and the reflections of the X-rays are recorded on film. Franklin began to use the technique on carbons and biological molecules, although it was traditionally used on simpler crystals.[3]

Weill found Franklin a job in Paris after World War II ended. After the war years in England, during which she was often terrified by air raids on her bicycle trip to and from work, she began to travel freely around the continent with her co-workers both male and female. Franklin disliked formal occasions, although she loved small gatherings, where she sparkled with wit. She enjoyed gossiping about love affairs, shopping at street markets, and playing French word games.[4]

She spent three years at the Laboratoire Central des Services Chimiques de l'État before returning to England in 1950. She was offered a fellowship at King's College in the University of London by John Randall to analyze DNA using her expertise in X-ray crystallography. At the time, DNA was known to carry the genetic code from one generation to another, but its structure was a mystery.

Franklin drew the DNA fibers apart and bundled them in parallel, because a single fiber was too thin to create a useful picture. As she continued her experiments, she found humidity to be a key component in producing a quality photograph. At 75 percent humidity in the air, Franklin produced pictures of DNA she called the dry A-form; good, detailed photographs were produced. At 95 percent humidity, however, the wet B-form of DNA showed fewer details, but a definite cross shape was manifested. The simple cross is the characteristic shape produced in X-ray crystallography by a helical structure. Franklin concluded that the phosphate sugars known to be in DNA were located on the outside of the molecule because the molecule could absorb water and desiccate so quickly.

Maurice Wilkins was second in command in the department and also was interested in DNA. He and Franklin got along at first, but their relationship deteriorated during the fall of 1951. The lab group generally worked as a team and published together, but Franklin felt she was being treated as a highly paid technical assistant and refused to share her data with others for analysis. In November 1951, Franklin gave a presentation of her work at King's College, which was attended by James Watson. The American Watson was also interested in the structure of DNA and was collaborating with Francis Crick, an English graduate student at Cambridge. In marked contrast to her gaiety in France, on her return to England she was described by Watson as being without warmth or frivolity.[5]

Watson took no notes at the presentation but, along with Crick, produced a model of the DNA molecule using parameters he remembered somewhat incorrectly from the lecture. Franklin spotted several mistakes when she viewed the model. A few months later, a government report by Franklin that summarized her colloquium was passed to Watson and Crick by Max Perutz, a member of the government agency's review committee. Crick realized after reading the report that one of the chains of the DNA helix must go up and the other down. The correct water content and location of phosphate sugars were also noted in the report.

During this period, Franklin was struggling to solve the structure of DNA through a Patterson analysis that helped her obtain accurate parameters for the unit cell but did not, in the end, give appropriate information for solving the structure. Franklin considered and discarded cylinders, double sheets, and figure eight structures. Only a few weeks after she finally turned her attention to single and multiple helices as possible structures did she learn that Watson and Crick had built a model that solved the structure of the B form of DNA. She expanded and revised a draft paper written with Raymond Gosling to appear in the same issue of *Nature* as did the Watson and Crick paper on their model.

Even before these papers were published, Franklin left King's College to work for John Desmond Bernal at Birkbeck College in the University of London. Bernal and Randall agreed that Franklin could bring her fellowship and become head of her own research group, but could not work on DNA. Despite this, Franklin finished her work on DNA at Birkbeck and began a study of tobacco mosaic virus (TMV).

Over the following years, Franklin became good friends with Crick and his wife Odile and traveled with them through Spain. During the summer of 1956 she was diagnosed with ovarian cancer, and she went through three operations and experimental chemotherapy during the next two years. At one point she convalesced with the Cricks, although she gave them no details about her illness. Only her close family and research group were told.

Franklin was a very private person, not given to easily discussing her personal problems—even with friends. Her cancer had not been diagnosed until it was far advanced because of her propensity to ignore pain, although ovarian cancer can be intensely painful. As a child, her parents had sent her to a convalescent boarding school to recover following severe bouts of flu. Franklin was very unhappy at the school, and this episode may have influenced her in her efforts to ignore painful health problems. She once walked several hours to a hospital in extreme pain when a needle became stuck in her knee.[6]

Rosalind Elsie Franklin died on April 16, 1958, within a few minutes of the time her last paper was to be read at the Faraday Society.[7] She was 37 years old. Four years later the Nobel Prize for medicine was awarded to Francis Crick, James Watson, and Maurice Wilkins for their work on DNA. Nobels are given only to living persons, and no more than three winners can share each award. None of the men's Nobel lectures cite references to Franklin, and only Wilkins included her in the acknowledgments.

Although she was a strong experimentalist, Franklin favored the inductive approach to science and was critical of speculation. Thus she did not make great leaps of imagination; but through her keen observations and employment of very precise techniques, she was able to make crucial contributions to one of the great discoveries of the twentieth century and help lay the foundations for the science of structural molecular biology.

Notes

1. Aaron Klug, "Rosalind Franklin and the Discovery of the Structure of DNA," *Nature* 219 (Aug. 24, 1968): 808–10, 843–44.

2. Rosalind Franklin and R.G. Goslin, "Evidence for a 2-Chain Helix in Crystalline Structure of Sodium Deoxyribonucleate," *Nature* 172 (1953): 156–57.

3. Rosalind Franklin, "Crystallite Growth in Graphitizing and Nongraphitizing Carbons," *Proceedings of the Royal Society* 209A (1951): 154.

4. Anne Sayre, *Rosalind Franklin and DNA* (New York: Norton, 1975).

5. James Watson, *The Double Helix* (New York: New American Library, 1968).

6. Sharon Bertsch McGrayne, *Nobel Prize Women in Science: Their Lives, Struggles, and Momentous Discoveries* (Secaucus, N.J.: Carol Pub. Group, 1993).

7. Rosalind E. Franklin and A. Klug, "Order-Disorder Transitions in Structures Containing Helical Molecules," *Discussions of the Faraday Society* 25 (1958): 104–10.

Bibliography

Bernal, J.D. "Dr. Rosalind Franklin." *Nature* 182 (July 19, 1958): 154.

Jevons, Frederick Raphael. *Winner Take All: Case Study of the Double Helix.* Waurn Ponds, Vic.: Deakin University, 1987.

Judson, Horace Freeland. "Annals of Science: The Legend of Rosalind Franklin." *Science Digest* 94 (Jan. 1986): 56–59.

———. *The Eighth Day of Creation.* New York: Simon and Schuster, 1969.

Kass-Simon, G., Patricia Farnes, and Deborah Nash, eds. *Women of Science: Righting the Record.* Bloomington: Indiana University Press, 1990.

Klug, Aaron. "Rosalind Franklin and the Discovery of the Structure of DNA." *Nature* 219 (Aug. 24, 1968): 808–10, 843–44.

———. "Rosalind Franklin and the Double Helix." *Nature* 248 (Apr. 26, 1974): 787–88.

McGrayne, Sharon Bertsch. *Nobel Prize Women in Science: Their Lives, Struggles, and Momentous Discoveries.* Secaucus, N.J.: Carol Pub. Group, 1993.

Sayre, Anne. *Rosalind Franklin and DNA.* New York: Norton, 1975.

MARGARET SYLVIA

CHARLOTTE FRIEND

(1921–1987)

Microbiologist

Birth	March 11, 1921
1944	B.A., Hunter College
1946–66	Associate Member, Sloan-Kettering Institute
1950	Ph.D., Yale University
1952–66	Associate Professor of Microbiology, Sloan-Kettering
1954	Alfred Sloan Award
1957	Alfred Sloan Award

1962	Alfred Sloan Award in Cancer Research; American Cancer Society Award
1966–87	Professor and Director, Center for Experimental Cell Biology, Mt. Sinai School of Medicine
1970	Presidential Medal Centennial Award, Hunter College
1974	Virus-Cancer Program Award, NIH
1979	Prix Griffuel
1986	Honorary Ph.D., Brandeis University
Death	January 13, 1987

Charlotte Friend opened new doors in the study of leukemia by discovering a virus that caused the disease in mice. Although her research was questioned, she remained firm and continued her work in that area. The citation for the Mayor's Award of Honor in Science that she received in 1985 noted that "her irrefutable data and persistence led to a fundamental rethinking of cancer research."[1]

Charlotte was born in lower Manhattan, New York City, in 1921, the daughter of emigrants from Russia. Her mother was a pharmacist but chose to stay home and raise their children. Her father was a successful businessman. As a child, Charlotte often visited the local public library, where she enjoyed the biographies of scientists such as Louis Pasteur, Paul Ehrlich, and Robert Koch. She was challenged by the stories of their discoveries of disease-causing bacteria as well as discoveries of vaccinations and immunizations for disease. Additionally, the death of her father when she was 3 years old, caused by bacterial endocarditis, may have motivated her to work in the field of microbiology.[2] After the death of Charlotte's father, the family moved to the Bronx to be closer to other relatives. They were forced to receive "home relief" after the money in her father's estate was lost in the 1929 stock market crash. This also had an enduring influence on Charlotte.[3]

She never married, but she served as the matriarch and leader of her family. She was devoted to her siblings and their children as well as to her large extended family. Her family was very supportive of her and very important to her, both in her childhood and in her adult life. She did not like sports, but in addition to her work Charlotte loved the theater, music, ballet, opera, art, and reading. In high school she frequently skipped classes to attend the opera and often visited the opera house during rehearsals.[4] She also loved to travel but always enjoyed returning to New York City. Her sense of humor was recognized by all acquainted with her.

Charlotte attended Hunter College High School, a tuition-free school for gifted students. She worked in a doctor's office during the day and attended college at night. After college she joined the Navy in 1944, where her first assignment was in the hematology laboratory at the naval hospital in Shoemaker, California. Working in the lab confirmed her childhood dream of becoming a scientist.[5] Charlotte went to work in the Public Health Service after her discharge from the Navy and finally earned her Ph.D. from Yale in 1950, using the G.I. Bill of Rights for financial support. At Yale she chose to take classes with the medical students for her first two years, thereby gaining a good foundation in anatomy, pathology, and other subjects that would be important to her research.

In 1946, Charlotte began work at the Sloan-Kettering Institute for Cancer Research under the mentorship of Cornelius P. Rhoads, the director. There she began her cancer research. Her first project at Sloan-Kettering involved the development of drugs that would assist cancer patients. Charlotte stayed at Sloan-Kettering for 20 years, leaving that institution to become professor and director of the Department of Cell Biology at Mount Sinai Hospital. She loved working in the lab but was not fond of giving lectures.[6] In fact, her appointment at Mount Sinai included no teaching responsibilities; this allowed her to focus on research, but she was required to raise the money needed to support the work of the lab. In later years the involvement in procuring grants became a burden for her, as it is for many researchers.[7] While serving in that position at Mount Sinai (which she held until her death), she helped to mold the educational and research philosophy of the young medical school. Upon her death Dr. Nathan Kase, dean of the Mount Sinai School of Medicine, noted that "her presence was a major factor in establishing at the fledgling medical school a balance between emphasis on clinical care and on basic scientific research."[8]

Charlotte was a strong supporter of the women's movement, often ensuring that women were well represented at seminars and on committees. Although she was described as a warm, social, and somewhat shy person, she had definite beliefs and a forceful personality. She was not shy about expressing her opinion; in fact, she often wrote to newspapers. While at Mount Sinai she worked for faculty rights, sometimes serving on committees and sometimes working behind the scenes on issues such as tenure and the right of faculty to express their opinions.[9]

Sometimes Charlotte served as a mentor to new researchers, giving advice and encouragement to those around her and often writing letters in their support.[10] She had a genuine interest in people and instilled loyalty in her colleagues and friends, always being willing to listen and take an interest in what was happening in their lives.

In 1981, Charlotte was diagnosed with lymphoma. Very few people knew of the diagnosis because she did not want reviewers of grants or manuscripts she had submitted to be influenced.[11] While undergoing therapy she continued to conduct research in the lab, apply for grants, write papers, and carry on business as usual, including standing firm against others who opposed her.[12] She did all she could to keep the lab operational. At one of her last public appearances she received an honorary doctorate from Brandeis University. She died on January 13, 1987, at age 65.

Charlotte made several major contributions to the field of cancer research. In 1956 she reported research on a virus-like agent that was responsible for a malignant disease of the hematopoietic system in mice. Previous studies had reported similar results but had met with resistance and skepticism from the scientific community. Early reports in the 1950s used the phrase "virus-like" because the virologists were cautious of noting that they were working with tumor viruses. Although the head of the Sloan-Kettering Institute did not believe that viruses could cause cancer, he encouraged Charlotte in her research, for which samples of spleen tissue were retrieved from mice with leukemia. From this tissue, cell-free filtrates were prepared and injected into healthy mice of the same genetic background. When the mice developed leukemia, Charlotte realized that it must be caused not by the cells but by a microscopic organism, a virus. With the aid of an electron microscope, she was able to view and photograph the virus. In 1956 the *New York Times* reported she had "found that a filterable virus in mice would produce leukemia in infants, weanlings and adults of both sexes." It was noted that 90 to 100 percent of all mice inoculated with the virus developed leukemia.[13] When Charlotte presented her research at the 1956 annual meeting of the American Association for Cancer Research, she was belittled and rejected.[14] She responded with composure and tenacity to the questions posed to her.[15] When she tried to publish her research results, she again faced opposition from a male-dominated group. Ludwik Gross, a scientist who had worked with the first and only other described mouse leukemia virus at that time, helped her to have the paper published.[16] Another scientist, Peyton Rous, assisted her in writing and editing the article. Rous had previously conducted research on a chicken tumor that was caused by a virus, one of the first cancer-causing viruses discovered, and he was then serving as co-editor of the *Journal of Experimental Medicine,* in which the research was first reported.

In the ensuing years Charlotte continued work in this area and, with other scientists (including a highly respected pathologist), worked on the genetic makeup of the host animal in the viral pathogenesis. The virus would not infect all animals into which it was injected. Charlotte made another important contribution to cancer research by developing a vaccine that was used to immunize mice against the live leukemia virus. In 1976, at the annual meeting of the group that had humiliated her in 1956, Charlotte was the presiding officer. During her presidential address she recalled the events of that earlier meeting. It may have been her experience at the 1956 meeting that made Charlotte a reluctant public speaker. She always carefully wrote down exactly what she wanted to say before each presentation.[17]

The latter part of Charlotte's career was spent studying cell progression, from cancerous to normal, when the cell is exposed to chemical stimuli. Again she made another major contribution to the field. The work involved the Friend leukemia virus and used dimethyl sulfoxide. She theorized that this would lead to a new type of cancer treatment that would focus on growing normal cells rather than destroying cancerous cells. In 1972 she directed a study in which it was discovered that a malignant red cell precursor could be made to differentiate in vitro or perform the special task for which it was intended.

Among the many awards and honors that recognized Charlotte's important contributions to scientific research was the Alfred Sloan Award in Cancer Research in 1962. The award allowed recipients to focus on cancer research for the award period with financial support. In addition to their regular compensation, they were given funds for travel and other expenses. Charlotte used the monetary award to travel throughout the world, spending three-month periods in various laboratories. It was during this travel that she was able to carry out a dream she had cultivated since childhood—visiting and working in the laboratory of the Pasteur Institute.[18] The greatest recognition by her peers came in 1976 when she was elected to the National Academy of Sciences. Yet one of her most prized awards was the Great Heart Award, given in 1970 by the Pennsylvania chapter of the Variety Club, a national organization that focused on children's diseases. The citation was presented "in recognition of her distinguished career in the fields of virology and immunology, as an outstanding scientist and a brilliant researcher, attaining new achievements in the study and control of leukemia, which is the most common malignant disease of children." It further recognized Charlotte as "a true friend of the physically afflicted child."[19]

In the late 1970s Charlotte was concerned with the trend to channel funds to patient care rather than to focus them on basic research. She was concerned that young people would not be encouraged to go into scientific research because of the lack of funds. She was also afraid that graduate students who had prepared educationally would not have their projects funded. She felt that the lack of funds would lead to a lack of discovery of new treatment methods in cancer research. Remarking on a quote by Pasteur, "It is the prepared mind that sees new things," she realized that the mind must be prepared both by education and by participating in research. Without funding, that research is stifled.[20]

The "Friend virus," since its discovery, has been basic to study of the relationship between viruses and animal cancers. It has been the subject of many papers and texts. The role of viruses as a cause of cancer in humans continues to be studied. Some viruses have been implicated as the cause of leukemia and cervical cancer. The research conducted by Charlotte Friend opened avenues for work that continues in this area. Her work has had and continues to have a major influence on cancer research.

Notes

1. W. Sullivan, "Five Will Get City Science Award," *New York Times* (May 19, 1985).

2. L. Diamond and S.R. Wolman, "Charlotte Friend, Ph.D., 1921–1987. A Scientist's Life," *Annals of the New York Academy of Sciences* 567 (1989): 1.

3. L. Diamond, "Charlotte Friend (1921–1987)," *Biographical Memoirs of the National Academy of Sciences* 63 (1994): 128.

4. W. Scher, telephone interviews by author, Mount Sinai Medical School, August 8 and August 25, 1994.

5. Diamond, "Charlotte Friend (1921–1987)," p. 129.

6. Scher, telephone interviews.

7. Diamond, "Charlotte Friend (1921–1987)," p. 135.

8. H.M. Schmeck, "Charlotte Friend Dies at 65; Researched Cancer Viruses," in *New York Times Biographical Service* (New York: Arno Press, 1987), p. 36.

9. Diamond and Wolman, "Charlotte Friend, Ph.D., 1921–1987," p. 2; Scher, telephone interviews.

10. Scher, telephone interviews.

11. Diamond, "Charlotte Friend (1921–1987)," p. 137.

12. Diamond and Wolman, "Charlotte Friend, Ph.D., 1921–1987," p. 2.

13. "Cancer Scientist Studies Proteins," *New York Times* (Apr. 14, 1957): 14.

14. Iris Noble, "Charlotte Friend: Medical Microbiologist," in *Contemporary Women Scientists of America* (New York: Messner, 1979), p. 68.

15. Diamond, "Charlotte Friend (1921–1987)," p. 131.

16. Scher, telephone interviews.

17. Diamond, "Charlotte Friend (1921–1987)," p. 131.

18. Diamond and Wolman, "Charlotte Friend, Ph.D., 1921–1987," p. 2.

19. Noble, "Charlotte Friend: Medical Microbiologist," p. 78.

20. *Ibid.*, pp. 76–77.

Bibliography

"Cancer (Medicine)." *McGraw-Hill Encyclopedia of Science and Technology,* 7th ed. New York: McGraw-Hill, 1992.

Diamond, L. "Charlotte Friend (1921–1987)." *Nature* 326 (1987): 748.

————. "Charlotte Friend (1921–1987)." *Biographical Memoirs of the National Academy of Sciences* 63 (1994): 127–48.

Diamond, L., and S.R. Wolman. "Charlotte Friend, Ph.D., 1921–1987; A Scientist's Life." *Annals of the New York Academy of Sciences* 567 (1989): 1–13.

Friend, Charlotte. "The Coming of Age of Tumor Virology: Presidential Address." *Cancer Research* 37 (1977): 1255–63.

Noble, Iris. "Charlotte Friend: Medical Microbiologist." In *Contemporary Women Scientists of America*. New York: Messner, 1979.

Rapp, F. "The Friend Legacy: From Mouse to Man." *Annals of the New York Academy of Sciences* 567 (1989): 349–53.

Schmeck, H.M. "Charlotte Friend Dies at 65; Researched Cancer Viruses." In *New York Times Biographical Service*. New York: Arno Press, 1987.

Who's Who in America, 40th ed. Chicago: Marquis, 1978.

Who's Who in American Women, 14th ed. Chicago: Marquis, 1985.

Who's Who in Frontier Science and Technology, 1st ed. Chicago: Marquis, 1984.

JUDY F. BURNHAM

SUSANNA PHELPS GAGE

(1857–1915)

Embryologist, Comparative Anatomist

Birth	December 26, 1857
1880	Ph.B., Cornell University
1881	Married S.H. Gage on December 15
1910	"Starred" in first and second editions of *American Men of Science*; Fellow, American Association for the Advancement of Science
Death	October 15, 1915

"The brilliant but unemployed Susanna Phelps Gage" (as Margaret W. Rossiter called her in *Women Scientists in America*) presents a stereotype of the fate of women scientists in the late nineteenth and early twentieth century.[1] As a faculty wife, Gage never held a position of her own after her marriage to Professor Simon Gage of Cornell University. Like most women scientists who married scientists, her work was seen as an extension of her husband's and her academic and scientific accomplishments were subordinated to her social position as his wife. *American Men of Science* listed her name as Mrs. S.H. Gage, and her husband's obituary in the *New York Times* merely said that he had married Miss Susanna Phelps. Her own death 29 years earlier had not been deemed important enough for an obituary in the *Times*.

Susanna Stewart Phelps Gage was born in 1857 in Morrisville, New York. Her father, Henry S. Gage, was a businessman; her mother, Mary

Austin Gage, a former schoolteacher. After early education in the Morrisville Union School and the Cazenovia Seminary in Cazenovia, New York, Susanna earned a Ph.B. degree from Cornell in 1880. (The notice of her death in *Science*, October 15, 1915, calls this "the degree of doctor of philosophy.") She engaged in independent research in embryology and comparative anatomy before pursuing neurological studies at Johns Hopkins and Harvard medical schools in 1904 and 1905. (The same death notice in *Science* refers to her as a neurologist, but there appears to be no record that she held a medical degree.)

Married on December 15, 1881, to Simon Henry Gage, who taught histology and embryology at Cornell from 1878 to 1919, Susanna Phelps Gage was the mother of one child, Henry Phelps Gage, born on October 4, 1886. Her son received his own Ph.D. from Cornell in 1911.

Susanna was not a prolific author, but her published works in comparative morphology of the brain and of muscle fibers display a talent for clear and accurate observation. An artist as well as a writer, she illustrated scientific papers for her husband and for Dr. Burt G. Wilder.

Only 25 women's entries were starred in the first two editions of *American Men of Science*. Susanna Phelps Gage was awarded her star in the second edition at the same time as her husband. They were one of eight couples who both received the honor. Susanna Phelps Gage was elected a fellow of the American Association for the Advancement of Science and a member of the Association of American Anatomists at a time when most learned societies either did not accept women members or severely restricted their numbers. In contrast, her husband's *Who's Who* biography listed five such societies to which his wife did not also belong.

Susanna died on October 15, 1915. Two years later her husband and son gave $10,000 to establish the Susanna Phelps Gage Fund for Research in Physics at Cornell University. According to the Deed of Gift that created the fund, the money was to be used to advance knowledge in physics. Although Susanna Phelps Gage had been a biologist, physics was chosen in recognition of her status as the first woman to take laboratory work in physics at Cornell.

Note

1. Margaret W. Rossiter, *Women Scientists in America* (Baltimore and London: Johns Hopkins University Press, 1982), p. 89.

Bibliography

American Men of Science, 1st and 2nd eds. New York: Science Press, 1906, 1910.
Bailey, Martha J. *American Women in Science*. Denver: ABC-CLIO, 1994.
Barr, Ernest S. *An Index to Biographical Fragments in Unspecialized Scientific Journals*. University: University of Alabama Press, 1973.

Ogilvie, Marilyn B. *Women in Science.* Cambridge, Mass.: MIT Press, 1974.

Rossiter, Margaret W. *Women Scientists in America.* Baltimore and London: Johns Hopkins University Press, 1982.

Visher, Stephen S. *Scientists Starred 1903–1943 in "American Men of Science."* Baltimore: Johns Hopkins University Press, 1947.

LEE MCDAVID

ALESSANDRA GILIANI

(1307–1326)

Anatomist

Birth	1307
Death	March 26, 1326

Alessandra Giliani was the first woman prosector recorded by historians.[1] (A prosector prepares dissections for anatomical demonstration.) She also became a pioneer and master in the art of anatomical injection.

Giliani was a student at the University of Bologna. She studied under Mondino dei Luzzi, the father of modern anatomy, and acted as his assistant. Initially she studied philosophy, like all university students of the period. Later, Mondino taught her anatomy. She became so skilled in the art of dissection that Mondino made her his prosector. In this position, Giliani was responsible for preparing cadavers for Mondino's class lectures and demonstrations. She developed a method of draining blood from the veins and arteries of the corpse and replacing it with colored liquids that quickly solidified. With her new method and her deft skill as a dissector, she was able to mark the most minute blood vessels. As a result, she could highlight any portion of the circulatory system that was needed for discussion. Her remarkable presentations helped to bolster the reputation and renown of Mondino.[2]

Alessandra Giliani's life was short. She died at age 19. Her death so devastated her lover, Otto Agenius, who was also an assistant to Mondino, that he died a short time later.[3] In the Church of San Pietro e Marcellino of the Hospital of Santa Maria de Mareto in Florence, Agenius erected a tablet memorializing Alessandra Giliani's life. It reads as follows:

In this urn enclosed, the ashes of the body of Alexandra Giliani, a maiden of Periceto, skillful with the brush in anatomical demonstrations and a disciple, equalled by few, of the most noted physician, Mondinus of Luzzi, await the resurrection. She lived nineteen years; she died consumed by her labours March 26, in the year of grace 1326, Otto Agenius Lustrulanus, by her loss deprived of his better part, his excellent companion deserving of the best, has erected this tablet.[4]

Notes

1. Murial Joy Hughes, *Women Healers in Medieval Life and Literature* (Freeport, N.Y.: Books for Libraries, 1968), p. 87.
2. Michele Medici, *Compendio Storico della Scuola Anatomica di Bologna* (Bologna: 1857), p. 29, as quoted in Hughes, *Women Healers*, p. 87.
3. Kate Campbell Hurd-Mead, *A History of Women in Medicine* (New York: AMS Press, 1977), p. 225.
4. Medici, *Compendio*, p. 30, as quoted in Hughes, *Women Healers*, p. 87.

Bibliography

Alic, Margaret. *Hypatia's Heritage.* Boston: Beacon Press, 1986.
Hughes, Murial Joy. *Women Healers in Medieval Life and Literature.* Freeport, N.Y.: Books for Libraries, 1968.
Hurd-Mead, Kate Campbell. *A History of Women in Medicine.* New York: AMS Press, 1977.
Mozans, H.J. *Woman in Science.* Cambridge, Mass.: MIT Press, 1913.

MICHAEL WEBER

MARY JANE GUTHRIE
(1895–1975)
Zoologist, Cancer Researcher

Birth	December 13, 1895
1916	A.B., zoology, University of Missouri, Columbia
1918	A.M., University of Missouri
1918–20	Demonstrator in Biology, Bryn Mawr College
1920–21	Instructor in Biology, Bryn Mawr College

1921–22	Fellowship in Biology, Bryn Mawr College
1922	Ph.D., Bryn Mawr College
1922–27	Assistant Professor of Zoology, University of Missouri
1925	Published *Laboratory Directions in Zoology*
1927	Published *Textbook of General Zoology*
1927–37	Associate Professor of Zoology, University of Missouri
1937–51	Professor of Zoology, University of Missouri
1939–50	Chair, Dept. of Zoology, University of Missouri
1944–47	Member, Editorial Board, *Journal of Morphology*
1950–61	Faculty, Dept. of Biology, Wayne State University, Pontiac, MI
1951–61	Research Associate, Detroit Institute of Cancer Research
1961	Retired
1962–65	Member and Chair of various committees, Women's Auxiliary, Pontiac (MI) General Hospital
1965–67	President, Women's Auxiliary, Pontiac General Hospital
1966–70	Member, State Board, Michigan Association of Hospital Auxiliaries
1968–69	President-Elect, Michigan Association of Hospital Auxiliaries, Southeast District
1969–70	President, Michigan Association of Hospital Auxiliaries
Death	February 22, 1975

Mary Jane Guthrie, a biologist, teacher, and cancer researcher, was born in 1895, in New Bloomfield, Missouri. Her working life spanned a period of nearly 40 years, during which time she worked to further the understanding of the causes of cancer. Even though she retired from active research in 1961, her work is still cited today.

Mary Jane was the daughter of George Robert and Lula Ella (Lloyd) Guthrie. New Bloomfield, where she was born, is a small town in central Missouri near Columbia. After graduating from Columbia High School, Mary Jane attended the University of Missouri, Columbia, receiving an A.B. degree in zoology in 1916 and an A.M. degree in 1918.

After finishing her master's degree, Guthrie continued her studies at Bryn Mawr College in Pennsylvania. There she worked as a demonstrator in biology and an instructor before receiving a fellowship for her last year. She received her Ph.D. from Bryn Mawr in 1922. Years later, when Dr. Guthrie was asked by an interviewer conducting a survey of Bryn Mawr alumnae why she had chosen to study at Bryn Mawr, she answered simply, "Peebles."[1] Dr. Florence Peebles, an influential teacher and researcher in marine biology, was teaching there at the time Guthrie attended Bryn Mawr.

After receiving her Ph.D., Dr. Guthrie returned to the University of Missouri to teach zoology. From 1922 to 1927 she was an assistant professor; from 1927 to 1937, an associate professor; and in 1937 she became a full professor. Two years later, in 1939, she became chair of the Department of Zoology, a position she held until 1950. While at the University of Missouri, Dr. Guthrie co-authored a textbook, *General Zoology*, published from 1927 to 1957, and an accompanying manual, *Laboratory Directions in General Zoology*, published from 1925 to 1958. During this time she was also a member of the editorial board of the *Journal of Morphology* (from 1944 to 1947).

Dr. Guthrie was a noted scientist, being listed in a number of editions of *American Men of Science* (later *American Men and Women of Science*). Her name was starred in the sixth and seventh editions, indicating continuing growth as a scientist.

Although Dr. Guthrie never married, pursuing her career with the single-minded dedication of her male colleagues, she experienced difficulty in obtaining grants because of her gender. "In 1934, an official of the Rockefeller Foundation explained to her that, although she might be an outstanding scientist, as a woman she had to present extra proof of her excellence in order to receive a grant (women were not officially excluded from the Rockefeller fellowship program)."[2]

In 1950, Dr. Guthrie took a leave of absence from the University of Missouri, where she had taught for almost three decades, and went to Pontiac, Michigan. She began teaching biology at Wayne State University. In November 1950 she was hired as a "visiting scientist" by the Detroit Institute of Cancer Research (now the Michigan Cancer Foundation). Her term of employment at the Institute was initially for a period of seven months, from February to August 1951, when her leave of absence from the University of Missouri was to end.[3] However, in May 1951 it was recommended that her term of employment be extended. She became a salaried researcher at the Institute while continuing to teach at Wayne State University until her retirement ten years later.[4]

Near the end of her stay at the Detroit Institute for Cancer Research, Dr. Guthrie's work was commended in a report by the Institute's Scientific Advisory Committee:

Dr. Guthrie is approaching the end of a productive career. Having completed a study of experimental ovarian tumorigenesis, she has redirected her efforts towards definition of the origin of renal tumors induced by estrogens in hamsters—a practical problem with an attainable objective. At the same time, Dr. Guthrie continues her long-term interest in cultivation of ovaries *in vitro*. The Scientific

Advisory Committee points to Dr. Guthrie's *modus operandi* as an excellent model to be studied by younger members of the staff. An extremely important long range objective of extraordinarily great technical difficulty is coupled with eminently practical problems of more limited significance which can be solved in reasonably short periods of time.[5]

Dr. Guthrie's working life as well as her life in retirement after 1961 showed active concern for the plight of humanity. She became a volunteer of the Women's Auxiliary of the Pontiac General Hospital, serving on various committees and even chairing some. From 1965 to 1967 she served as president of the Auxiliary, and from 1966 to 1970 she was a member of the State Board of the Michigan Association of Hospital Auxiliaries. From 1968 to 1969 she served as president-elect of the Association, and from 1969 to 1970 as its president.[6] In 1970, when asked about her plans for the coming years, she said she would "continue volunteer work with hospital auxiliaries as long as I can get around."[7] She was 75 years old at that time.

Throughout her life Dr. Guthrie found time to pursue interests in other fields. She had a broad range of personal interests including music, art, the theater, and sports; but her particular hobbies were gardening, stamp collecting, and breeding dogs.[8] She died on February 22, 1975, at age 79 and was buried in New Bloomfield, Missouri.

Notes

1. *Bryn Mawr College Survey of the Alumnae and the Alumni, 1970.* Mariam Coffin Canaday Library, Bryn Mawr College, Pa.

2. Martha J. Bailey, *American Women in Science: A Biographical Dictionary* (Santa Barbara, Calif.: ABC-CLIO, 1994), p. 145.

3. Detroit Institute of Cancer Research, Executive Committee Meeting, November 20, 1950.

4. Detroit Institute of Cancer Research, Executive Committee Meeting, May 16, 1951.

5. Detroit Institute of Cancer Research, Minutes of the Scientific Advisory Committee Meeting, February 20, 1959.

6. *Bryn Mawr College Survey.*

7. *Ibid.*

8. *Ibid.*

Bibliography

American Men of Science: A Biographical Directory, 6th ed. New York: Science Press, 1938.

Bailey, Martha J. *American Women in Science: A Biographical Dictionary.* Santa Barbara, Calif.: ABC-CLIO, 1994.

Bryn Mawr College Alumnae Journal (Fall 1975).
The Detroit Institute of Cancer Research, 1948–1958. Detroit: The Institute, 1958.
Who Was Who in America with World Notables, Vol. 6, 1974–1976. Chicago: Marquis
    Who's Who, 1976–.

CARA KENDRIC

ALICE HAMILTON
(1869–1970)
Physician

Birth	February 27, 1869
1893	Medical degree, University of Michigan Medical School
1895–96	Studied pathology and bacteriology at Universities of Leipzig and Munich, Germany
1896–97	Studied pathology and bacteriology, Johns Hopkins University
1897–1902	Professor of Pathology and Director of Histological and Pathological Laboratories, Women's Medical College of Northwestern University
1897–1919	Resident, Hull House, Chicago
1908	Appointed by governor of Illinois to Illinois Commission on Occupational Diseases
1910–19	Investigator of industrial poisons for U.S. Department of Labor
1915	Became involved with Jane Addams and others in a "proposal for a neutral commission to end the war"
1919	Visited Germany to survey the effects of postwar starvation
1919–35	Assistant Professor of Industrial Medicine, School of Public Health, Harvard University—first female faculty member at Harvard
1935–70	Professor Emerita—Industrial Medicine, School of Public Health, Harvard University
1947	Lasker Award
Death	September 22, 1970

Alice Hamilton. Photo courtesy of the Library of Congress.

Alice Hamilton was a physician and a pioneer. Her major contribution was to discover and publicize the dangers of working in many of our nation's major industries. She was effective in bringing these dangers to the attention of the proper authorities so that laws could be enacted requiring safer practices in the workplace. In her ultimate vocation, what is now known as industrial medicine, Alice Hamilton saved countless lives—more than she could have as a practicing physician. She truly had a revolutionary impact on industrial practices in the United States.

Alice Hamilton was an extraordinary woman from an extraordinary family. She was one of five children, four daughters and one son, of a prominent family from Fort Wayne, Indiana. Intellectual achievement and contribution to the community were expected from all members of the family. Her sister, Edith Hamilton, became well known as headmistress of the Bryn Mawr School, a prestigious school for girls, and was the author of now-classic volumes on Greek and Roman culture and mythology. Her sister Margaret was an educator, also at Bryn Mawr School. The youngest sister, Norah, was an artist. Norah's sketches enhance Alice Hamilton's autobiography, *Exploring the Dangerous Trades.*

Quint, the only boy in the family, became a respected professor of foreign languages at the University of Illinois.

The Hamilton children were educated at home, where they were taught by both parents with supplementary instruction in German from the housemaids and a teacher at a local Lutheran school. Science and mathematics were neglected because Mr. Hamilton thought them unimportant. Alice later spent two years at Miss Porter's School in Farmington, Connecticut, where many fellow students were from prominent families. These contacts later became important in her work. When Alice decided to become a physician, she had to study with tutors to repair her deficiencies in science and mathematics.

Two major themes inform Alice Hamilton's life, affecting her choice of careers and life style: (1) the need to find work that would benefit people, particularly the poor and weak, and (2) the need to exercise her intellect. At one time she seriously considered missionary work but felt that she was not good enough to pursue that as a career. The practice of medicine seemed to fulfill both requirements and, also important, to permit financial independence.

Hamilton studied at the University of Michigan Medical School, receiving her medical degree at age 24. Two brief internships convinced her that clinical practice did not suit her. She decided to pursue graduate study in bacteriology and pathology, although doctoral degrees in those fields were not granted to women by most universities at the time. She spent some time studying in Michigan under her former medical school professors. But because the most prominent pathologists and bacteriologists of the time were in Germany, as were many of the greatest scholars in all fields, she and her sister Edith went together to Germany to further their studies. Women in universities in Germany were very rare. In one case Alice had to be escorted to a chair on the lecture platform before the male students entered the lecture hall. After the lecture she was required to remain until all the students had left, then discreetly exit escorted by the professor who had daringly given her permission to attend. It was even suggested that she and other women who wished to attend lectures be hidden behind a curtain so as not to distract the male students. In spite of these obstacles, the Hamilton sisters enjoyed their time in Germany and always maintained a love for that country and its people.

During her medical training in Michigan, Alice had become aware of the severe deprivation and medical problems of immigrant laborers and their families. She wanted to do something to help these hard-working people. Upon returning from Germany, she attended a lecture by Jane Addams, founder of Hull House. This settlement house in the Chicago slums was famous throughout the world for the work its residents did in assisting poor immigrants in their adjustment to life in America. The settlement house provided English lessons, a day nursery, and a playground, as well as assistance to the sick. Alice was inspired and determined to work at Hull House. She was accepted as a resident in 1897.

Like most other Hull House residents, Hamilton pursued full-time professional responsibilities while participating in the work of the House. She was a professor and researcher at Women's Medical College of Northwestern University. She also conducted research in pathology and bacteriology at the University of Chicago. Her work at Hull House included establishing a well-baby clinic and counseling new mothers on principles of nutrition and cleanliness. During this time she formed an enduring friendship with Jane Addams, with whom she later traveled and worked devotedly in the cause of peace.

From visiting families in the slum tenements, she began to see the correlation between the occurrence of certain diseases and membership in certain occupations. She also saw the devastating effects on a family when the breadwinner was incapacitated owing to accidental injury or acute disease acquired on the job. Children often worked 12-hour days in factories. Weary women had to care for their children and sick husbands while trying to earn enough to put food on the table. Alice began to read widely in the medical literature to find out as much as was known about occupational disease and what could be done to avoid it.

She discussed her findings at Hull House, where she met many of the nation's prominent politicians, labor organizers, and economic theorists. These progressive leaders flocked to Hull House for good conversation and firsthand education in the social realities of the time. They learned of Hamilton's interest in the "dangerous trades." So when the governor of Illinois was establishing a commission to study the extent of occupational disease in the state, Alice Hamilton was a natural choice for director and chief medical investigator.

Dividing the task of gathering information with other members of the commission, Hamilton concentrated her investigations on the study of lead poisoning in the state. She determined the number of cases of lead poisoning and found that a majority of these were indeed of occupational origin. She visited lead factories to see how the men were exposed to the lead. She became convinced that inhalation of lead particles was the origin of most occupational lead poisoning. Owners of factories preferred to blame the disease on lead ingested by workers who were too lazy to wash their hands before lunch. They knew that changing manufacturing techniques, installing ventilation systems, and providing intensive worker training to protect workers from airborne lead particles would be expensive.

Thus began Alice Hamilton's official entry into the field of industrial hygiene, a field that hardly existed in the United States before that time. Although many European nations, including Germany and England, had laws and practices for protecting workers in industrial settings, the United States lagged in that area. Those espousing laws and programs limiting working hours or exposure to health hazards were often labeled as radicals and their arguments dismissed. Industrialists were certain that because their plants were newer and more modern than those in Europe, workers could not be coming to harm from working in them.

Having completed the Commission's study for the state of Illinois, Hamilton was invited often to participate in government-sponsored research into occupational diseases. She studied the effects of mercury on felt workers in Connecticut, the famous "mad hatters," and "quicksilver" miners in the West. During the war she studied the safety of workers in the rapidly expanding explosives industry. She always was meticulous in her research and painstakingly collected data to establish the connection between the working conditions in factories or mines and the diseases suffered by the workers. She went directly into the mines or factories she was studying, observing the working conditions, even trying the equipment, looking for the dangerous elements. She visited homes and hospitals to determine the extent of disease in the community. Finally, she would offer suggestions for alternative techniques or safety equipment that could be used to make the activity safer. The public was receptive to the concept of requiring a safe working environment. Lawmakers began to follow through with regulations and legislation establishing safety standards and setting up programs such as workmen's compensation.

But Alice Hamilton was surely not a bureaucrat. Although she was conscientious about making her reports to the agencies commissioning the surveys, she did not want to wait for the results of governmental actions. Often she went straight to the factory owners to tell them what she had learned of the dangers in their plants, offering her recommendations to minimize the dangers. They often took her advice at great cost to their companies, but with greater benefit to their employees.

In 1919 Alice left Chicago and Hull House to become the first woman faculty member at Harvard University. As an assistant professor in the School of Public Health, she arranged to teach and do research for half the year, freeing the other six months for fieldwork and travel. This appointment added to her professional stature and increased her already substantial influence on policy in the field of occupational health and safety.

Because of her impeccable social credentials, Alice Hamilton was able to make inroads where others could not. When she visited a factory or mine, she could often meet with the head of the company, a social equal, who would be forced by good manners to listen politely to her arguments. Her charm and demeanor alone would not convince them. It was the logic and force of the facts behind her arguments that convinced many of these industrial leaders that there was, indeed, some cause and effect involved in the premature illnesses and deaths of many of their workers. She also had to convince them, however, that the expense of correcting the situations to promote worker safety would be cost-effective in the long run. When she could not convince the owners and managers, she could still wield great influence on lawmakers. The first half of the twentieth century saw an explosion of protective laws and regulations, both state and federal. Alice Hamilton's influence and contributions to these efforts undoubtedly saved many more lives than she could have in years of medical practice.

Hamilton's commitment to pacifism should not go unmentioned. During World War I she visited Europe with Jane Addams, working with world leaders and laymen in the cause of peace. She campaigned to stop the blockade of Germany at the end of the war, having seen firsthand the near-starvation of many German civilians. She visited Russia when it was politically unwise to do so. She was open-minded enough to be willing to accept innovations the communists had made in the cause of worker safety, while rejecting the political ideology behind them. She never flinched from her ideals in spite of the criticism to which she was subjected for her unpopular views.

Hamilton's book *Industrial Toxicology* became a classic and was published in several editions. Her fascinating autobiography, *Exploring the Dangerous Trades*, covers her life until the early 1940s. She lived to the advanced age of 102, a loved and honored celebrity, recognized by presidents and the public alike as a pioneer in her field, to whom all workers owe a debt of thanks.

Bibliography

"Alice Hamilton." *National Cyclopaedia of American Biography, Current Volume G,* pp. 107–8. New York: James T. White and Co., 1946.

Grant, Madeleine P. *Alice Hamilton, Pioneer Doctor in Industrial Medicine.* New York: Abelard-Schuman, 1967.

Hamilton, Alice. "Edith and Alice Hamilton, Students in Germany." *Atlantic Monthly* 215 (Mar. 1965): 129–32.

———. *Exploring the Dangerous Trades: The Autobiography of Alice Hamilton.* Boston: Little, Brown and Co., 1943.

"The Lasker Awards for 1947." *American Journal of Public Health* 37 (1947): 1612–16.

Sicherman, Barbara. *Alice Hamilton: A Life in Letters.* Cambridge, Mass.: Harvard University Press, 1984.

HELEN HOFFMAN

ETHEL NICHOLSON BROWNE HARVEY

(1885–1965)

Zoologist

Birth	December 14, 1885
1906	A.B., Woman's College of Baltimore
1907	A.M., Columbia University
1909	Described induction of a differentiated animal by a specific tissue
1913	Ph.D., Columbia University
1913–14	Instructor in Biology, Dana Hall School, Wellesley, MA
1914–15	Sarah Berliner Fellow, Hopkins Marine Station, University of California
1915–16	Assistant in Histology, Cornell Medical College
1916–27	Part-time researcher
1928–31	Instructor in Biology, Washington Square College, New York University
1931–59	Independent researcher, Princeton University
1935	Performed centrifugal force experiments on nuclei
1938	"Starred" in sixth edition of *American Men of Science*
1941	Developed a method for determining the sex of sea urchins
1950–58	Trustee, Marine Biology Laboratory, Woods Hole, MA
Death	September 2, 1965

Ethel Browne Harvey's early experiments with *Hydra* and later experiments using centrifugal force on egg nuclei helped to redefine the very meaning of life. Her work earned her a coveted star in the sixth edition of *American Men of Science* (which meant that her fellow scientists considered her one of America's top researchers), but her work also caused controversy in the popular press.

Jennie and Bennet Browne did not follow the norm of the late 1800s when it came to the education of their children. They felt that their three daughters, as well as their two sons, should have the opportunity of good educations. Consequently their last child, Ethel, born in 1885 in Baltimore, attended and graduated from the first solely preparatory girls' school in the United States, the Bryn Mawr School. She attended Woman's College and received her bachelor's degree in 1906. Between her graduation from Woman's College and her enrollment at Columbia University, she spent her first summer at the Marine Biological Laboratory at Woods Hole, Massachusetts. Thus began a long and profitable association between Ethel and the laboratory.

As do many graduate students, Ethel supported her graduate work with a series of jobs and fellowships. She taught science and math at the Bennett School for Girls from 1908 to 1911 and biology at Dana Hall School in 1913–1914. She was awarded a fellowship from Goucher for 1906–1907 and one from the Society for the Promotion of University Education for Women for 1911–1912. She was also awarded the Sarah Berliner Fellowship, which she used to attend the Hopkins Marine Station, University of California, for 1914–1915.

As a predoctoral fellow at Columbia University, Ethel worked with T.H. Morgan, who suggested that she research the induction of new hydranths in *Hydra*. In 1909 she published the results of her experiments in a paper entitled "The Production of New Hydranths in *Hydra* by the Insertion of Small Grafts."[1] This article describes her experiments with implanting a tentacle cut from the base of one hydra into the column of another hydra. Her results, the formation of new hydranth by the original hydra, show that induction of a differentiated animal by specific tissue can occur. Even though this work occurred almost 12 years before the induction experiments of Hans Spemann and Hilde Mangold, it is not found in reviews that outline and document the development of the concept of induction.[2] For unknown reasons Ethel's early work seems to have gone unnoticed by the embryological researchers of the time, but she did not continue in this line of research. Like many of her contemporaries, her research emphasis shifted toward an interest in the role of the nucleus and cytoplasm in inheritance and development. This shift may have been owing to the influence of her doctoral thesis advisor, cell biologist Edmund Beecher Wilson. During this period she studied the male germ cells of the aquatic carnivorous insect genus *Notonecta*.

Ethel Browne married Edmund Newton Harvey, an assistant professor and physiologist at Princeton, on March 12, 1916. Later that same year their first son, Edmund Newton Harvey, Jr., was born. Their second son, Richard Bennet Harvey, was born six years later. Even though Ethel had shown unusual potential as a researcher, her career changed to part-time status following the birth of her first child. As with many women scientists of her generation, obtaining support for her work was difficult and inconsistent. Having a husband and children appears to have made continuing in her chosen career very difficult. Owing to the assistance given to her by a mother's helper, a maid, and governesses, however, she did not have to give up her career completely. She was able to continue her research, spend summers at Woods Hole, and attend l'Institut Océanographique in Monaco (1920–1921) and the Stazione Zoologica in Naples (1925–1926). She became an instructor in biology at Washington Square College in 1928 and held the position until 1931, when she was given office space in Edwin Conklin's laboratory in the biology department at Princeton. Princeton also agreed to pay her summer fees at the Marine Biological Laboratory. She continued to work in Conklin's laboratory and share summer space with her husband at Woods Hole until 1962. It is interesting to note that it was the laboratory at Woods Hole, not a university, that provided her with a stable professional base. Except for a grant from the American Philosophical Society in 1937, her work, although internationally recognized, was unsupported. Neither was Ethel ever appointed to a faculty position at Princeton, whereas her husband was tenured and promoted to professor in 1920.

Ethel's significant contributions were made during the 30 years that she worked at Princeton. Among these was the application of centrifugal force to the study of developing eggs. In her article entitled "Cleavage without Nuclei," she described the use of centrifugal force to redistribute various parts of the egg's content. She concluded that "cleavage can therefore occur in eggs without either maternal or paternal nucleus."[3] She first used the term "parthenogenic merogony" to describe cleavage and development of an unfertilized enucleated egg, and her research suggested that cytoplasm had the ability to develop itself into early embryonic stages of life without the input from a nucleus. This work was important in furthering research on the mechanism of cell division. In addition, her work led her to develop a method by which the sex of the sea urchin can be determined. This in itself does not seem remarkable; but for scientists who need to obtain eggs and sperm with which to study the process of differentiation and development, this was a major contribution. She also published *The American Arbacia and Other Sea Urchins* (1956), which has become a standard reference for scientists working in the field of sea urchin embryology.

Harvey's work drew some criticism from the popular press because some people perceived her work as an "intervention" in the life process. Her work was seen as "a creation of life without parents."[4] Her obituary noted that she "stirred the scientific community in the nineteen thirties with her experiments in the chemical creation of life." Many people had feared that this type of research would lead to the creation of organisms that had been "fathered by a chemical and brought into the world in a glass jar."[5]

Notes

1. See Ethel Browne, "The Production of New Hydranths in *Hydra* by the Insertion of Small Grafts," *Journal of Experimental Zoology* 7, no. 1 (1909): 1–23.

2. G. Kass-Simon, "Biology Is Destiny," in G. Kass-Simon and P. Farnes, eds.,

Women of Science: Righting the Record (Bloomington: Indiana University Press, 1993).

3. Ethel B. Harvey, "Cleavage without Nuclei," *Science* 82 (1935): 277.

4. D.J. Haraway, "Harvey, Ethel Browne," in *Notable American Women: The Modern Period* (Cambridge, Mass.: Harvard University Press, 1980).

5. "Dr. Ethel Harvey, Biologist, Was 79," *New York Times* (September 3, 1965): 27.

Bibliography

Browne, E. "The Production of New Hydranths in Hydra by the Insertion of Small Grafts." *Journal of Experimental Zoology* 7, no. 1 (1909): 1–23.

"Dr. Ethel Harvey, Biologist, Was 79." *New York Times* (September 3, 1965): 27.

Haraway, D.J. "Harvey, Ethel Browne." In *Notable American Women: The Modern Period.* Cambridge, Mass.: Harvard University Press, 1980.

Harvey, E.B. *The American Arbacia and Other Sea Urchins.* Princeton: Princeton University Press, 1956.

———. "Cleavage without Nuclei." *Science* 82 (1935): 277.

Kass-Simon, G. "Biology Is Destiny." In *Women of Science: Righting the Record,* eds. G. Kass-Simon and P. Farnes. Bloomington: Indiana University Press, 1993.

MOLLY WEINBURGH

ELIZABETH DEXTER HAY

(1927–)

Cell Biologist

Birth	April 2, 1927
1948	A.B., *summa cum laude*, Smith College; Phi Beta Kappa
1952	M.D., Johns Hopkins University School of Medicine
1953–56	Instructor in Anatomy, Johns Hopkins University School of Medicine
1957–60	Assistant Professor of Anatomy, Cornell Medical College
1960–64	Assistant Professor of Anatomy, Harvard Medical School
1964	Honorary M.A., Harvard University
1969–	Louise Foote Pfeiffer Professor of Embryology, Harvard Medical School
1973	Honorary Sc.D., Smith College

Elizabeth Dexter Hay. Photo courtesy of Elizabeth Dexter
Hay.

1975	Member, American Academy of Arts and Sciences
1975–93	Chair and Professor, Dept. of Anatomy and Cellular Biology, Harvard Medical School
1982	Johns Hopkins Society of Scholars
1984	Member, National Academy of Sciences
1988	Alcon Award for Vision Research
1989	E.B. Wilson Medal, American Society for Cell Biology; Honorary Sc.D., Trinity College
1990	Honorary Sc.D., Johns Hopkins University; Excellence in Science Award, Federation of American Societies for Experimental Biology
1991	Salute to Contemporary Women Scientists Award, New York Academy of Sciences

| 1992 | Henry Gray Award, American Association of Anatomists |
| 1993– | Professor of Cell Biology, Harvard Medical School |

Elizabeth Dexter Hay is a cell biologist, embryologist, and anatomist. She is currently a professor of cell biology at Harvard University School of Medicine. Her work is significant because her research on cellular mechanisms aids the understanding of the metastasis of cancer cells, birth defects such as cleft palate, and childhood diseases in which a cell is unable to make or respond to an extracellular matrix.

Elizabeth Hay was born in St. Augustine, Florida, in 1927. One of three children, she attended public schools and graduated from Melbourne High School in Florida in 1944. During World War II her father, a physician, enlisted in the Army Medical Corps and Elizabeth spent her junior year and part of her senior year in Biloxi, Mississippi, and Hayes, Kansas. In 1948 she graduated *summa cum laude* from Smith College and in 1952 received an M.D. degree from Johns Hopkins University School of Medicine. She was interested in a career in biology during her undergraduate years: "I took a course in mammalian anatomy and physiology my first year in college given by a wonderful teacher that convinced me I wanted to learn all I could about biology."[1] Her mentor, Dr. S. Meryl Rose, convinced her to pursue an M.D. degree rather than a Ph.D. because "at that time women with Ph.Ds in biology got stuck at small women's colleges like Smith. There was far more motility with an M.D."[2]

Her father had wanted her twin brother, Jack, who died in 1942, to pursue a medical career; and although Elizabeth was delighted to tell her father that she was becoming a doctor, it was Dr. Rose's advice that was the deciding factor in her career choice. Dr. Hay did research under Dr. Rose during her four years at Smith and, after medical school, continued to work with him on limb regeneration during the summers at the Marine Biological Laboratory in Woods Hole, Massachusetts.

Dr. Hay entered academic medicine in 1953 as an instructor in the anatomy department at Johns Hopkins Medical School. She was interested in salamanders and their ability to grow new limbs. At about the same time, the electron microscope was first being used in the study of biological structure. Hay said, "I became an excellent microscopist, and that is the biological instrument that I still love most to use, because I am a visual person (in high school I wanted to be an artist) and electron microscopy is all about looking at cell ultrastructure."[3]

In 1957 she moved to Cornell Medical College to conduct research with Don Fawcett, one of the foremost electron microscopists in the world. When Fawcett was offered the chair of the Department of Anatomy at Harvard Medical School in 1960, Hay followed him to Boston and became an assistant professor in the Department of Anatomy. During these early years of her career, she published several important studies on limb regeneration and on the process of dedifferentiation (the process or state of a neoplasm characterized by the loss of normal cellular differentiation).

At Harvard, she and Jean-Paul Revel were among the first to successfully apply autoradiography (the technique of recording on a photographic emulsion the radiations emitted by radioactive material in the object being studied) to the electron microscope. Hay and Revel published a series of papers using electron microscopic autoradiography to localize metabolic activities in cells. In 1961, Hay and Revel demonstrated DNA synthesis in the nucleolus (a dense spherical accumulation of fibers and granules found in the nucleus of most eukaryotic cells) long before the widespread acceptance of the idea that the nucleolus contains DNA. Their studies of organelle function suggested that epidermis secreted collagen, a very heretical theory at that time. In 1969, Hay published a monograph with Revel on the fine structure of the developing avian cornea. This work has become a classic in the field.

In order to culture isolated epithelium, Dr. Hay turned to the embryonic avian cornea. Since that time, studies of eye tissues and the functions of collagen and other extracellular matrix molecules have dominated her research. In a series of experiments over the course of 20 years, she demonstrated that the early corneal matrix stimulates and possibly determines the differentiation of the corneal epithelium. Her research on the development of critical in vitro systems proceeded to metabolic and functions studies involving all members of the extracellular matrix armamentarium (collagens, proteoglyceans, laminin, and fibronectin) and went on to examine the role of matrix receptors.

A research theme that occupied Dr. Hay for many years and that has had a profound impact on cell and developmental biology is epithelium-mesenchyme transformation and related subjects, such as cell shape, cell polarity, cell motility, and cytoskeletal dynamics. Mesenchyme is the part of the embryonic mesoderm from which connective tissue and the circulating and lymphatic systems develop; epithelial-mesenchymal transformations are involved in many morphogenetic events. Hay's 1975 publication with J.B.L. Bard, which describes the different shape and behavior of motile corneal fibroblast in vivo and in vitro, has been widely cited in the scientific literature. Other related studies have culminated in convincing new theories concerning the cascade of intracellular events resulting from the interaction of matrix molecules with cell surface receptors.

Dr. Hay's contributions to her profession are numerous. She has served as the president of four societies—the Society for Developmental Biology; the American Society for Cell Biology; the American Association of Anatomists; and the Association of Anatomy Chairmen. From 1971 to

1975 she was the editor of the journal *Developmental Biology*. She has served on the councils of three institutes at the National Institutes of Health and has authored 130 publications, including four books. Dr. Hay has also received honorary doctorates from Smith College, Johns Hopkins University, and Trinity College. She is a member of the American Academy of Arts and Sciences, as well as the National Academy of Sciences and its Institute of Medicine, and has been the recipient of numerous other honors.

Elizabeth Hay has been successful not only as a scientific investigator but also as an effective administrator. She was the first woman chair of an academic department at Harvard Medical School. For over 18 years she served as chair of the Department of Anatomy and Cellular Biology—a department that typically had no more than two other women members at the same time. She said, "I really did not want to be chair, but they would have gone outside if I had not accepted it."[4] She said further that "as chair, I have enjoyed the most helping the junior faculty succeed in their careers. Being at Harvard has helped me hire the best young people."[5]

Dr. Hay has taught histology, developmental biology, and embryology. She runs a small research group in the Hay Lab at Harvard, working with three or four postdoctoral fellows. Using techniques such as genetic labeling and video microscopy, the lab analyzes the complex interactions between cells and their surroundings that allow cells to reach their ultimate destination. Many of her publications were written with students. Dr. Hay's students remember her dedication to excellence, her meticulous supervision, and her collegiality. She never married, not necessarily by choice but because "the people who wanted to marry me were not interested in my career. At that time, men were not willing to marry a woman who wanted a career."[6] She was, perhaps, a victim of the times: "I never married. The opportunity for women to do so while continuing full-time careers in biomedicine essentially did not exist then."[7] Elizabeth Hay chose her career instead of marriage, and she has had a prodigious one.

Notes

1. E.D. Hay, interview with Kathy Cook, Springfield, Mo., March 1, 1994.
2. E.D. Hay, telephone interview with Mary Ann McFarland, St. Louis, Mo., October 11, 1994.
3. E.D. Hay, remarks to Fiftieth High School Reunion, Melbourne, Fla., April 14, 1994.
4. E.D. Hay, telephone interview, October 11, 1994.
5. E.D. Hay, interview, March 1, 1994.
6. E.D. Hay, telephone interview, October 11, 1994.
7. E.D. Hay, remarks, April 14, 1994.

Bibliography

American Men and Women of Science, 18th ed. New Providence, N.J.: R.R. Bowker, 1992.

Pollard, T.D. "E. B. Wilson Medalist, 1988 [Elizabeth Hay]." *Journal of Cell Biology* 108 (1989): 2245.

Who's Who in America, 48th ed. New Providence, N.J.: Marquis Who's Who, 1994.

MARY ANN MCFARLAND

ELIZABETH LEE HAZEN

(1885–1975)

Mycologist

Birth	August 24, 1885
1910	B.S., Mississippi Industrial Institute and College
1917	M.S., biology, Columbia University
1918–23	Worked in U.S. Army diagnostic laboratories and directed Clinical and Bacteriological Laboratory of Cook Hospital, Fairmont, WV
1927	Ph.D., microbiology, Dept. of Bacteriology and Immunology, College of Physicians and Surgeons, Columbia University
1931	Director, Bacterial Diagnosis Laboratory, Division of Laboratories and Research, New York State Department of Health (NYC Branch)
1945–48	Determined nutritional requirements of *Microsporum audouini,* the cause of ringworm; applied knowledge to help children afflicted with the disease
1948	With Dr. Rachel Brown, discovered fungicin (later renamed nystatin), the first safe, effective antifungal antibiotic
1954	Food and Drug Administration approved nystatin; E.R. Squibb and Sons, Inc., began commercial production of Mycostatin
1954–73	Continued medical mycological work at Division of Laboratories and Research (until 1960), Albany Medical

Elizabeth Lee Hazen. Photo reprinted by permission of the John Clayton Fant Memorial Library, Mississippi University for Women.

	College (1958–60), and Columbia University Mycology Laboratory (1960–73)
1955	Squibb Award in Chemotherapy (with Rachel Brown)
1957	Secured U.S. patent for nystatin (with Rachel Brown)
1968	Honorary doctorates from Hobart College and William Smith College
1972	Rhoda Benham Award, Medical Mycological Society of the Americas (with Rachel Brown)

1975	Chemical Pioneer Award, American Institute of Chemists (with Rachel Brown)
Death	June 24, 1975
1994	Posthumously inducted into National Inventors Hall of Fame (with Rachel Brown)

Fungus diseases—the mycoses—are costly in human and monetary terms. Dr. Elizabeth Hazen's co-discovery of nystatin, "the first broadly effective antifungal antibiotic available to the medical profession," has saved countless lives, and the funding program endowed by royalties from her invention has ensured that future generations of researchers will continue the fight against these devastating diseases.[1]

Elizabeth Lee Hazen was born in 1885 in Rich, Mississippi. When her parents, cotton farmers William Edgar Hazen and Maggie (Harper) Hazen, died before she was 4 years old, Elizabeth and her sister and brother (who died at age 4) were adopted by their aunt and uncle. Hazen attended the rural public schools of Lula, Mississippi, and graduated in 1910 from the Mississippi Industrial Institute and College (today Mississippi University for Women) with a B.S. degree and a certificate in dressmaking.[2] An entry in her senior yearbook notes the following:

> If anyone ever passed through Senior Hall without seeing "Friz" in a sprightly jig or hearing her begging "Dock" for something to eat, it must have been while she was doing Physics with her fourth year Normal in the laboratory. B.S. is rightly attached to her name, for she has a most scientific mind. . . . Lee is never despondent before the outside world, even over her debts, and her quick repartee and good cheer win her many friends, who predict great things for her.[3]

Hazen taught high school physics and biology in Jackson, Mississippi, for six years and persevered with her studies in bacteriology, earning her M.S. degree from Columbia University in 1917. After working in Army diagnostic laboratories and directing the bacteriological laboratory of a West Virginia hospital, she completed her studies at Columbia and received a doctorate in microbiology in 1927 at age 42.[4]

In 1931, Hazen joined the Division of Laboratories and Research of the New York State Department of Health, whose record for hiring and promoting women scientists outshone that of most state and federal science agencies.[5] She greatly enjoyed working at the New York City Branch Laboratory, analyzing vaccines and serums and conducting microbiological research on a wide range of infectious diseases.[6] Hazen traced outbreaks of the animal diseases anthrax and tularemia to their sources and was the first in North America to implicate *Clostridium botulinum* Type E toxin in deaths owing to imported canned fish. Her collaboration with Dr. Ruth Gilbert in research on the fungal infection moniliasis (candidiasis) sparked her interest in the mycotic diseases.[7]

Penicillin was ineffective against fungus infections, which afflicted many children and members of the armed forces, and the prevailing antifungal treatments were too toxic for human use.[8] Penicillin actually promotes monilial infections by destroying the normal intestinal bacteria that keep the *Candida* fungi in check. Hazen was determined to find a naturally occurring antifungal antibiotic that would be safe for human therapy.[9]

With the encouragement of Dr. Rhoda Benham, co-founder of the Medical Mycology Laboratory of Columbia University, Hazen built a collection of fungal cultures and slides for teaching, reference, and diagnostic and consulting purposes.[10] Working tirelessly, she screened soil samples for actinomycetes (microorganisms that had yielded antibacterial antibiotics and demonstrated antifungal activity) and tested their action against two pathogenic fungi, *Candida albicans* and *Cryptococcus neoformans*. The division director, Dr. Gilbert Dalldorf, introduced her to chemist Dr. Rachel Fuller Brown. This meeting inaugurated a long, fruitful collaboration and warm friendship between the two women. Working at the Division's Central Laboratory in Albany, Brown extracted, isolated, and purified the antifungal substances within the cultures and returned them to Hazen for further testing.[11]

In 1948, Hazen collected a soil sample near Warrenton, Virginia, from which she isolated a streptomyces culture that yielded two antifungal antibiotics. The first, obtained from the culture's broth, was active against *C. neoformans* but was too toxic to test animals. It was also identical to an existing agent, cycloheximide, which was used to inhibit fungus on golf course greens. The second substance, extracted from the surface growth, was a previously undiscovered antibiotic that prolonged survival in mice infected with *C. neoformans*. It was only mildly toxic to the test animals and was potent against *C. albicans* and 14 other pathogenic and nonpathogenic fungi. They called the substance "fungicidin," and Hazen named the microorganism from which it was derived *Streptomyces noursei* in honor of the Nourse family on whose property she had obtained the successful sample.[12]

Brown reported their findings at the October 1950 meeting of the National Academy of Sciences. Hazen, who cherished her privacy, avoided the photographers there, but she was equally proud of their discovery. She cautioned an overzealous science reporter from the *New York Times*, "Don't you dare say that it's fit for human use!" She then explained that the antibiotic could not be considered safe for human therapy because it had been tested only in animals.[13]

E.R. Squibb and Sons, Inc., was the first of many pharmaceutical firms to contact the inventors, who renamed the antibiotic "nystatin" for the New York State Division of Laboratories and Research and consulted on its production process. Hazen declared, "It's my baby, and I'm going to look after it."[14] Dalldorf, who had backed their collaboration to the fullest, shepherded the invention through its 1957 patenting with Research Corporation of New York. Following its FDA approval, commercial production of Mycostatin, Squibb's trademark for the drug, commenced in 1954.[15]

Nystatin is highly effective as a specific for oral, vaginal, and skin infections caused by *Candida* species. It combats monilial infections of the intestinal tract when combined with the antibacterial antibiotics that allow the opportunistic fungi to proliferate, and prevents the development of monilial infections following intestinal surgery. Nystatin has controlled fungus infections in chickens and turkeys, salvaged trees afflicted with Dutch elm disease, aided in wine preservation, protected stored bananas during shipment, and restored priceless paintings and frescoes mildewed by the 1966 flooding in Florence, Italy.[16]

Hazen and Brown both received the Squibb Award in Chemotherapy in 1955. They shared honorary degrees conferred by Hobart College and William Smith College in 1968 and the Rhoda Benham Award bestowed in 1972 by the Medical Mycological Society of the Americas. In 1975 the American Institute of Chemists awarded Hazen and Brown the Chemical Pioneer Award, which had never before been given to a woman or to a scientist who was not a chemist.[17] In 1994, Hazen and Brown became the second and third women to be inducted into the National Inventors Hall of Fame.[18]

Nystatin generated over $13 million in royalties over the 17-year life of its patent. Its co-discoverers stipulated that Research Corporation allocate one-half of all royalties for support of its own grant programs and place the other half in a special fund for support of basic research in biochemistry, microbiology, and immunology.[19] Brown-Hazen grants strengthened academic programs in the biological and medical sciences, giving faculty, staff, and students fresh impetus to undertake independent research and motivating students to pursue careers in science. The Brown-Hazen Fund allocated scholarships in the inventors' names to the women's colleges they had each attended and supported programs advancing women's participation in the sciences.[20]

Characteristically, Hazen encouraged research directed toward the alleviation of human suffering: "We have some of the finest and best-equipped mycologic laboratories in the world, and we have highly trained and gifted mycologists. . . . But what we have not done is to improve the treatment of the infected person. That is the crying need in mycology." For a time, the Brown-Hazen Fund, which totaled $6.7 million and lasted until 1976, was the single most important source of support for research and training in medical mycology in the United States.[21]

Determined to "earn her keep," as she put it, Hazen accepted an associate professorship at Albany Medical College in 1958 and maintained her collaboration with Brown as well as her demanding schedule at the Division. She resumed a long-term project, research on the fungus *Microsporum audouini*, culprit of the disease ringworm of the scalp; and, with colleague Frank Curtis Reed, she produced a classic reference text, *Laboratory Identification of Pathogenic Fungi Simplified*.[22] In 1960 she became guest investigator in the Columbia University Medical Mycology Laboratory, where, recalled Dr. Margarita Silva-Hutner, then director of the lab, " 'she contributed scholarly lectures to our mycology courses, carefully written and documented from the most recent literature. . . . She also provided benchside instruction in and out of the classroom, giving friendly or stern advice as the occasion required, but always imparting invaluable details of technique, and providing standards of ethics for the young.' "[23]

Hazen always enjoyed the company of family and friends and her many pastimes, which included reading, attending the theater, and discussing politics and world affairs.[24] Her modesty prevented her from speaking of her accomplishments, but she wanted her hard-won professional status to be respected. When a young physician once addressed her as "Miss Hazen," she promptly reminded him that in academic processions M.D.s precede Ph.D.s because the doctoral degree represents a higher level of specialization.[25]

Hazen interrupted her work in 1973 to visit her ailing sister in Seattle and died there on June 24, 1975.[26]

Notes

1. Richard S. Baldwin, *The Fungus Fighters: Two Women Scientists and Their Discovery* (Ithaca: Cornell University Press, 1981), p. 100.

2. Lewis P. Rubin, "Elizabeth Lee Hazen," in *Notable American Women: The Modern Period*, eds. Barbara Sicherman and Carol Hurd Green (Cambridge, Mass.: Harvard University Press, 1980), p. 326; Baldwin, *Fungus Fighters*, pp. 37–39.

3. Erin Hearon et al., *Meh Lady* (Columbus: Mississippi Industrial Institute and College, 1910), p. 30.

4. W. Stevenson Bacon, "Elizabeth Lee Hazen, 1885–1975," *Mycologia* 68 (1976): 962; Baldwin, *Fungus Fighters*, p. 40; Elizabeth Lee Hazen, "General and Local Immunity to Ricin," *Journal of Immunology* 13 (1927): 171.

5. Elizabeth Moot O'Hern, "Elizabeth Hazen (1895–1975)," in Elizabeth Moot O'Hern, *Profiles of Pioneer Women Scientists* (Washington, D.C.: Acropolis Books, 1985), p. 96; Margaret W. Rossiter, *Women Scientists in America: Struggles and Strategies to 1940* (Baltimore: John Hopkins University Press, 1982), pp. 241–43.

6. Baldwin, *Fungus Fighters*, p. 43.

7. O'Hern, *Profiles*, p. 96.

8. W. Stevenson Bacon, "Elizabeth Lee Hazen, 1885–1975," *Mycologia* 68 (1976): 964; O'Hern, *Profiles*, p. 98; Albert Schatz and Elizabeth L. Hazen, "The Distribution of Soil Microorganisms Antagonistic to Fungi Pathogenic for Man," *Mycologia* 40 (1948): 462.

9. Baldwin, *Fungus Fighters*, p. 28, 73.

10. O'Hern, *Profiles*, p. 96. Benham was a gifted teacher as well as a brilliant researcher. After her death, Hazen and Benham's successor, Dr. Margarita Silva (later Silva-Hutner) paid tribute to their beloved mentor in a profile of her distinguished career; see Margarita Silva and Elizabeth L. Hazen, "Rhoda Williams Benham, 1894–1957," *Mycologia* 49 (1957): 596.

11. Edna Yost, "Rachel Fuller Brown (1898–   )," in *Women of Modern Science*, ed. Edna Yost (New York: Dodd, Mead, 1959), pp. 74–75. Brown, a New Englander and a graduate of Mount Holyoke College, was 13 years Hazen's junior; she too had pursued her goal of a scientific education with unswerving determination and distinguished herself early in her career.

12. Baldwin, *Fungus Fighters*, pp. 77–78; Elizabeth L. Hazen and Rachel Brown, "Two Antifungal Agents Produced by a Soil Actinomycete," *Science* 112 (1950): 423.

13. Margarita Silva-Hutner, telephone interview, September 23, 1994.

14. Baldwin, *Fungus Fighters*, pp. 98–101.

15. Research Corporation, *Gilbert Dalldorf and the Dalldorf Fellowship in Medical Mycology* (Tuscon: Research Corporation, 1991), pp. 5–6; Baldwin, *Fungus Fighters*, pp. 99–100.

16. Baldwin, *Fungus Fighters*, pp. 100–101, 178–82.

17. *Ibid.*, pp. 107, 109.

18. Margarita Silva-Hutner, telephone interview, September 23, 1994; Autumn Stanley, *Mothers and Daughters of Invention: Notes for a Revised History of Technology* (Metuchen, N.J.: Scarecrow Press, 1993), p.196; Anne L. Macdonald, *Feminine Ingenuity: Women and Invention in America* (New York: Ballantine Books, 1992), pp. 357–61. The first woman to be inducted into the National Inventors Hall of Fame was Gertrude Belle Elion.

19. Bacon, "Elizabeth Lee Hazen," pp. 967–69.

20. Baldwin, *Fungus Fighters*, pp. 123–31, 169.

21. *Ibid.*, pp. 103, 134.

22. Bacon, "Elizabeth Lee Hazen," p. 967; Elizabeth Hazen and Frank Curtis Reed, *Laboratory Identification of Pathogenic Fungi Simplified*, 2nd ed. (Springfield, Ill.: Charles C. Thomas, 1960).

23. Baldwin, *Fungus Fighters*, p. 106.

24. *Ibid.*, pp. 110–11.

25. Margarita Silva-Hutner, telephone interview, September 23, 1994.

26. Rubin, "Elizabeth Lee Hazen," p. 327. I am especially grateful to Ms. Frieda Davison, Director of Library Services at the John Clayton Fant Memorial Library, Mississippi University for Women, for generously taking the time from her busy schedule to search through the University's archives and locate very valuable material relating to Dr. Hazen. A special thank you to Dr. Margarita Silva-Hutner for allowing me to interview her about Dr. Hazen, her close friend

and colleague, and for providing fascinating insights into Dr. Hazen's personality, contributions, and honors.

Bibliography

Bacon, W. Stevenson. "Elizabeth Lee Hazen, 1885–1975." *Mycologia* 68 (1976): 961–69.

Baldwin, Richard S. *The Fungus Fighters: Two Women Scientists and Their Discovery.* Ithaca: Cornell University Press, 1981.

Gilbert Dalldorf and the Dalldorf Fellowship in Medical Mycology. Tuscon: Research Corporation, 1991.

Hazen, Elizabeth Lee. "General and Local Immunity to Ricin." *Journal of Immunology* 13 (1927): 171–218.

Hazen, Elizabeth, and Frank Curtis Reed. *Laboratory Identification of Pathogenic Fungi Simplified,* 1st ed. Springfield, Ill.: Charles C. Thomas, 1955.

Hazen, Elizabeth L., and Rachel Brown. "Two Antifungal Agents Produced by a Soil Actinomycete." *Science* 112 (1950): 423.

Macdonald, Anne L. *Feminine Ingenuity: Women and Invention in America.* New York: Ballantine Books, 1992.

New York State Association of Public Health Laboratories. "The Discovery, Development, and Application of Nystatin." *New York State Association of Public Health Laboratories, Proceedings* (1955): 58–62.

O'Hern, Elizabeth Moot. "Elizabeth Hazen (1895–1975)." In *Profiles of Pioneer Women Scientists,* with introduction by Morris Schaeffer, pp. 95–102. Washington, D.C.: Acropolis Books, 1985.

Rivett, Francis. "The Case of the Florentine Fungus." *Health News* (New York State Department of Health) 44 (1967): 18–19.

Rubin, Lewis P. "Elizabeth Lee Hazen." In *Notable American Women: The Modern Period,* eds. Barbara Sicherman and Carol Hurd Green, pp. 326–28. Cambridge, Mass.: Harvard University Press, 1980.

Siegel, Patricia Joan, and Kay Thomas Finley. "Elizabeth Lee Hazen," in *Women in the Scientific Search: An American Bio-Bibliography,* pp. 248–50. Metuchen, N.J.: Scarecrow Press, 1985.

Silva, Margarita, and Elizabeth L. Hazen. "Rhoda Williams Benham, 1894–1957." *Mycologia* 49 (1957): 596–603.

Stanley, Autumn. *Mothers and Daughters of Invention: Notes for a Revised History of Technology.* Metuchen, N.J.: Scarecrow Press, 1993.

Vare, Ethlie Ann, and Greg Ptacek. *Mothers of Invention. From the Bra to the Bomb: Forgotten Women and Their Unforgettable Ideas.* New York: William Morrow, 1987.

MARIA CHIARA

HELOISE

(ca. 1098–1164)

Natural Philosopher

Birth	ca. 1098
ca. 1118	Began further education with Peter Abelard
Death	May 15, 1164

Heloise, abbess of the convent of the Paraclete in twelfth-century France, is perhaps better known to posterity because of the circumstances of her tragic romance with Peter Abelard, a great medieval theologian, than for her own intellectual gifts, which were considerable. Heloise was born in France ca. 1098 (some sources give her year of birth as 1101), and by the time she was 17 she was living in the home of her uncle Fulbert, a canon of the cathedral of Notre Dame in Paris. Here she met Peter Abelard, whom Fulbert hired to continue the education of his remarkable niece.

By the time Heloise came under the tutelage of Abelard she already had a reputation as a scholar, and Abelard continued her education. Abelard himself was perhaps the most notorious teacher of the twelfth century. His fame at the time he became involved with Heloise was extensive, although he was even then a subject of controversy.

The two became lovers, and Heloise soon became pregnant. She retired to a convent in Brittany to give birth to the couple's son, whom they named Astrolabe. After she returned from Brittany, Heloise and Abelard were secretly married. When Heloise's relatives discovered what had happened, they were outraged; in revenge, they had Abelard castrated. Abelard retired to the monastery of St. Denis just outside Paris, and Heloise was sent to the convent at Argenteuil.

The enemies of both Abelard and Heloise were many. The convent at Argenteuil was dispersed, and Heloise went to the community of the Paraclete, near Nogent-sur-Seine, where Abelard gave Heloise and her nuns property he owned. Heloise thus became abbess of the convent, where she remained until her death in 1164.

According to tradition, Abelard instructed Heloise in medicine, along with the other traditional subjects of medieval higher education. At least one source gives Heloise quite a reputation as a physician in her role as abbess of the Paraclete, but there is little hard evidence to substantiate these claims. Heloise was certainly one of the most intellectual and best-

Heloise. Illustration courtesy of the Library of Congress.

educated women of the Middle Ages. Her surviving letters to Peter Abelard give full evidence of the power and depth of her mind, although some misogynistic writers have attempted to prove that Heloise herself did not write the letters.

Heloise was buried beside Peter Abelard at the Paraclete, but in the nineteenth century the remains of both were removed to the Père-Lachaise cemetery in Paris, where they lie together forever.

Bibliography

Alic, Margaret. *Hypatia's Heritage: A History of Women in Science from Antiquity through the Nineteenth Century.* Boston: Beacon Press, 1986.

Dronke, Peter. *Women Writers of the Middle Ages: A Critical Study of Texts from Perpetua (+203) to Marguerite Porete (+1310).* Cambridge: Cambridge University Press, 1984.

Echols, Anne, and Marty Williams. *An Annotated Index of Medieval Women.* New York: Markus Wiener Publishing, 1992.

Gilson, Etienne. *Heloise and Abelard.* Ann Arbor: University of Michigan Press, 1960.

Hurd-Mead, Kate Campbell. *A History of Women in Medicine: From Earliest Times to the Beginning of the Nineteenth Century.* Boston: Milford House, 1973 (first printed in 1938).

Klapisch-Zuber, Christiane, ed. *Silences of the Middle Ages. A History of Women in the West,* Vol. 2. Cambridge, Mass.: Belknap Press, 1992.

DEAN JAMES

HERRAD OF LANDSBERG, ALSO CALLED HERRADE AND HERRAD OF HOHENBURG
(Fl. 1150s–1195)
Natural Philosopher

Birth	Fl. 1150s
1160–70	Began writing encyclopedia, *Hortus deliciarum*
Death	1195

Like many educated women of the Middle Ages, Herrad of Landsberg was a nun. Historians have been unable to determine the year in which

Herrad was born, but by 1167 she was noted as a teacher when she became abbess of the convent of Hohenburg on Mount St. Odile in Alsace. Herrad had been sent to the abbey as a child, and she remained there until her death in 1195.

Herrad was perhaps the first woman ever to compile an encyclopedia, any number of which had been written by men before, during, and after the Middle Ages. She wrote the greater part of her *Hortus deliciarum* (or *Garden of Delights*) between 1160 and 1170, but she continued to add to it until at least 1190. The *Hortus deliciarum* is an encyclopedia of religion, history, astronomy, geography, philosophy, natural history, and medical botany. Technical terms are given in both Latin and German, which meant that the encyclopedia could serve as a teaching tool for the novices and nuns of the convent of Hohenburg. Herrad is considered by many historians to have illustrated the *Hortus* herself with some 636 illustrations; indeed, the manuscript is of great artistic value, aside from its value to historians of medieval science and medicine. The only manuscript copy of the *Hortus* was destroyed, except for a few pages, in a siege of the city of Strasbourg in 1870, but fortunately an early nineteenth-century scholar had copied large sections of it.

In compiling the *Hortus deliciarum*, Herrad relied primarily on biblical sources and her own knowledge and experience of herbal medicine, although she was perhaps also familiar with the work of the school of medicine at the university in Salerno and the work on drugs of Isidore of Seville. There are striking similarities between the work of Herrad and that of possibly the most famous woman physician-scientist of the Middle Ages, **Hildegard of Bingen** (1098–1179). As far as historians can determine, however, there was no direct contact between the two women, even though they lived at approximately the same time. Both women apparently put their practical experience with the medicine of the day into use in their own communities. For example, in 1187 Herrad had a large hospital built on the grounds of the convent of Hohenburg, and here she worked as chief physician until her death in 1195. This role as provider of healthcare for the community, both secular and cloistered, was characteristic of monastic foundations in the Middle Ages.

In addition to the many subjects of study that the *Hortus* covers, it offers glimpses of the inner workings of the convent of Hohenburg and the daily lives of the nuns and novices there. Another important component of the *Hortus*, at least in practical terms, was the table that Herrad worked out for the computing of medieval feast days. Feast days were very important in the life of the medieval church, and Herrad's table was considered one of the best. It provided the dates for Easter and the day of the week for Christmas for a cycle of 532 years, from 1175 to 1706. As a book of medicine, the *Hortus* describes plants and their medicinal uses.

Herrad, along with her peers **Hildegard of Bingen** and **Heloise** of the Paraclete, was one of the last great scholar-abbesses of the Middle Ages. Trends in monasticism with regard to women and their education in the convent had begun to change by the late twelfth century, and thus we see few figures like these three women in the late Middle Ages.

Bibliography

Alic, Margaret. *Hypatia's Heritage: A History of Women in Science from Antiquity through the Nineteenth Century.* Boston: Beacon Press, 1986.

Carr, Anne Marie Weyl. "Women Artists in the Middle Ages." *Feminist Art Journal* 5 (Spring 1976): 5–9.

Dronke, Peter. *Women Writers of the Middle Ages: A Critical Study of Texts from Perpetua (+203) to Marguerite Porete (+1310).* Cambridge: Cambridge University Press, 1984.

Echols, Anne, and Marty Williams. *An Annotated Index of Medieval Women.* New York: Markus Wiener Publishing, 1992.

Hurd-Mead, Kate Campbell. *A History of Women in Medicine: From Earliest Times to the Beginning of the Nineteenth Century.* Boston: Milford House, 1973 (first printed in 1938).

Katzenellbogen, Adolf. *Allegories of the Virtues and Vices in Medieval Art: From Early Christian Times to the Thirteenth Century.* Medieval Academy Reprints for Teaching, No. 4. Toronto: University of Toronto Press, 1989.

Klapisch-Zuber, Christiane, ed. *Silences of the Middle Ages. A History of Women in the West,* Vol. 2. Cambridge, Mass.: Belknap Press, 1992.

DEAN JAMES

HOPE HIBBARD

(1893–1988)

Zoologist

Birth	1893
1916	B.A., University of Missouri
1918	M.A., University of Missouri
1919–20	Demonstrator in Biology, Bryn Mawr College
1921	Ph.D., zoology, The Sorbonne
1921–25	Associate Professor, Elmira College

1925–26	Berliner Fellow, The Sorbonne
1927–28	International Education Board Fellow, The Sorbonne
1928–30	Assistant Professor, Oberlin College
1930–33	Associate Professor, Oberlin College
1933–61	Professor, Oberlin College
1944	"Starred" in seventh edition of *American Men of Science*
1954–58	Chair, Dept. of Zoology, Oberlin College
1961–88	Professor Emerita, Oberlin College
1987	Honorary Life Member, American Association of University Women
Death	1988

Hope Hibbard was best known for her seminal work on the Golgi apparatus, which was done before the invention of the electron microscope. Because the Golgi apparatus is a cytoplasmic component so fine in structure that its details and function in the cell were not apparent without an electron microscope, Hibbard was said to be intensely frustrated by her inability to research it fully.

Hope Hibbard was born in Altoona, Pennsylvania, in 1893 to Mary (Scofield) Hibbard and Herbert Wade Hibbard. Two years later, Mary Scofield Hibbard died. Herbert Wade Hibbard then married Mary Davis and moved the family to Minneapolis, where he worked as a professor of mechanical engineering on the faculties of the University of Minnesota and Cornell University. He joined the engineering faculty of the University of Missouri in 1909. It was at this university that Hope Hibbard received a baccalaureate degree in 1916 and then a master's degree in 1918. In 1921 she went to Bryn Mawr College for the first of her two doctorates.

Hibbard's dissertation at Bryn Mawr College was entitled "Cytoplasmic Inclusions in the Egg of Echinarachnius Parma." It was this dissertation, a 27-page study of fertilization in sea urchin eggs, that she submitted to earn her Ph.D. from Bryn Mawr College in 1921. Dr. Hibbard then took up a teaching job from 1921 to 1925 at Elmira College in New York, but she became restless living in a small, rural environment and wanted to experience much more of the world of science.[1] She applied for a fellowship from the American Association of University Women to do research in Paris. After being awarded the fellowship, she began her studies at the Laboratory of Anatomy and Comparative Histology at the Sorbonne, where she began a study of oogenesis in frog eggs. This fellowship was extended for another year, and Hibbard was able to stay for a third year on a small stipend from a teaching assistantship. Her second doctorate was awarded by the Sorbonne at the end of this third year.

Hibbard then returned to the United States to take a teaching position at Oberlin College. She was granted tenure and promoted to associate professor only two years after coming to Oberlin. Professor David Egloff says, "Hope came to Oberlin, not because of its location, but because of the good colleagues and students who spend the winter here." Oberlin had strong connections with the Marine Biological Laboratory in Massachusetts. Hibbard had spent a summer there in 1917 and enjoyed it very much. Because of the laboratory's strong history of encouraging women to do science, its innovation in science, and its links with Oberlin College, Hope Hibbard came to Oberlin to work and study after receiving her second doctorate from the Sorbonne.[2]

Hibbard's research at Oberlin in the 1930s was in histology. She published papers on studies of the tissues and organs of limpets, earthworms, squid, and silkworms. Her last published studies were done in the 1940s, when she began working on the Golgi apparatus. Her definitive review of the Golgi apparatus continues to be cited in the literature.

Hope Hibbard is well remembered on Oberlin's campus for her excellent teaching skills and the inspiration she lent her students. Her dedication to encouraging her students to achieve their full potential is mentioned more than once in correspondence from Oberlin College.[3] Egloff mentions that her strength of character was such that less than five weeks after removal of a lung for cancer, she gave the introductory biology lecture in Hall Auditorium.[4]

Egloff, who worked with Hibbard, describes the less-than-ideal conditions for biological research present during Hibbard's tenure at Oberlin. It was not until after she had retired in 1961 that the biology department moved into the new science building on Oberlin's campus. Approximately 12 years later, Oberlin acquired a transmission electron microscope, which was dedicated to Hibbard in 1973. Egloff says, "At last Oberlin biologists had the means to clearly see the Golgi apparatus."[5] Because Dr. Hibbard had worked so exhaustively on the Golgi apparatus, this dedication seems very fitting. For 32 of her 33 years at Oberlin, she was the only woman faculty member of the Department of Zoology.

After Hibbard's retirement in 1961, she cared for her second mother, Mary Davis Hibbard. After Mary Hibbard died in 1967, Hope began to travel extensively. Professor Egloff mentions Hawaii, Florence, Italy, Australia, New Zealand, South America, Africa, and Europe as being among her destinations. At age 84 she visited Israel, Egypt, and Yugoslavia. One notable event in her life occurred at the age of 88, when she traveled two nights by bus from Oberlin to the United Nations to attend a nuclear disarmament rally, stating that she would give her life to achieve disarmament.[6]

Other achievements in Hibbard's life include her position as chair of the Department of Zoology from 1954 to 1958 and her election to the Adelia A. Field Johnson Professorship, which was established in 1898 by contributors who desired "that Oberlin College should have a woman professor in its faculty of instruction." Professor Hibbard received this honor in 1952; when she retired in 1961, the professorship was retired with her. She was also included in the fourth through eleventh editions of *American Men of Science*, having achieved the accolade of a star next to her name (signifying those scientists considered to be eminent in their fields and among the top 1,000 scientists in America). Only 27 women were starred in *American Men of Science* from 1921 through 1943.[7] Hibbard also founded the Oberlin Chapter of the League of Women Voters and served in the American Association of University Women, in which she was an honorary life member. She died on May 12, 1988, at age 94.

Notes

1. David Egloff, "Memorial Minute: Hope Hibbard," *Oberlin Alumni Magazine* (Fall 1988): 54.

2. *Ibid.*

3. *Ibid.*, p. 55.

4. *Ibid.*, p. 54.

5. *Ibid.*

6. *Ibid.*

7. Margaret W. Rossiter, *Women Scientists in America: Struggles and Strategies to 1940* (Baltimore: Johns Hopkins University Press, 1982), pp. 291, 293.

Bibliography

American Men and Women of Science, eds. 1–14, Cumulative Index. Compiled by Jaques Cattell Press. New York: R.R. Bowker Co., 1983.

Egloff, David. "Memorial Minute: Hope Hibbard." *Oberlin Alumni Magazine* (Fall 1988): 54–55.

Rossiter, Margaret W. *Women Scientists in America: Struggles and Strategies to 1940*. Baltimore: Johns Hopkins University Press, 1982.

KIMBERLY J. LAIRD

HILDEGARD OF BINGEN

(1098–1179)

Physician

Birth	1098 at Bockelheim, Mainz, Germany
1136–79	Abbess of Disibodenberg; later moved near Bingen
1145	Moved the convent to its location near Bingen
1151–58	Wrote *Book of Simple Medicine* and *Book of Compound Medicine*
1158–63	Wrote *Book of Life's Merits*
Death	1179 at Rupertsberg, near Bingen

Known as the "Sibyl of the Rhine," Hildegard was perhaps the foremost writer on science in the twelfth century. A visionary and a theologian, she gained fame as a healer as well. Although she was never officially canonized by the Catholic Church, she is listed in the *Acta sanctorum* and is thus known as St. Hildegard of Bingen.

Hildegard was born in 1098 to parents of the minor German nobility. Other than the name Mechtilde, nothing is known of Hildegard's mother; but her father, Hildebert, was a knight in attendance upon Meginhard, count of Sponheim. By age 5 Hildegard had begun receiving visions from God, and at age 7 she was sent to nearby Disibodenberg to the tutelage of her aunt Jutta, an anchoress whose cell was attached to the Benedictine monastery there. Here it was that Hildegard apparently began her education, and soon Jutta began to attract more followers. After Jutta's death in 1136, Hildegard succeeded her as abbess of the convent, which had grown out of the anchorite's cell. By 1145, Hildegard and her nuns had outgrown their facilities. Hildegard sought a new location, moving with her 18 nuns to a new convent on the Rupertsberg, near Bingen.

Hildegard continued to enjoy the visions from God that had made her famous, and she used her growing fame as a springboard from which to involve herself in the political issues of the day. She corresponded with popes and secular rulers alike, not hesitating to castigate the formidable Frederick Barbarossa, the Holy Roman Emperor, when his actions displeased her. Hildegard had many enemies. In fact, her later years were somewhat troubled as a result of her enemies' machinations against her. She died in 1179 at age 81, revered by many and well known throughout Latin Christendom.

Hildegard's reputation as a scientist and physician rests on several of her writings. Much of what Hildegard wrote was perforce influenced by her milieu and the theology of the Catholic Church. To understand her properly, one must consider her context, both spiritual and temporal. Hildegard did demonstrate in her writings the accepted characteristics of the scientist. Although she might have dwelled on the spiritual and its implications, she supplied theoretical explanations for what she had witnessed in the physical world. This approach was perfectly acceptable to the medieval audience for which she wrote. For them, there was a fundamental connection between the spiritual and physical, and science was a necessary bridge between the two.

Two of Hildegard's five main works belong more properly to the realm of science than to theology. The first of these is known as the *Physica*—or, in English, *Natural History* or *Book of Simple Medicine*. Written from 1151 to 1158, the *Physica* contains references to nearly 500 plants, metals, stones, and animals and explains their medicinal value to human beings. The second work, written during the same period, is known as the *Causae et curae*—or, in English, *Causes and Cures* or *Book of Compound Medicine*. This work consists of five sections, the first of which covers the creation of the world, both cosmology and cosmography. The second places humans in this context, expounds on the medieval theory of the humors, and then lists the diseases to which human beings are subject. The next two sections present remedies for the diseases in the second section. The final, fifth section contains a mixture of topics, including signs of life and death and astrological prognostications according to the phase of the moon at the time of conception.

Hildegard's influence was tremendous. Her name was renowned throughout medieval Europe, and spurious prophecies and writings circulated for a long time. Unlike her contemporary, **Herrad of Landsberg** (d. 1195), Hildegard's works were copied over and over and were widely disseminated. One historian, George Sarton, went so far as to call Hildegard the most distinguished naturalist and the most original twelfth-century philosopher in Western Europe.[1] Of all medieval women scientists, Hildegard of Bingen is perhaps justly the most famous.

Note

1. George Sarton, *Introduction to the History of Science* (London: Bailliere, Tindall & Cox, 1927–48), Vol. 2, Pt. 1, p. 310.

Bibliography

Alic, Margaret. *Hypatia's Heritage: A History of Women in Science from Antiquity through the Nineteenth Century*. Boston: Beacon Press, 1986.
Dronke, Peter. *Women Writers of the Middle Ages: A Critical Study of Texts from*

Perpetua (+203) to Marguerite Porete (+1310). Cambridge: Cambridge University Press, 1984.

Echols, Anne, and Marty Williams. *An Annotated Index of Medieval Women*. New York: Markus Wiener Publishing, 1992.

Flanagan, Sabina. *Hildegard of Bingen, 1098–1179: A Visionary Life*. London: Routledge, 1989.

Hurd-Mead, Kate Campbell. *A History of Women in Medicine: From Earliest Times to the Beginning of the Nineteenth Century*. Boston: Milford House, 1973 (first printed in 1938).

Klapisch-Zuber, Christiane, ed. *Silences of the Middle Ages. A History of Women in the West*, Vol. 2. Cambridge, Mass.: Belknap Press, 1992.

Newman, Barbara. *Sister of Wisdom: St. Hildegard's Theology of the Feminine*. Berkeley: University of California Press, 1987.

Ogilvie, Marilyn Bailey. *Women in Science, Antiquity through the Nineteenth Century: A Biographical Dictionary with Annotated Bibliography*. Cambridge, Mass.: MIT Press, 1986.

Sarton, George. *Introduction to the History of Science*. 3 vols. London: Bailliere, Tindall, & Cox, 1927–1948.

Thorndike, Lynn. *A History of Magic and Experimental Science*. 8 vols. New York: Columbia University Press, 1923–1958.

DEAN JAMES

ARIEL CAHILL HOLLINSHEAD

(1929–   )

Pharmacologist, Cancer Researcher

Birth	August 24, 1929
1951	A.B., Ohio University
1955	M.A., George Washington University
1957	Ph.D., pharmacology, George Washington University
1958	Married Montgomery Hollinshead
1958–59	Assistant Professor of Virology and Epidemiology, Baylor University Medical Center
1959–61	Assistant Professor of Pharmacology, George Washington University
1961–73	Associate Professor of Pharmacology, George Washington University

1964–89	Established and developed the Laboratory for Virus and Cancer Research
1974–	Professor of Medicine, George Washington Medical Center
1977	Honorary D.Sc., Ohio University; Chair, Review Board of Oncology, Veterans Administration
1980	Star of Europe Medal
1985	Distinguished Scientist Award, Society for Experimental Biology and Medicine
1990	Silver Medal, Scholar Speciale Medicina, Italy

Ariel Cahill was born in 1929 to an educated Pennsylvania family. Her father was a Quaker with an engineering degree from Lehigh. Her mother had been both president and valedictorian of her class at Barnard College. Ariel's parents brought her up to feel equal to her older brother. At age 15, while her brother served in the Navy in World War II, Ariel read Paul De Kruif's *The Microbe Hunters*. This book inspired her to pursue an investigative career in the sciences, becoming the first to identify animal and human antigens in cancerous tumors.

Ariel was fortunate to be able to complete her studies uninterrupted to become a medical researcher, studying first at Swarthmore College in Pennsylvania and then receiving her bachelor's degree from Ohio University in 1951, where she served as an instructor during her senior year. During this same period her older brother was serving in Korea. Ariel went on to George Washington University to earn a master's degree in 1955 and a Ph.D. in pharmacology in 1957. In 1958 she married attorney Montgomery Hollinshead. They have two children, William and Christopher.

After receiving her doctorate, Dr. Hollinshead accepted a postdoctoral fellowship in virology at Baylor Medical School. She enjoyed her work there with Dr. Joseph Melnick but returned to Washington, D.C., to be with her new husband. She became an assistant professor at George Washington University in 1959 and was promoted to associate professor just two years later. From 1964 through 1989 she worked to establish the Laboratory for Virus and Cancer Research, starting as its head and then serving as its director. She now is its president.

While she was establishing herself as a researcher, Ariel was beginning her family. She notes wryly some of her experiences fulfilling her "expected" role: ironing seven shirts each week; planning, buying, and preparing meals. "I had a little chart to convert teaspoons and pounds into milliliters and grams, since I had never cooked and had many chemistry courses."[1] Most of her salary went to pay a nanny before the boys were old enough to attend Sidwell Friends, a private school. There was no maternity leave, and two weeks of annual vacation did not go far. She often took her babies to the laboratory in the evenings and on weekends.

Their company did not distract her from her work. Dr. Hollinshead is a prolific researcher, having written over 260 scientific papers. Her research includes basic and clinical work in many kinds of cancers throughout the body. She has been involved in the development of immunotherapies for larynx, breast, lung, and gastric cancers. She has studied chemotherapy in animal viral diseases and cancers. Hollinshead is also interested in the way viruses can trigger neoplasms. Investigating the nucleoprotein chemistry of viruses, she was the first to isolate, purify, and identify animal, and then human, tumor-associated antigens (TAA). She developed epitopes for study and use in human cancers. Her other research interests include cancer immunogenetics and human-human hybridoma research. With T.H.M. Stewart, she is the co-discoverer of a specific immunotherapy cancer treatment using TAA immunogens to produce a long-lasting cell-mediated immunity. In 1994, Dr. Hollinshead worked alone on the "development of a breast cancer drug sensitivity test and . . . a possible combination immunochemotherapeutic approach to the treatment of AIDS."[2]

In 1977 this top-notch researcher became the first woman appointed to chair the Review Board of Oncology for the Veterans Administration. She was presented with the prestigious Medical Woman of the Year Award by the Board of American Medical Colleges. Her other honors include Distinguished Cancer Scientist in the United States, awarded by the American Association for the Advancement of Science; a 1988 award as Outstanding Woman of America; and the 1980 Star of Europe. Dr. Hollinshead is a member of the Society for Experimental Biology and Medicine, winning its 1985 Distinguished Scientist Award; the International Agency for Research on Cancer; and the American Society of Clinical Oncology. She is also a member of a number of honor and service organizations, including Phi Beta Kappa, the National Honor Society, and Graduate Women in Science/Sigma Delta Epsilon.

Hollinshead has said of being female in medical research, "You get knocked plenty, but you just get up and fight."[3] She was an early advocate of mentoring and networking for women in the sciences and has established prizes for papers on this topic. "I . . . spent time I did not have on these things as well as giving much time to Graduate Women in Science/Sigma Delta Epsilon."[4]

Four volumes of Dr. Hollinshead's papers are already housed in the archives of the Medical College of Pennsylvania. Recently her family has grown, as both sons married women in the same fields as theirs. William married Nancy, a scientist, and Christopher married Maria, a lawyer. Ariel enjoys spending time with her family, playing the piano, oil painting, and swimming. She reads many books, including those about the lives of members of the Society of Friends.

Notes

1. Ariel Hollinshead, personal letter to Kelly Hensley, November 1994.
2. *Ibid.*
3. Lois Decker O'Neill, *The Women's Book of World Records and Achievements* (Garden City, N.Y.: Anchor Press/Doubleday, 1979), p. 223.
4. Ariel Hollinshead, personal letter to Kelly Hensley, November 1994.

Bibliography

American Men and Women of Science 1992–1993, 18th ed. New Providence, N.J.: R.R. Bowker, 1992.
De Kruif, P. *The Microbe Hunters.* New York: Harcourt, Brace, 1926.
O'Neill, Lois Decker. *The Women's Book of World Records and Achievements*, pp. 222–23. Garden City, N.Y.: Anchor Press/Doubleday, 1979.
Who's Who in Health Care, 1st ed. New York: Hanover Publications, 1977.

KELLY HENSLEY

RUTH HUBBARD
(1924–   )
Biologist

Birth	March 3, 1924
1944	A.B., Radcliffe College
1950	Ph.D., Radcliffe College
1950–52	Research Fellow, Harvard University
1952–53	Guggenheim Fellow, Carlsberg Laboratory, Copenhagen
1954–58	Research Fellow, Harvard University
1958–74	Research Associate, Harvard University
1967	Paul Karrer Medal (with George Wald), University of Zürich
1974–90	Professor of Biology, Harvard University
1982–	Board of Directors, Council for Responsible Genetics

1985	Peace and Freedom Award, Women's International League for Peace and Freedom
1990–	Professor Emerita, Harvard University; Board of Trustees, Trustee Emerita, Marine Biological Laboratory, Woods Hole, MA
1991	D.Sc., Macalester College; D.H.L., Southern Illinois University, Edwardsville; D.Sc., University of Toronto; Feminist Marathoner Award, National Organization for Women, Boston Chapter; Board of Directors, Civil Liberties Union of Massachusetts
1992	Distinguished Service Award, American Institute of Biological Sciences

Ruth Hubbard—scientist, educator, women's health advocate, and activist—is a woman with impeccable credentials who does not want to leave the future to the technocrats. She wants to bring science to the people.

Ruth Hubbard was born in 1924 in Vienna, Austria. Her parents were both doctors, and as a girl, Hubbard sometimes followed her father on his home-visits.

> Once I stopped wanting to be a nursemaid (a role model for little girls in my socioeconomic class equivalent to the fireman for boys), I assumed that I, too, would be a physician. When I moved [to America] . . . I was too much of an outsider to the social life and expectations of American adolescents to realize that here this was a rather non-traditional ambition for a girl.[1]

When Hitler annexed Austria in 1938, Hubbard's family, which was Jewish, emigrated to America. Hubbard attended her final two years of high school in Brookline, Massachusetts. In 1941 she enrolled at Radcliffe College to pursue the standard premedical course, Biochemical Sciences, and married Frank Hubbard, who later pioneered the modern harpsichord revival. Originally planning for a career as a clinician, she applied to medical school and was accepted at the Woman's Medical College of Pennsylvania. But after her undergraduate years at a women's college, the prospect of spending several more years at a women's school no longer appealed to her. Rejecting the offer of admission, she instead embarked on a career in research in a coeducational setting. Hubbard worked in both research and clinical laboratories, at Harvard (1944–1945 and 1946–1948), at the Tennessee Public Health Service (1945–1946), and at the University College Hospital Medical School in London (1948–1949). "I loved the atmosphere of the laboratory, the discussions, the socializing, the give and take," she explained.[2]

During the late 1940s Hubbard published several articles from her research. Her first published piece, a collaboration with D.R. Griffin and George Wald (the man who later became her husband), was entitled "The Sensitivity of the Human Eye to Infra-Red Radiation." This was the first of many scientific papers and review articles Hubbard would author, both independently and with colleagues, on the biochemistry and photochemistry of vision in vertebrates and invertebrates.

Hubbard returned to Cambridge to do research in Wald's Harvard laboratory and to begin formal work toward a Ph.D. in biology from Radcliffe, which she completed in 1950. She later reflected on her career's trajectory:

> I had done some teaching while in graduate school, but decided that in a choice of teaching or research, I wanted to do research. Now why did I think I had to make that choice? Because though I wasn't really thinking about it in a conscious way, I realized that as a woman, I had to choose. The point is that the large research universities . . . where young faculty members were expected to excel in research while also teaching, were not hiring my kind. They were hiring young men. The colleges that were hiring women had little or no research facilities or support.[3]

Hubbard stayed on in Cambridge, Massachusetts, as a research fellow at Wald's laboratory until 1952, when she traveled to Copenhagen as a Guggenheim fellow at the Carlsberg Laboratory. After her year abroad, Hubbard returned to Harvard as a research fellow in Wald's laboratory and remained there, moving up the ranks to positions that combined research and teaching: from research fellow to research associate, to lecturer, and then in 1974 to professor, through 1990, when she became professor emerita. During this time she and Wald published a dozen articles in collaboration, married, and had two children. Elijah was born in 1959 and Deborah Hannah in 1961.

From the late 1940s through the early 1970s, Hubbard's work focused on the biochemistry and photochemistry of vision in vertebrates and invertebrates, an area in which she published many articles and reviews. External events in the late 1960s and early 1970s precipitated an important shift in her career's focus. Two major political upheavals, the Vietnam War and the women's movement, made her realize that science is neither impersonal nor objective, and "recognize that the ways scientists work are strongly affected by their social and personal backgrounds."[4] With this realization, she wrote:

> I could not be satisfied with continuing the kinds of research I had been doing during the 1950s and 1960s. I became more and more interested in exploring the basic assumptions that make scientists

believe that we can understand the world, by asking the kinds of questions about it that I had been asking at the laboratory bench. ... I finally began to notice the misogynist ways in which male biologists and physicians have described women and the effects this has had on biological theory and worse than that, on women's health.[5]

Recognizing how society's attitudes toward women and science had affected her own opportunities as a scientist, she reflected on her early career:

I genuinely believed that Harvard was doing me a favor to let me work there. Though I knew that my work was good, that I was getting international recognition for it and that recognition accrued to the greater glory of the institution which "allowed" me to work within its ivied walls, it did not occur to me that I ought to be offered a real job. I also did not understand, not consciously at any rate, that my male contemporaries were going on to bigger and better jobs that gave them more power and control over their own lives, while I was dead-ended in a situation that could, in fact, end if "my professor" were to leave.[6]

When she received tenure in 1973, she wrote:

A reporter asked me whether I thought that my being a woman had something to do with my being appointed. I answered him then, as I would now, that I didn't know whether my being a woman had something to do with my getting tenure in 1973, but that I was sure that my being a woman had something to do with my not even having been considered before.[7]

Hubbard's new perspective on scientific practice changed the way she chose to present her own work. Rather than concentrating on working in university laboratories and classrooms or writing in scholarly journals, she decided to reach out proactively to educate a far broader audience. Hubbard began to publish articles on women's health in more popular, less academic journals and magazines. Notably, she worked as a consultant to the Boston Women's Healthbook Collective, the group that published the classic book on women's health, *Our Bodies, Ourselves*. Beyond her writings, Hubbard worked as a counselor at the Pregnancy Counseling Service of Boston in 1972 and served as a board member of several organizations, including the National Women's Health Network, Science for the People, the Council for Responsible Genetics, and the Massachusetts chapter of the American Civil Liberties Union.

Hubbard is a scientist who has not been afraid to examine her own discipline critically, even when it has meant recognizing personal limitations:

> When I as a feminist write and think about science it is not because I think that I . . . can operate in the "free spirit" of unbiased inquiry. Rather, I think that it is my job to try to understand what my own limitations are—imposed by my class, race, sex, ethnicity, age, sexual orientation and the many other constraints and opportunities that define the range of my experiences—and try to compensate for them by working together with people whose experiences differ from mine. When we do that, we still have a limited view, since we are members of a particular culture living in a particular time and place. But, at least, we have a broader base from which to decide what questions we should be asking, how we should ask them, and what answers are relevant and meaningful. I believe that some such, more representative process of inquiry is important no matter what types of questions scientists ask. It becomes crucial when we ask questions about human biology, behavior and health. Women have far too long accepted the edicts of "experts" even when what we are told contradicts our own experience. It is important that women and other kinds of people (women or men) who have been underrepresented among the social groups involved in generating knowledge, recognize that our absence invalidates the traditional descriptions of the world and the actions that are based on such representations.[8]

Whereas in the 1970s and 1980s Hubbard concentrated on women's health concerns, in the 1990s she spreads her energies more widely. She continues with women's health advocacy and, in addition, has taken on the campaign to educate the public about the social implications of biotechnology and genetic engineering. An educator and activist, she writes with the intention to inspire and motivate her readers to take action. In *Exploding the Gene Myth* she writes, "This is a crucial time in the development of genetics and biotechnology. . . . It is crucial that we, as citizens, not leave this process in the hands of the 'experts.' "[9]

Notes

1. Ruth Hubbard, "Reflections on My Life as a Scientist," *Radical Teacher* (Jan. 1986): 3.
2. *Ibid.*
3. *Ibid.*, p. 4.
4. *Ibid.*, p. 6.
5. *Ibid.*, p. 4.

6. *Ibid.*
7. *Ibid.*
8. *Ibid.*, p. 7.
9. Ruth Hubbard and Elijah Wald, *Exploding the Gene Myth* (Boston: Beacon Press, 1993), p. xiii.

Bibliography

Hubbard, Ruth. *The Politics of Women's Biology.* New Brunswick, N.J.: Rutgers University Press, 1990.

———. *Profitable Promises: Essays on Women, Science and Health.* Monroe, Maine: Common Courage Press, 1994

———. "Reflections on My Life as a Scientist." *Radical Teacher* (Jan. 1986): 3–7.

Hubbard, Ruth, and Elijah Wald. *Exploding the Gene Myth.* Boston: Beacon Press, 1993.

Hubbard, Ruth, and Lynda Birke, eds. *Reinventing Biology: Respect for Life and the Creation of Knowledge.* Bloomington: Indiana University Press, 1995.

Hubbard, Ruth, and Margaret Randall. *The Shape of Red: Insider/Outsider Reflections.* Pittsburgh and San Francisco: Cleis Press, 1988.

Hubbard, Ruth, and Marian Lowe, eds. *Genes and Gender II: Pitfalls in Research on Sex and Gender.* New York: Gordian Press, 1979.

———. *Woman's Nature: Rationalizations of Inequality.* New York: Pergamon, 1983.

Hubbard, Ruth, et al., eds. *Biological Woman: The Convenient Myth.* Cambridge, Mass.: Schenkman, 1982.

———. *Women Look at Biology Looking at Women.* Cambridge, Mass.: Schenkman, 1989.

JENNIFER LIGHT

IDA HENRIETTA HYDE
(1857–1945)
Physiologist

Birth	September 8, 1857
1891	B.S., Cornell University
1893–94	European Fellow, Association of Collegiate Alumnae
1894	Petitioned for German universities to be opened to women

1896	Ph.D., physiology, University of Heidelberg; conducted research in marine biology at the Heidelberg Table, Naples Zoological Station; founded the American Women's Naples Table Association (for promoting scientific research by women)
1897	Researcher, Harvard Medical School; first woman to conduct research in the laboratories
1898	Associate Professor of Physiology, University of Kansas
1902	First woman member of the American Physiological Society
1905	Professor of Physiology and Head of newly separate Dept. of Physiology, University of Kansas
1918–19	Invented the first microelectrode for intracellular work
1920	Retired from University of Kansas
1927	Endowed Ida H. Hyde Scholarship
1945	Established Ida H. Hyde Woman's International Fellowship, American Association of University Women
Death	August 22, 1945

The career of Ida Henrietta Hyde can be characterized as one of many obstacles and many firsts. She was known as a great professor and researcher. She concentrated in the field of physiology but also conducted research and taught in zoology and medicine. Her research focused on the circulation, respiratory, and nervous systems of animals, with a special interest in embryological development. She "showed originality, breadth of interest, and admirable scientific precision."[1] As did many scholars in her field at the time, she tended to analyze a system with respect to its evolutionary development. She also was one of the first researchers to study at the micro-level by stimulating an individual cell and examining it.

Hyde was born in 1857 in Davenport, Iowa, to Mayer (or Meyer) H. and Babette (Loewenthal) Heidenheimer. The family name was later changed to Hyde. Her parents were originally from Württemberg, Germany. Her father was a merchant according to the 1860 census, and Ida had an older sister, a younger sister, and a younger brother.[2]

Between 1873 and 1880, Hyde apprenticed in a millinery in Chicago and took classes at the Chicago Athenaeum, a school for working people. She attended the University of Illinois for the year 1881–1882 but had to leave for lack of money. For the next seven years she taught in the Chicago public schools. When her finances improved, she resumed her education at Cornell University, where she earned her undergraduate degree in 1891. Hyde then conducted research for physiologist and embryologist Jacques Loeb as well as for zoologist Thomas Hunt Morgan. Morgan won a Nobel Prize in 1933 for research on the role of chromosomes in heredity. This work earned Hyde the Bryn Mawr Biology Fellowship in 1892. In the following year, she became an instructor in biology at Bryn Mawr.

Meanwhile, two European professors were in a hot dispute to answer the very research question on which Hyde was working. When the one whose beliefs were corroborated by Hyde's research discovered her work, he invited her to his department. She attained the European Fellowship for the year 1893–1894 from the Association of Collegiate Alumnae, which later became the American Association of University Women. Thus Hyde journeyed to the University of Strassburg (now Université de Strasbourg) to conduct research with Professor Alexander Wilhelm Goette in zoology and physiology. Many professors felt her work deserved a Ph.D. However, her request to take the Ph.D exam caused such a stir—because she was a woman—that she withdrew.[3]

Hyde moved on to Heidelberg. She had heard that conditions might prove more favorable there. Surprisingly, many of the faculty and students were quite supportive of her work at Heidelberg. The exception was Professor Wilhelm Friedrich Kuhne, the department head for physiology, Hyde's favorite subject. He placed many obstacles in her path, such as not allowing her to attend his classes. Hyde persevered through this trying time by studying a great deal on her own. She also had to work her way through a complex bureaucracy to allow women to be admitted to the university on equal terms with men; the effort culminated in petitioning royalty. In 1894 the Grand Duke of Baden granted favorable judgment, thus placing Hyde as one of the major figures in helping to open German universities to women.

In 1896, Hyde had met all the requirements for the doctorate and performed splendidly on her examination. Most of the professors present agreed that Hyde deserved a *summa cum laude* on her exam; however, Kuhne refused to grant a woman this highest honor.[4] The group agreed on a new wording, *multa cum laude superavit* (greater than high honors). Although disappointed that she did not receive the praise she deserved, Hyde was pleased to be the first woman Ph.D. from Heidelberg. She was also proud that her degree was in physiology, despite all the obstacles for women in this particular department.

Having an apparent change of heart, Kuhne was instrumental in helping Hyde secure her next appointment. She conducted research in marine biology at the Heidelberg Table of the Naples Zoological Station, arguably the most advanced laboratory of its type at the time.[5] Hyde loved her experience in Naples so much that she wanted to help other women scientists. When she returned to the United States later that year, she founded the American Women's Naples Table Association (for promoting scientific research by women) with other well-known women scholars M. Carey Thomas and Ellen Swallow Richards. Hyde also received help from Kuhne in obtaining her next assignment at the University of Berne conducting research in physiology.

Hyde next became the Irwin Research Fellow in biology at Radcliffe College. In 1897 she began conducting research at the Harvard Medical School, the first woman to conduct research in the laboratories there. Over the next few years she was an instructor in a few of the Cambridge preparatory schools and also director of the biology department at the State Teachers College in Hyannis, Massachusetts. She spent her summers in 1897–1899 and 1901–1907 as professor of physiology and researcher at the Woods Hole Marine Biological Station.

In 1898, Hyde became associate professor of physiology at the University of Kansas; between the years 1900 and 1913 she was on the faculty at the School of Medicine there. She conducted research at the University of Liverpool in 1904. In 1905 she became full professor of physiology and head of the newly separate Department of Physiology at the University of Kansas. She maintained an "outstanding reputation as a teacher."[6] The summers from 1908 to 1912 were spent at Rush Medical College, University of Chicago, studying for the M.D. degree. Hyde completed all but one semester of the requirements. The year 1912 brought Hyde another first, that of being the first woman member in the American Physiological Society. She was the only one until 1914. During World War I she chaired the Women's Commission of Health and Sanitation of the State Council of National Defense, and in 1918 she organized a unit of the Women's Land Army to work the farms of Kansas.

Another first for Hyde occurred sometime during the years 1918–1919. She "invented the first microelectrode for intracellular work."[7] This can be considered the "single most useful and powerful tool in electrophysiology" and still is used today.[8] However, Hyde was not given credit for this innovation; but it is described and drawn in her 1921 article in *Biological Bulletin.*[9] Due to the fact that her discovery was either not widely known or not given credence, the microelectrode was reinvented in 1933 by someone else.

In 1920, Hyde retired from the University of Kansas, but she still conducted research. For example, between 1922 and 1923 she returned to the University of Heidelberg to research the biological effects of radium. She then relocated to San Diego and later moved to Berkeley. Her religious affiliation was with the Society for Ethical Culture, a nondenominational group that believed in the importance of ethics instead of faith in a supernatural being. Hyde was interested in social problems, especially women's issues such as the suffrage movement. She also was concerned with women and academic standards, serving on an accreditation board of the Association of Collegiate Alumnae to monitor the qualifications of women wanting to study abroad. She continued to help women scientists financially. She endowed a scholarship at Cornell University as well as the Ida H. Hyde Scholarship at the University of Kansas. In 1945 she established the Ida H. Hyde Woman's International Fellowship of the American Association of University Women. Later that year, on August 22, Hyde died in Berkeley at age 87 of a cerebral hemorrhage and arteriosclerosis.

Notes

1. *Notable American Women: 1607–1950. A Biographical Dictionary* (Cambridge, Mass.: Belknap, 1971), p. 248.

2. *Ibid.*, p. 248.

3. Leon Stein, "Before Women Were Human Beings: Adventures of an American Fellow in German Universities of the '90s," *Journal of the American Association of University Women* 31 (1938): 228.

4. *Ibid.*, p. 234.

5. *Ibid.*, p. 235.

6. Marilyn Bailey Ogilvie, *Women in Science: Antiquity through the Nineteenth Century* (Cambridge, Mass.: MIT Press, 1986), p. 103.

7. G. Kass-Simon., ed., *Women of Science: Righting the Record* (Bloomington: Indiana University Press, 1990), p. 241.

8. *Ibid.*

9. *Ibid.*, p. 242.

Bibliography

Kass-Simon, G., ed. *Women of Science: Righting the Record.* Bloomington: Indiana University Press, 1990.

Leonard, John William, ed. *Woman's Who's Who of America.* Detroit: Gale, 1976.

National Cyclopedia of American Biography, Current series. Vol. B, pp. 146–47. New York: J.T. White and Co., 1927.

Notable American Women, 1607–1950: A Biographical Dictionary. Cambridge, Mass.: Belknap Press, 1971.

Ogilvie, Marilyn Bailey. *Women in Science: Antiquity through the Nineteenth Century.* Cambridge, Mass.: MIT Press, 1986.

Rossiter, Margaret W. *Women Scientists in America.* Baltimore: Johns Hopkins University Press, 1982.

Stein, Leon. "Before Women Were Human Beings: Adventures of an American Fellow in German Universities of the '90s." *Journal of the American Association of University Women* 31 (1938): 226–36.

JILL HOLMAN

LIBBIE HENRIETTA HYMAN
(1888–1969)
Zoologist

Birth	December 6, 1888
1910	B.S., zoology, University of Chicago
1915	Ph.D., University of Chicago
1915–31	Research Assistant, University of Chicago
1940	Published first of six volumes of *The Invertebrates*
1940–67	Research Associate, Dept. of Living Invertebrates, American Museum of Natural History
1941	Honorary Sc.D., University of Chicago
1953	Vice President, American Society of Zoologists
1955	Daniel Giraud Medal, National Academy of Sciences
1958	Honorary Sc.D., Goucher College
1959	President, Society of Systematic Zoology; Honorary Sc.D., Coe College
1959–63	Editor, *Systematic Zoology*
1960	Gold Medal, Linnean Society of London
1963	Honorary LL.D., Upsala College, NJ
1969	Gold Medal, American Museum of Natural History
Death	August 20, 1969

Libbie Henrietta Hyman was a Phi Beta Kappa scientist who devoted her life to the study of invertebrates, determined to bring to English-language readers a knowledge of and her love for invertebrate life.

Libbie Hyman was born in 1888 in Des Moines, Iowa, the third child and only daughter of Joseph and Sabina Hyman. Her parents were Jewish immigrants. Her father escaped Russian Poland at age 14, making his way to London and eventually to the United States. In 1884 he married Sabina Neumann, a woman 20 years his junior from Stettin, Germany, who worked for his sister's family.

Libbie grew up in Fort Dodge, Iowa, where her father had a clothing store. The store was unprofitable, making Libbie's childhood less than idyllic—her father was constantly worried by financial troubles and her mother's domineering personality. However, Libbie showed an early interest in biology, collecting and identifying butterflies and plants from the woods around Fort Dodge. She attended the public school system and graduated as class valedictorian in 1905.

Libbie continued her education by taking additional courses in science and advanced German at the local high school. But she soon became ineligible for the high school courses and found a job pasting labels onto boxes at a Mother's Rolled Oats factory. Shortly after Hyman began work, Mary Crawford, her German instructor, discovered what she was doing. Horrified at the waste of her talent, Crawford helped her acquire a scholarship to the University of Chicago. Hyman entered the university in the fall of 1906.

The scholarship paid her tuition at Chicago, but Hyman still had to find a way to finance her room and board. She worked as a cashier at the Women's Commons and lived with her aunt and uncle until her father's death in 1907. After that, her mother and three brothers moved to Chicago. She lived with them until her mother died in 1929. Her family did not approve of the career she had chosen. Libbie's mother believed that daughters should stay home and do housework.

Despite this, Hyman continued to perform high quality schoolwork and maintained her scholarship. She was a botany major during her first year at Chicago, but she perceived anti-Semitism in the department and switched to chemistry, only to discover that she found the study of living things more interesting. She switched to zoology in her third year.

As a zoology major, Libbie received encouragement from Mary Blount, the Ph.D. candidate who ran the elementary zoology laboratory. At her urging, Hyman took Charles Manning Child's invertebrate zoology course during her senior year. Dr. Child was impressed by the young woman and suggested that she enter the University of Chicago's graduate school. In 1910, after receiving her B.S. in zoology, Hyman became Child's graduate student.

Libbie replaced Mary Blount as the laboratory assistant in elementary zoology and comparative vertebrate anatomy. Her experience in teaching these labs led to the publication of her first two books: *A Laboratory Manual for Elementary Zoology* (1919) and *A Laboratory Manual for Comparative Vertebrate Anatomy* (1922). These early books show that she had already developed the frank, wry writing style that characterized her later work. For example, in the introduction of her second book she stated: "Our experience with laboratory manuals of the type in which the burden of discovery is left to the student is that the student becomes highly dissatisfied and that the instructors are brought into a state of irritation and fatigue by the continuous demands for assistance with which they are bombarded."[1]

These textbooks were extremely successful. The substantial royalties from them and their subsequent revisions made Hyman financially independent. By 1960, however, she had tired of vertebrate anatomy and lost interest in making any further revisions.

After receiving her Ph.D. in 1915, Hyman remained at the University of Chicago as Dr. Child's research assistant. Over the next 16 years she experimented on planarians and hydroids, trying to produce results to support Child's ideas about physiological gradients. She never thought this work was very important and discovered that she preferred taxonomic work to experimental research.

During this period "Libbie realized that in English there was no major monographic series comparable to Bronn's *Klassen und Ordnungen des Tierreichs*, the beautifully illustrated *Traité de Zoologie Concrete* of Delage and Herouard, or the Kukenthal-Krumbach *Handbuch der Zoologie*."[2] It became her goal to rectify this with a grand monograph on the invertebrates.

In 1931, comfortable with the royalties from her laboratory manuals, Hyman resigned her research assistantship and spent 15 months touring European scientific centers. When she returned, she moved to an apartment in New York near the American Museum of Natural History and began work on the first volume of *The Invertebrates*. She worked at home until 1937, when she received a research appointment to the American Museum. The museum did not pay her a salary but provided an office, laboratory space, and library access.

For the next 30 years Hyman devoted much of her time to her monograph. She never used an assistant or technician, relying on her photographic memory and familiarity with European languages to translate, process, and organize scientific information before writing it as text. She made all her own histological preparations and illustrations for the books, spending several summers at Woods Hole and other marine labs to make drawings from living and prepared animals. Ultimately, each volume of *The Invertebrates* integrated the anatomy, embryology, physiology, and ecology of the animals described, but was "more than a compilation; it involves incisive analysis, judicious evaluation, and masterly integration of information."[3]

In addition to her monographic work, Dr. Hyman continued to study the morphology, physiology, development, systematics, and bionomics of lower invertebrates. She was particularly interested in the taxonomic identification of free-living flatworms. She published some 145 scientific papers during her lifetime and always took time to encourage young investigators in her field.

Dr. Hyman was a member of many scientific societies, served as the vice president of the American Society of Zoologists in 1953, and became president of the Society of Systematic Zoology in 1959. She edited *Systematic Zoology* from 1959 to 1963, taking over the editorship while president of the Society of Systematic Zoology, because the editor of the journal was falling behind in his deadlines. According to Richard Black-welder, "Most presidents would have consulted the other officers, and perhaps written some letters, but when Miss Hyman made up her mind that something had to be done, she didn't fool around with such red-tape. She got onto the train to New Haven, walked into the editor's office, and said 'I'm here to take over the editorship.' "[4]

She received many honors for her work, including honorary doctorates from the University of Chicago, Goucher College, Coe College, and Up-sala College. She was the first woman to be awarded the Daniel Giraud Medal from the National Academy of Sciences in 1955. She later received the Gold Medal of the Linnean Society of London and the Gold Medal of the American Museum of Natural History.

Libbie Hyman was unable to finish *The Invertebrates* before her death in 1969. Parkinson's disease slowed her progress on the sixth volume of the monograph; and after its publication in 1967, Hyman discontinued her work. During the last two years of her life Hyman was confined to a wheelchair, unable to move, and had to be nursed around the clock.

After her death Libbie Hyman was honored by many scientists, but none did so as uniquely as Masaharu Kawakatsu: "The memory of my friendship and the token of my gratitude lives on the muddy bottom of beautiful Lake Tahoe, a rare species of planarians, *Dendrocoelopsis hymanae.*"[5]

Notes

1. Libbie H. Hyman, *A Laboratory Manual for Comparative Vertebrate Anatomy* (Chicago: University of Chicago Press, 1922), p. viii.
2. Horace W. Stunkard, "In Memoriam, Libbie Henrietta Hyman, 1888–1969," in *Biology of the Turbellaria* (Libbie H. Hyman Memorial Volume) (New York: McGraw-Hill, 1974), p. xi.
3. *Ibid.*
4. Richard E. Blackwelder, "In Memoriam . . . Libbie Henrietta Hyman, 1888–1969, Her Life . . . ," *Journal of Biological Psychology* 12 (1970): 10.
5. *Ibid.*, p. 15.

Bibliography

Blackwelder, Richard E. "In Memoriam . . . Libbie Henrietta Hyman, 1888–1969, Her Life . . ." *Journal of Biological Psychology* 12 (1970): 3–15.
Hyman, Libbie H. *Comparative Vertebrate Anatomy.* Chicago: University of Chicago Press, 1942.
———. *The Invertebrates*, 6 vols. New York: McGraw-Hill, 1940–1967.
———. *A Laboratory Manual for Comparative Vertebrate Anatomy.* Chicago: University of Chicago Press, 1922.
———. *A Laboratory Manual for Elementary Zoology.* Chicago: University of Chicago Press, 1919.

Stunkard, Horace W. "In Memoriam, Libbie Henrietta Hyman, 1888–1969." In *Biology of the Turbellaria* (Libbie H. Hyman Memorial Volume), pp. ix–xii. New York: McGraw-Hill, 1974.
Yost, Edna. *American Women of Science.* Philadelphia: Frederick A. Stukio, 1943.

DIANE A. KELLY

ALETTA HENRIETTE JACOBS
(1854–1929)
Physician

Birth	February 1854
1870	Passed examination to become an apothecary; became apprenticed to her brother
1871	Received permission to begin medical studies at Groningen University
1879	Completed thesis to obtain M.D. degree at Amsterdam University; opened practice in Amsterdam
1882	Began prescribing diaphragms to patients at the world's first birth control clinic
1892	Married Dr. Carel Victor Gerritsen
1893	Appointed to the first committee to draft statutes for the Association for Women's Suffrage
1904	Retired from the practice of medicine
Death	1929

Born in the Province of Groningen, the Netherlands, Aletta was the eighth child in a Jewish family of eleven. From age 6, Aletta wanted to become a physician like her father and older brothers. Women did not pursue such ambitions at that time, but her father kindly arranged for her to learn Greek and Latin in addition to the secondary education that was customary. He also let his daughter help prepare medicines. Aletta spent a brief, unhappy stint at a finishing school but found that it did nothing to further her ambition to become a doctor. "Why should a girl be taught to lower her eyes if she passes a man on the street?"[1]

Aletta used her time in high school to study the current laws and found that there were no laws forbidding women from attending universities. At age 16 she took the examination to become an apothecary, passed, and apprenticed to one of her brothers who was already an apothecary. But this was not enough for Aletta.

In early 1871 she wrote a letter to Dutch liberal Prime Minister Thorbeck requesting his permission to attend medical lectures for one year and take examinations at the University of Groningen. Thorbeck replied that he would consent if her father would also give permission. On April 29, 1871, at age 17, Aletta entered the University of Groningen as the first female in Holland to attend university classes. She had to reapply in the next year to continue to study at the university, after which universities nationwide were opened to women. She considered this among her most important accomplishments.

Jacobs studied at the University of Groningen as the only female student until 1874. Her behavior on and off campus was closely scrutinized, but she found at least some of her fellow students and professors to be very supportive. She wrote of one instance in which an outsider tried to sour her college experience:

> On one occasion, when a student at Leiden wrote an article which was published in the students' weekly journal, containing insinuations against me, advising the students at Groningen to make my life there so unbearable that I would be obligated to leave the university, and thus deter other women from following my example, one of my fellow students answered the article, closing with these words: "Miss Jacobs trusted the students at Groningen, and we thank her for it. That trust has not been abused, and the students at Groningen may be proud of it. Therefore, since your words insult us, I have taken up the gauntlet for her."[2]

Not every student at Groningen was so kind, however. Her own brother Johan, an upperclassman, declared that he would rather see her dead than in medical school.

Because her studies were frequently interrupted by chronic malaria, Jacobs finished her studies at the University of Amsterdam. She had a serious bout with typhoid fever after completing all her coursework, delaying her state examination until April 1878. Two of the professors administering the examination in Utrecht were vehemently opposed to the idea of a woman doctor. Jacobs wrote that "the treatment they accorded me was more than uncivil, and it was only because of the fair attitude of my other professors, no less than the kindness of my Amsterdam teachers, that I did not withdraw from the examinations."[3] At the end of the three-week test, she was given the right to practice medicine. It was important to her to receive the degree of Doctor of Medicine, however, so she immediately began work on a thesis, "Localization of Physiological and Pathological Phenomena in the Brain." She defended the thesis before the medical faculty at Groningen, and her degree was granted in April 1879. She had become the first woman physician in the Netherlands.

Aletta began her practice of medicine in Amsterdam in September 1879. Hostile male doctors advised her to limit her work to midwifery, advice that she ignored. Although she was successful from the very start, Aletta faced problems her male colleagues did not. For example, it was conventional for unmarried women of better social classes to be escorted everywhere, even at midday, because this was the time when prostitutes walked the streets. Each time Dr. Jacobs encountered a problem, she notified the police and requested protection. She deliberately asked a civil agency to do this so that the problem would be dealt with "in such a way that other women would not be hampered in the future."[4] Jacobs soon had enough patients so that she could limit her practice to women and children. She began spending two mornings a week in her clinic for the poor in 1878.

The symptoms her patients presented her with were often the result of social conditions rather than disease. Dr. Jacobs began pushing for reforms. After treating many shop clerks for ailments related to the long hours spent on their feet, she approached her male colleagues for help in getting legislation passed for shorter working hours as well as provisions for employees to sit down when they were not busy. These physicians ridiculed her medical opinion and the campaign, sometimes publicly. Jacobs then took her efforts outside of the medical community and urged Dutch women to fight for the law, which eventually was passed.

Far more delicate was the issue of birth control. Many of her patients were worn from too many pregnancies. Jacobs soon began prescribing diaphragms to reduce fertilization, thus earning the distinction of establishing the world's first birth control clinic. Of her experience in the clinic she wrote:

> I was gratified, early in 1882, to obtain a pessarium occlusivium, and as soon as convinced that its use, if properly adjusted, would not have harmful results, did not hesitate to prescribe it in my practice among the poor, as well as in all other cases where conception was not desirable. Then the entire medical and clerical world rose against me! But the gratitude of the mothers was my compensation, and I continued my work. For many years now every adult woman in Holland has known how to control the number of her children.[5]

Dr. Jacobs's concern for her patients drew her again into a firestorm when she began campaigning against brothels. She treated both prostitutes and respectable women with venereal diseases, passed from one to the other through unfaithful husbands. Jacobs wrote articles against prostitution and was again vocally attacked by physicians, the press, and even other women. Eventually Holland became the first country in the world to legally abolish prostitution.

During the late 1880s Aletta met and fell in love with Carel Victor Gerritsen, a radical journalist and politician who held many of the same ideas about social reform and even attended women's suffrage conferences with her. They tried for several years to have a "free union," because they were both unimpressed with marriage laws, but they eventually married in 1892. Although they wanted a family, Jacobs and Gerritsen had only one child, who lived for one day.

After successfully campaigning for medical reforms, Dr. Jacobs broadened her activist horizons to include the reform of voting, penal, and women's rights laws. She had noticed, as with the laws regarding university attendance, that there was nothing to prohibit women from voting and tried as early as 1883 to have her name added to the voting list. The word "male" was promptly added to the law. In 1893 she helped to draft the statutes for the Association for Women's Suffrage, and within ten years she was the president of the Dutch association. The right for women to vote was acquired in 1919, and Jacobs said, "Now that we have got the vote, we are no longer powerless. Now it is up to us to see to it that all injustice disappears from our laws and from our society."[6]

Dr. Jacobs retired from medicine in 1904 after practicing for 25 years. Her husband died in 1905. Aletta became a relentless campaigner for reform, throwing herself full-time into the fight. During World War I she worked for peace with Jane Addams through the women's suffrage movement, even calling together a convention of women from hostile nations. Aletta Jacobs was well aware of the role she played in the history of medicine and of women's rights. She noted that "the political and economic independence of women in our country has come to pass with my assistance."[7] And she was clearly pleased that "I have lived to see women doctors needed and accepted as a necessity in every country in the world and in every community."[8]

Notes

1. Beatrice Levin, *Women and Medicine* (Lincoln, Neb.: Media Publishing, 1988), p. 43.

2. Aletta H. Jacobs, "Holland's Pioneer Woman Doctor," *Medical Woman's Journal* 35 (Sept. 1928): 257–58.

3. *Ibid.*, p. 258.

4. *Ibid.*

5. *Ibid.*

6. "Facts and Figures about Dutch Women in Medicine," *Journal of the American Medical Women's Association* 13 (June 1958): 253.

7. Cornelia de Lange, "Pioneer Medical Women in the Netherlands," *Journal of the American Medical Women's Association* 7 (Mar. 1952): 99.

8. Jacobs, "Holland's Pioneer Woman Doctor," p. 259.

Bibliography

de Lange, Cornelia. "Pioneer Medical Women in the Netherlands." *Journal of the American Medical Women's Association* 7 (Mar. 1952): 99–101.

"Dr. Aletta Jacobs." *Medical Woman's Journal* 45 (June 1938): 175.

"Facts and Figures about Dutch Women in Medicine." *Journal of the American Medical Women's Association* 13 (June 1958): 251–53.

Jacobs, Aletta H. "Holland's Pioneer Woman Doctor." *Medical Woman's Journal* 35 (Sept. 1928): 257–59.

Levin, Beatrice. *Women and Medicine*, pp. 43–45. Lincoln, Neb.: Media Publishing, 1988.

"A Memorial to Dr. Aletta Jacobs." *Medical Woman's Journal* 38 (Sept. 1931): 235.

Potter, Ada. "The History of Dutch Medical Women." *Medical Woman's Journal* 30 (Jan. 1923): 5–6.

Uglow, Jennifer S. *The Continuum Dictionary of Women's Biography*, p. 279. New York: Continuum Pub., 1982, 1989.

KELLY HENSLEY

SOPHIA JEX-BLAKE

(1840–1912)

Physician

Birth	January 21, 1840
1858	Student, Queen's College, London
1862	Helped Elizabeth Garrett try to enter Edinburgh University; became a teacher at the Grand Ducal Institute, Mannheim, Germany
1864	Planned a girls' school at Manchester

1865	Arrived in Boston and met Dr. Lucy Sewall; toured American schools and colleges
1868	Visited New York and met Dr. Elizabeth Blackwell
1869	One of the first five women students to matriculate at Edinburgh University
1877	M.D., Berne, Switzerland; obtained license to become a registered physician
1878	Returned to Edinburgh and began medical practice
1886	Published *Medical Women*; founded the Edinburgh School of Medicine for Women
1899	Retired from practice and moved to Sussex
Death	1912

Sophia Jex-Blake led the "battle of Edinburgh." The war was not a military one, but a major event in the history of medicine. When Sophia applied to the medical school at the University of Edinburgh in 1869, she knew that seven years earlier her attempts to help her friend Elizabeth Garrett enter had been unsuccessful. At that time, university authorities had implied that the decision might be reversed at some future time. Her letter to the dean of the Faculty of Medicine gives clues to the character of Sophia Jex-Blake:

SIR, As I understand that the statutes of the University of Edinburgh do not in any way prohibit the admission of women, and as the Universities of Paris and Zurich have already been thrown open to them, I venture earnestly to request from you and the other gentlemen of the Medical Faculty permission to attend the lectures in your Medical School during the ensuing session.

I beg to signify my willingness to accede to any such conditions, or agree to any such reservations as may seem desirable to you, and indeed to withdraw my application altogether if, after due and sufficient trial, it should be found impracticable to grant me a continuance of the favour which I now request. You, Sir, must be well aware of the almost insuperable difficulty of pursuing the study of Medicine under any conditions but those which can be commanded by large colleges only; and, in view of the increasing demand for the medical service of women among their own sex, I am sure that you will concede the great importance of providing for the adequate instruction of such as desire thoroughly to qualify themselves to fulfil the duties of the medical profession.

Earnestly commending my request to the favourable con-
sideration of yourself and your colleagues,

I am, Sir,
Yours obediently,
Sophia Jex-Blake[1]

Sophia had not reached this momentous decision at age 29 without
having already accomplished many unusual goals for a woman of her
era. She was born in Brighton, England, in 1840, the third and youngest
child of a wealthy family. Her father was already retired from his work
as a proctor of Doctors' Commons. She joined an 8-year-old brother,
Thomas William, who grew up to become headmaster of Rugby and then
dean of Wells; and a 6-year-old sister, Caroline. Both parents were descended from well-known families. They were very religious, conservative, and serious. Devoted to their children's upbringing, they attempted
to give Sophia the best education for girls at that time. Unlike her siblings, she was a very energetic, sometimes naughty child. Sophia out-
smarted her governesses and boarding school teachers. By age 19, as she
studied at Queen's College, London, she was offered a job as a mathematics tutor. Because her father disapproved of her being paid, she volunteered. But her desire to earn her own living led her to further study
in Edinburgh and a teaching job in a girls' school in Germany.

At first Sophia wanted to start a school for girls. She traveled to America in 1865. The story of her tour, *A Visit to Some American Schools and
Colleges*, was published in 1867. In her plan for better education for
women, she was advised by such famous people as Ralph Waldo Emerson and Dr. Thomas Hill, the president of Harvard. She had said before
leaving England that she felt a new life would open to her in the New
World. In Boston, she found her new life and an area of service for
herself and for women to come.

Sophia became associated with Dr. Lucy E. Sewall and a group of
young women physicians at the New England Hospital for Women and
Children. As a patient of Dr. Sewall's, she overcame an illness and
gained strength. As a volunteer in the hospital, she found a cause even
more appealing than women's education—women's healthcare. Having
been rejected at Harvard, she became the first student to register at the
Women's Medical College of the New York Infirmary for Women and
Children in 1868. But the illness and death of her father forced her return
to England. When she departed, her friend and mentor, Dr. Lucy Sewall,
prophesied that Sophia would open the medical profession to women in
England. Already Sophia had contributed an essay, "Medicine as a Profession for Women," to Josephine Butler's book, *Woman's Work and Woman's Culture*. She asserted in the essay that

So far from there being no demand for women as physicians, I believe that there is at this moment a large amount of work actually awaiting them; that a large amount of suffering exists among women which never comes under the notice of medical men at all, and which will remain unmitigated till women are ready in sufficient numbers to attend medically to those of their own sex who need them, and this in all parts of the world.[2]

In spite of the protests of eminent physicians and professors, Sophia and four other women received enough support from other regarded teachers to enroll in the University of Edinburgh in 1869. This was the first British university to admit women as medical students. Later two other women joined the group, and they became known as the "seven against Edinburgh." This negative description and another, "the battle of Edinburgh," came from the women's long struggle in attempting to receive a medical education. Even the initial plan for their instruction was a compromise. They had to take separate classes, which involved extra expense. Sophia wrote in 1875:

During the first session all went smoothly; the women received the same instruction, and passed the same examinations, as the other students; taking, by-the-bye, more than their share of honours in the class-lists. . . . But they lost several friends in the Medical Faculty by death or resignation, and the hostile element became strengthened by the consequent changes.[3]

At one point the women were met by nearly 200 male students as they tried to enter the hospital for an exam. "The Riot at Surgeons' Hall" was unsuccessful, and the women managed to reach the classroom. However, this small battle won was followed by too many that were lost. In spite of Sophia's and the other women's struggle, citizens' petitions on their behalf, and newspapers' support, lawsuits and other persecution followed. By 1873 the university had won its legal appeal and was once again closed to women.

Sophia refused to give up her dream of bringing medical education to women in England. By October 1874 she helped to open the London School of Medicine for Women. Unable to finish her degree in Scotland or England, she and Edith Pechey, another of the "seven," went to Switzerland. By January 1877, at age 36, she completed her M.D. at the University of Berne. Her final personal goal was to be officially licensed to practice in England. In May 1877 she and Edith became licentiates of the Irish College of Physicians, and their names were entered on the Medical Register of England. Only the American physician **Elizabeth Blackwell** and **Elizabeth Garrett Anderson** preceded them.

When the key administrative position with the London School was given to another, Sophia decided to return to Edinburgh to practice medicine. By June 1878, Scotland had its first woman doctor. She was especially helpful to working-class and poor women. In a letter to her mother, she wrote:

[E]very one of my patients has done well. Several have left my hands practically recovered, and those who are still there are all going on satisfactorily. . . . I have had 23 patients (nearly 100 visits) at my private house, and about as many more at my Dispensary, which has only been open a fortnight; so I don't think there is much doubt about the "demand" nor about my prospects.[4]

By 1885 her outpatient clinic became the Edinburgh Hospital and Dispensary for Women, Scotland's first hospital for women staffed by women. The opening of the Scottish Royal College's examinations to women led Sophia to aid the formation of the Edinburgh School of Medicine for Women and to become its dean. A few years later, some of Sophia's former students founded a second medical school for women in Edinburgh. Because there was not enough need for two schools and the second was affiliated with the famous Edinburgh Royal Infirmary, Sophia's school closed in 1898.

Sophia Jex-Blake retired in 1899 and returned to England. She settled into a farm in a lovely Sussex village. Her home was always open not only to former classmates, colleagues, and students but also to young medical students and visiting physicians from all over the world. One woman doctor wrote: "Thinking it over, I see that the best new influence that came into my life during the last seven years was the Doctor's young fresh interest, her enthusiasm, her breadth of mind, her spiritual force and faith, and her strong original wisdom."[5]

Margaret Todd, one of Sophia's former students, quoted a newspaper report about her death in 1912: "She it was, more than anyone else, who compelled the gates of the medical profession to be opened to women. Through years of hostility and obloquy she never lost heart in her Cause; and, meeting violence with reason and coarseness with dignity, she won at last."[6]

Notes

1. Shirley Roberts, *Sophia Jex-Blake, A Woman Pioneer in Nineteenth-Century Medical Reform* (New York: Routledge, 1993), pp. 82–83.

2. Sophia Jex-Blake, *A Visit to Some American Schools and Colleges* (London: Macmillan, 1867), p. 104.

3. Sophia Jex-Blake, "The Practice of Medicine by Women," *Fortnightly Review* 17 (Mar. 1875): 398.

4. Roberts, *Sophia Jex-Blake*, pp. 165–66.

5. Margaret Todd, M.D., *The Life of Sophia Jex-Blake* (London: Macmillan and Company, 1918), p. 528.

6. *Ibid.*, pp. 541–42.

Bibliography

Allen, Maggie, and Michael Elder. *The Walls of Jericho: A Novel Based on the Life of Sophia Jex-Blake*. London: British Broadcasting Corporation, 1981. (Derived from the BBC-TV serial.)

Hume, Ruth Fox. *Great Women of Medicine*. New York: Random House, 1964.

Jex-Blake, Sophia. *Medical Women*. London: Macmillan, 1886.

———. "Medicine as a Profession for Women," in *Woman's Work and Woman's Culture*, ed. Josephine Butler, pp. 78–120. London: Macmillan, 1869.

Lovejoy, Esther Pohl. *Women Doctors of the World*. New York: Macmillan, 1957.

Lutzker, Edythe. *Women Gain a Place in Medicine*. New York: McGraw-Hill, 1969.

Roberts, Shirley. *Sophia Jex-Blake: A Woman Pioneer in Nineteenth-Century Medical Reform*. New York: Routledge, 1993.

Thorne, J.O., and T.C. Collocott, eds. *Chambers Biographical Dictionary*. Cambridge: Cambridge University Press, 1986.

Todd, Margaret, M.D. *The Life of Sophia Jex-Blake*. London: Macmillan and Co., Ltd., 1918.

Uglow, Jennifer S., ed. *The Continuum Dictionary of Women's Biography*. New York: Continuum, 1989.

CAROL BROOKS NORRIS

ANANDIBAI JOSHEE

(1865–1887)

Physician

Birth	March 31, 1865
1874	Married Gopal Vinayak Joshee on March 31
1883	Sailed from Calcutta, India, to America; reached Ellis Island on June 4
1886	M.D., Woman's College of Pennsylvania, March 11; accepted the position of Physician-in-Charge of the Female Ward, Albert Edward Hospital, Kolhapur, India, on June 16; sailed back to India to assume her job at Kolhapur
Death	February 26, 1887

Anandibai Joshee. Photo reprinted by permission of the Archives and Special Collections on Women in Medicine, Medical College of Pennsylvania.

Anandibai Joshee was a pioneer. In 1886 she was the first Indian woman to receive the M.D. degree from the Woman's College of Pennsylvania, now known as the Medical College of Pennsylvania and Hahnemann University School of Medicine. She was also the first Hindu Brahmin woman to break the barrier of the time to seek medical education in the Western Hemisphere. She inspired Indian women to pursue medical education.

Anandi was born in Poona (now known as Pune), located 120 miles south of Bombay, India. She was the daughter of Ganpatrao and Gungubai Joshee, a wealthy landlord family of Pune. Her given name at birth was Yamuna, which was changed to Anandi according to the Hindu custom, when she married Gopal. "Anandi" means joy. The suffix "bai" is added to female names as a sign of respect. Joshee is also spelled as Joshi.

Ganpatrao organized a school in one of the large rooms of his house to educate Anandi after he noticed her love for education. Gopal, a postal employee in Pune and a Sanskrit scholar, began teaching Sanskrit to Anandi. In 1874, at age 9, Anandi married Gopal, a widower 20 years older. Anandi gave birth to her first and only child, who survived for only 10 days, at age 14. She thought the child would have survived if there were female doctors and good hospitals. At this tragic point in her life, she decided to become a doctor to help Indian women and to save babies.

In 1880 an event took place that allowed Anandi's dreams to come true. Halfway around the world in Elizabeth, New Jersey, a Mrs. Carpenter of Roselle, New Jersey, read a letter in *Missionary Review* from Gopal to Reverend Dr. R.G. Wilder while visiting her dentist. Wilder was the editor of *Missionary Review*, which was published at Princeton University. In his letter Gopal had expressed Anandi's desire to visit America for a medical degree. Mrs. Carpenter was moved by the genuine appeal and felt Dr. Wilder's response was inhumane. It seems, she thought, that "he does not want any unconverted Hindu to come to America and he believes that his intelligent correspondence will be led to confess 'Christ.' . . . Her whole soul was roused to indignation by the brutal manner in which she thought this cry was repulsed."[1]

Anandi wanted to go to America with Gopal, as it was uncommon for a Hindu woman to travel alone abroad during the nineteenth century. Gopal's self-esteem, dignity, and work ethics were evident in his letter: "I do not want to live on charity, but I am very much in need of assistance in securing a place suitable to me."[2] Immediately, Mrs. Carpenter wrote to Gopal expressing her willingness to offer shelter to the ambitious Anandi in her own house. Mrs. Carpenter's encouraging letter came as a bolt from the blue to Anandi and was the starting point of her journey to America.

Anandi and Mrs. Carpenter began writing letters to each other. Anandi described Hindu culture, Hindu festivals, and food to Mrs. Carpenter, who admired Anandi's "remarkable mastery of English."[3] She also described her broad, liberal philosophy: "The whole universe is a lesson to me. I am required by duty to respect every creed and sect, and value its religion. I therefore read the Bible, with so much interest as I read my own religious book."[4]

Gopal was not able to come with Anandi owing to financial and family responsibilities. Because of her decision to travel alone without Gopal, there was a strong opposition by Bengalis at her Serampore residence. "The Christians, natives and Europeans did not want Anandi to go abroad until she would submit to baptism before she went."[5] Anandi herself, rather than her husband, decided to address the crowd at Serampore College Hall to explain her decision to go alone to America for a medical degree. It seems that Anandi was the first Indian woman to deliver a public address. Col. Hans Matison, American consul-general in India at the time, was one of the Europeans in the audience to hear Anandi's address.

In her famous address entitled "The Courage of Her Conviction," delivered on February 24, 1883, at Serampore College Hall in Calcutta, Anandi told the audience that she wanted to procure a thorough knowledge in medicine and open a medical college for the instruction of women. "In my humble opinion, there is a growing need for Hindu female doctors in India, and I volunteer myself for one."[6] She found male domination and inferior quality of instruction in midwifery education in colleges in India at that time. She described to the audience her torture, the derogatory remarks she had heard from natives, and the sad episodes of people spitting on her face and throwing stones at her when she walked on the streets with books. Anandi found it more difficult at the time for a Hindu woman to take education than to convert to Christianity. She made a heroic public pledge: "I will go as a Hindu and come back to live as a Hindu."[7] A question of excommunication or baptism would not arise.

H.E.M. Jones, the director-general of the Post Office in India, wrote to Gopal, "I was very glad that Mrs. Joshee has made her debut and has succeeded. Pray, give her my congratulations. I wish her every success. In recognition of her courage and public spirit, permit me to offer the enclosed check of 100 rupees which may be useful to her."[8] Besides his check, he collected 750 rupees from Anandi's well-wishers through an appeal and established the Jones Fund. Even the governor-general of India donated 200 rupees to the Jones Fund.

Anandi sold the gold bangles she had received from her parents during her wedding for her passage to America. She arrived in New York on the steamer *City of Calcutta* in the company of European ladies. Mr. and Mrs. Carpenter received her at New York. During her stay with the Carpenter family from June 4 to October 1, 1883, "she stole into the hearts of those who met her."[9] After her arrival in America she wrote a letter for admission to Mr. Alfred Jones, Woman's College of Pennsylvania, describing her educational and financial background in these words:

I may not meet all points, the requirements for entering the college . . . to render to my poor suffering country women the true medical aid they sadly stand . . . they would rather die for than to accept at the hands of a male physician. The voice of humanity is within me and I must not fail. My soul is moved to help the many who cannot help themselves and I feel sure that God who has me in his care will influence the many . . . and assistance as I may need.[10]

These words had an impact on Jones and Dean **Rachel Bodley**, who saw in them a strong motivation to pursue a medical degree. With the kindness of Jones and Bodley, Anandi was admitted in the medical program after matriculation. She was offered financial assistance. Dean Bodley felt such love for Anandi that she took her into her own home when Anandi said, "I feel very lonesome."[11]

Anandi began her semester in October 1884. She had weekly examinations on the subjects covered in the lectures. The Pennsylvania Hospital Library was available for study for a deposit of three dollars. At the end of her coursework, she submitted her thesis entitled "Obstetrics among the Aryan Hindoos" for her degree requirement; the thesis covered 50 pages and was considered the longest in her class. She received the Doctor of Medicine degree at the commencement ceremony held at the Academy of Music Hall, Philadelphia, on March 11, 1886. Queen Victoria sent a congratulatory message to Anandibai and complimented Dean Bodley for the medical program at the College to train women from Asia in the medical field.

As a college student Anandi had visited and delivered talks to various groups in Philadelphia, New Jersey, and Washington, D.C. Because of her strict diet and unfamiliarity with cold weather, she contracted tuberculosis. She was sent to Colorado Springs for recuperation, but the visit did not improve her health.

After her graduation Anandibai accepted a job offer to work as the physician-in-charge of the Female Ward at Albert Edward Hospital, Kolhapur, India, for a salary of 300 rupees, free lodging and board, and a one-way ticket from New York to Kolhapur. Dewan Meherjee Cooverjee of Kolhapur had written to Dean Bodley with the offer when he heard about Anandibai's graduation.

Anandibai went home with high hopes to take the new job and to help women and children in India. She received a heroine's welcome from her friends, family members, and relatives upon her return to India. But Anandibai became weaker and weaker from exhaustion and tuberculosis. The newspapers issued bulletins of her health. She breathed her last in her mother's arms at her birthplace in Pune on February 26, 1887. Her last words deserve to be remembered: "I have done all that I could."[12] Anandibai's ashes were sent to Mrs. Carpenter, who buried them in the Carpenter family cemetery under Anandibai's grave in Poughkeepsie, New York.

Anandibai struggled to demonstrate to Indian society and to the British colonizers that the Indian women were capable of being doctors. Special tributes were paid to Anandibai's undaunted heroism in obituaries in the local and national newspapers of India, including *Kesari* and *Dnyana Chakshu*, the prestigious newspapers of Pune:

It is wonderful that a Brahmin lady has proved to the world that the great qualities—perseverance, unselfishness, undaunted courage and an eager desire to serve one's country—do exist in the weaker sex. . . . Thus may the memory of the late distinguished lady be perpetuated.[13]

We think it will take a long time before we again see a woman like her in the country. We do not hesitate to say Dr. Joshee is worthy of high place on the roll of historic women who have striven to serve and to elevate their native land.[14]

Dean Rachel Bodley also demonstrated her love for Anandibai and her humanitarian cause in remembering her: "These lines are written with the deep emotions, the binding tears which fall upon the page are the saddest tears my eyes have ever wept."[15]

Notes

1. Caroline Healey Dall, *The Life of Anandibai Joshee, a Kinswoman of the Pundita Ramabai* (Boston: Roberts Brothers, 1888), pp. 33–35.

2. "Correspondence of Gopal and Rev. Wilder," *Missionary Review* (Jan. 1879): 47.

3. Dall, *The Life of Anandibai Joshee*, p. 36.

4. *Ibid.*, p. 52.

5. *Ibid.*, p. 80.

6. *Ibid.*, p. 84.

7. *Ibid.*, p. 87.

8. *Ibid.*, p. 92.

9. *Ibid.*, p. 95.

10. Anandibai Joshee's application letter to Mr. Alfred Jones, June 28, 1883, Pennsylvania College of Medicine (PCM) Archives, Women in Medicine.

11. Dall, *The Life of Anandibai Joshee*, p. 107.

12. *Ibid.*, p. 185.

13. Pundita Ramabai Sarswati, *The High Caste Hindu Woman*, Introduction by Rachel L. Bodley, reprint of 1889 ed. (New Delhi: Inter-India Publication, 1984), pp. v–vi.

14. *Ibid.*, p. vii.

15. *Ibid.*, p. i.

Bibliography

Baker, Frances J. *First Women Physicians to the Orient.* Boston: Woman's Foreign Missionary Society, Methodist Episcopal Church, 1904.

"Correspondence of Gopal and Rev. Wilder." *Missionary Review* (Jan. 1879): 47–50.

Dall, Caroline Healey. *The Life of Anandibai Joshee, a Kinswoman of the Pundita Ramabai.* Boston: Roberts Brothers, 1888.

"Dr. Anandibai Joshee." *Medical Missionary Record* 1, no. 12 (1887): 291.

"Dr. Anandibai Joshi: Time & Achievement." Produced and directed by Anjali Kirtane. 20 minutes. Bombay, India: Salil Chitra, 1993. Videocassette.

Joglekar, Ram. "AnandiGopal" (a play in the Marathi language). Bombay, India: Majestic Books, 1976.

Joshi, S.J. *AnandiGopal* (English). Translated and abridged by Asha Damle. Calcutta: Stree, 1992.

Pearce, Louise. "A Century of Medical Education for Women." *Independent Woman* 29, no. 4 (Apr. 1950): 104–6, 122.

Pundita Ramabai Sarswati. *The High Caste Hindu Woman,* Introduction by Rachel L. Bodley. Reprint of 1889 edition. New Delhi: Inter-India Publication, 1984.

Saxena, T.P. *Women in Indian History: A Biographical Dictionary.* New Delhi: Kalyani Publishers, 1979.

SAROJINI D. LOTLIKAR

EVELYN FOX KELLER

(1936–)

Biologist

Birth	March 20, 1936
1957	B.A., *magna cum laude,* Brandeis University
1957–61	National Science Foundation Fellowship
1959	M.A., Radcliffe College
1962–63	Instructor in Physics, New York University
1963	Ph.D., theoretical physics, Harvard University
1963–66	Assistant Research Scientist, New York University
1963–69	Assistant Professor, Cornell University Medical College
1964	Married Joseph Bishop Keller
1970–72	Associate Professor of Mathematical Biology, New York University
1972–74	Chair, Math Board of Study, SUNY
1972–82	Associate Professor of Natural Science, SUNY College, Purchase
1973	Vice Chair, Gordon Conference on Theoretical Biology
1974	Chair, Gordon Conference on Theoretical Biology

1976	Center for Policy Research
1979–80	Visiting Fellow, MIT Program in Science, Technology, and Society
1980–84	Visiting Scholar, MIT
1981–82	Visiting Professor, Northeastern University; Mina Shaughnessy Scholars Award, Fund for the Improvement of Post-Secondary Education
1982–	Professor of Humanities and Mathematics, Northeastern University; Organizer and coordinator of Boston Area Colloquium on Feminist Theory
1984	Mellon Fellowship, Wellesley College Center for Research on Women
1985	Published *Reflections on Gender and Science;* won visiting professorship for women, National Science Foundation; Visiting Professor, Northwestern University
1985–86	Visiting Professor, MIT; Rockefeller Humanities Fellowship
1986	Distinguished publication award, Association for Women in Psychology
1986–87	Senior Fellow, Society for the Humanities, Cornell University
1987–88	Member, Institute for Advanced Study, Princeton
1988–93	Professor of Rhetoric, Women's Studies, and History of Science, University of California, Berkeley
1992	MacArthur Foundation Fellowship Award; published *Secrets of Life, Secrets of Death*
1993–	Professor in Program in Science, Technology, and Society, MIT

Evelyn Fox Keller has worked in the fields of theoretical physics, molecular biology, and mathematical biology. She has worked on mathematical models of chemotaxis and pattern formation. She has also studied the psychological basis for scientific beliefs and given science a feminist critique. She wants to expand the thinking styles available to scientists, to offer them more than the traditional "masculine" ways. She also believes that paying attention to differences can lead to an understanding of a larger order, whereas scientists often ignore differences or try to assimilate them. In addition, Keller has worked to uncover other women scientists.

Evelyn Fox was born in 1936 in New York, New York. Her parents, Albert and Rachel Fox, were Russian Jewish immigrants; theirs was a working-class family. Evelyn was not attracted to science at an early age, but in 1947 her sister Frances piqued her interest in psychology by telling her what she had learned at college about the unconscious.[1] During her teen years her brother, Maurice, a biologist 11 years her senior, tried to no avail to interest his bright youngest sister in science through popular writers such as George Gamow and Isaac Asimov.

In 1953, at Queens College, Evelyn was getting C's in freshman composition. She started writing papers on Gamow in desperation and received great grades. She ended up writing on quantum mechanics because "[a]s far as she could see, it was the only way she could get a decent grade."[2] Although evidence of her talent appeared here, she still was not attracted to science. At the end of that year Evelyn wanted to transfer to Antioch or Reed, but her family could not afford those schools and wanted her close by. Conflict resulted, but Maurice saved the day by proposing the idea of Brandeis University. He also, as Evelyn puts it, " 'sicced [physicist] Leo Szilard on me' " in hopes of capturing her interest for physics.[3] She decided to major in physics but only to stick with it long enough to get to medical school.

These plans were altered, however, when she did her senior thesis on physicist Richard Feynman. She became hooked on science.[4] As she relates, "I fell in love, simultaneously and inextricably, with my professors, with a discipline of pure, precise, definitive thought, and with what I conceived of as its ambitions. I fell in love with the life of the mind. I also fell in love . . . with the image of myself striving and succeeding in an area where women had rarely ventured."[5] The year was 1957 and Evelyn graduated with a B.A., *magna cum laude.*

Evelyn won a National Science Foundation Fellowship for the years 1957–1961 and was lured to Harvard, where she experienced "two years of almost unmitigated provocation, insult, and denial."[6] Harvard colleagues told her that she couldn't possibly understand physics and that her "lack of fear was proof of [her] ignorance."[7] She was quite isolated because she intimidated most of the students. She also was disappointed in how physics was viewed at the time. She viewed physics not as a skill of calculation, as others did, but "as a vehicle for the deepest inquiry into nature."[8] To her, Einstein epitomized this noble goal. The emphasis on calculation and the lack of support from colleagues or friends finally affected her to the extent that she gave up physics after two years, intending to switch back to psychology.

Her brother, however, saved the day once again by inviting her to spend the summer in Cold Spring Harbor. Here was a colony of fascinating, intelligent biologists who treated her better. She worked in a lab and stumbled on an idea that would work for a physics thesis but was actually in the field of molecular biology. Back at Harvard she found a professor, Walter Gilbert, who was also making the switch to molecular biology. Gilbert (who later won the Nobel Prize in 1980) agreed to be her advisor. Evelyn pulled through and received her Ph.D. in theoretical physics in 1963.

She became an assistant research scientist at New York University, working for Joseph Bishop Keller, whom she married in 1964. A son, Jeffrey, was born in 1965, and a daughter, Sarah, in 1966. Evelyn then held a number of successive and concurrent positions, including assistant professor at Cornell University Medical College from 1963 to 1969; associate professor of mathematical biology at New York University from 1970 to 1972; associate professor of natural science at SUNY College, Purchase, from 1972 to 1982; chair of the mathematics board of study at SUNY from 1972 to 1974; vice chair of the Gordon Conference on Theoretical Biology in 1973; and chair of the same conference the following year.

During this period Evelyn experienced conflict with the roles of wife, mother, teacher, and scientist. As a result of a combination of the women's movement and therapy, she began to see that the personal was political and that the difficulties she had undergone in graduate school were not her fault. She transformed her shame to rage and eventually to political conscience.[9] She had to analyze and discard labels that did not seem to apply to her. In 1974 she was a special lecturer in mathematical biology at the University of Maryland. For her last lecture she talked about the absence of women in science, an unheard-of gesture. Soon after this lecture she wrote "The Anomaly of a Woman in Physics," which concludes with the following words: "I hope that the political awareness generated by the women's movement can and will support young women who today attempt to challenge the dogma, still very much alive, that certain kinds of thought are the prerogative of men."[10]

When "Anomaly" was published in 1977, it led to the next phase in Keller's career. Someone suggested she write about the geneticist **Barbara McClintock**. Keller says from the first conversation, "I was obliged to put everything aside and become the vehicle for Barbara McClintock."[11] Keller was a good person to do the work on McClintock as a result of her beliefs and vision evidenced in "Anomaly." However, it was a difficult task for Keller because "Barbara McClintock represented everything I was most afraid of—that becoming a scientist would mean I'd be left alone."[12]

In 1983 *A Feeling for the Organism*, the biography of Barbara McClintock, was published. McClintock saw scientists trying to impose answers on their subjects. "Precisely because nature's complexity exceeds our ability to understand it, McClintock believe[d] that scientists must 'listen to the material' and 'let the experiment tell you what to do.' Her major criticism of most contemporary research is based on what she sees as an inadequate humility."[13] In 1984 Keller continued to question the traditional way science is carried out in her essay entitled "Science and

Power for What?" She argued for a science without domination or chastity, but one of intimacy and identification with the subject.[14] The trend in her thinking continued in 1985 with the publication of *Reflections on Gender and Science,* which generated much controversy. Here she argued to expand the thinking styles available to scientists. She believed that intuition has a place and that scientists need to realize and admit that their beliefs and goals shape their work.

In 1986, Keller continued to uncover other women scientists with her article on Lynn Margulis, "One Woman and Her Theory." Keller happily portrayed this rebel woman scientist as sexy and energetic as well as gifted—she discovered that plant and animal cells resulted from symbiosis of different bacterial encounters over time. Margulis's life story must have been interesting to Keller because it contains several similarities to her own life, such as being a woman in science when there weren't many, struggling to balance a career with the roles of wife and mother, getting divorced, and being devoted to unaccepted biological theories.

Beginning in 1988, Keller became a professor of rhetoric, women's studies, and history of science at the University of California, Berkeley. In 1991 she was interviewed by Larry Casalino for an article entitled, "Decoding the Human Genome Project: An Interview with Evelyn Fox Keller." The Project aims at mapping human genes and determining their function. Keller affirms that it is exciting to study human genetics but warns, "We need a much larger, much better political, social, historical, and philosophical framework for understanding the consequences of what's happening. I don't think we're prepared at all."[15] She believes that people are motivated by a desire to control nature, not a simple desire to understand. This critique of the Human Genome Project leads Keller toward developmental molecular biology in her research interests. The field studies how outer stimuli affect genes.[16]

The year 1992 brought Keller a coveted MacArthur Award. These awards provide five years' income to individuals "to fulfill their promise by devoting themselves to their own endeavors at their own pace."[17] In that same year *Secrets of Life, Secrets of Death* was published. Keller believes that "sharing a language means sharing a conceptual universe."[18] Because people are subjective, she believes that examining language leads to understanding thoughts that are sometimes taken for granted. In 1993, Keller became a full professor in the Program in Science, Technology, and Society at MIT.

Notes

1. Beth Horning, "The Controversial Career of Evelyn Fox Keller," *Technology Review* 96 (Jan. 1993): 62.
    2. *Ibid.*

3. *Ibid.*, p. 63.

4. *Ibid.*

5. Evelyn Fox Keller, "The Anomaly of a Woman in Physics," in Sara Ruddick, ed., *Working It Out: 23 Women Writers, Artists, and Scholars Talk about Their Lives and Work* (New York: Pantheon, 1977), p. 78.

6. *Ibid.*, p. 81.

7. *Ibid.*, p. 82.

8. *Ibid.*, p. 83.

9. *Ibid.*, p. 80.

10. *Ibid.*, p. 91.

11. Horning, "Controversial Career," p. 65.

12. *Ibid.*

13. Evelyn Fox Keller, "Women and Basic Research: Respecting the Unexpected," *Technology Review* 87 (Nov./Dec. 1984): 46.

14. Evelyn Fox Keller, "Science and Power for What?" in *Nineteen Eighty-Four: Science between Utopia and Dystopia* (Dordrecht: D. Reidel, 1984), pp. 270–71.

15. Larry Casalino, "Decoding the Human Genome Project: An Interview with Evelyn Fox Keller," *Socialist Review* 21 (Apr. 1991): 115.

16. Horning, "Controversial Career," p. 68.

17. MacArthur Fellows Program Brochure (Chicago: MacArthur Foundation, n.d.), p. 1.

18. Horning, "Controversial Career," p. 67.

Bibliography

American Men and Women of Science. New York: Bowker, 1989/1990.

Casalino, Larry. "Decoding the Human Genome Project: An Interview with Evelyn Fox Keller." *Socialist Review* 21 (Apr. 1991): 111–28.

Contemporary Authors, Vol. 125. Detroit: Gale, 1989.

Hirsch, Marianne, and Evelyn Fox Keller, eds. *Conflicts in Feminism*. New York: Routledge, 1990.

Horning, Beth. "The Controversial Career of Evelyn Fox Keller." *Technology Review* 96 (Jan. 1993): 58–68.

Jacobus, Mary, Evelyn Fox Keller, and Sally Shuttleworth, eds. *Body/Politics: Women and the Discourses of Science*. New York: Routledge, 1990.

Keller, Evelyn Fox. "The Anomaly of a Woman in Physics," in *Working It Out: 23 Women Writers, Artists, and Scholars Talk about Their Lives and Work*, ed. Sara Ruddick. New York: Pantheon, 1977.

———. *A Feeling for the Organism: The Life and Work of Barbara McClintock*. New York: W.H. Freeman and Co., 1983.

———. "One Woman and Her Theory." *New Scientist* 111 (July 3, 1986): 46–50.

———. *Reflections on Gender and Science*. New Haven: Yale University Press, 1985.

———. "Science and Power for What?" in *Nineteen Eighty-Four: Science between Utopia and Dystopia*. Dordrecht: D. Reidel, 1984.

———. *Secrets of Life, Secrets of Death: Essays on Language, Gender and Science*. New York: Routledge, 1992.

———. "Women and Basic Research: Respecting the Unexpected." *Technology Review* 87 (Nov./Dec. 1984): 44–47.

Keller, Evelyn Fox, and Elisabeth A. Lloyd, eds. *Keywords in Evolutionary Biology.* Cambridge, Mass.: Harvard University Press, 1992.
Keller, Evelyn Fox, and Jane Flax. "Missing Relations in Psychoanalysis: A Feminist Critique of Traditional and Contemporary Accounts of Analytic Theory and Practice," in *Hermeneutics and Psychological Theory: Interpretive Perspectives on Personality, Psychotherapy, and Psychopathology.* New Brunswick, N.J.: Rutgers University Press, 1988.

JILL HOLMAN

FRANCES OLDHAM KELSEY

(1914–)

Pharmacologist

Birth	July 24, 1914
1934	B.Sc., McGill University
1935	M.Sc., McGill University
1936	Moved to the United States
1938	Ph.D., pharmacology, University of Chicago
1938–50	Instructor, then Assistant Professor of Pharmacology, University of Chicago
1943	Married Fremont Ellis Kelsey on December 6
1950	M.D., University of Chicago
1954–57	Associate Professor of Pharmacology, University of South Dakota School of Medicine
1956	Became naturalized U.S. citizen
1960–63	Medical Officer, FDA Bureau of Medicine, Washington, DC
1962	D.Sc., Hood College; President's Award, Distinguished Federal Civilian Service; Award from AMVETS and *Good Housekeeping* magazine
1963–66	Chief, Investigational Drugs Branch, FDA, Bureau of Medicine, Division of New Drugs, Washington, DC
1964	D.Sc., University of Nebraska; Western College for Women; University of New Brunswick
1966	D.Sc., Middlebury College

1966–67	Director, Division of Oncology and Radiopharmaceutical Drug Products, FDA, Bureau of Medicine, Office of New Drugs, Washington, DC
1967	D.Sc., Wilson College
1968–71	Assistant to the Director for Scientific Investigations, FDA, Bureau of Medicine, Office of Medical Support
1969	D.Sc., St. Mary's College
1971–	Director, Division of Scientific Investigations, FDA
1973	D.Sc., Drexel University
1976	Public Service Award, Association of Federal Investigators
1984	Department of Health and Human Services Distinguished Service Award (Scientific)

In 1962 the *New York Times* noted that Dr. Frances Kelsey usually "wears suits and low heels" with "short straight hair cut in a somewhat mannish style."[1] An article in the *Saturday Review* of the same year noted that her faith in the goodness of people sometimes caused moments of credulity, but that she could stand firm in her beliefs.[2] Also in 1962 a *Good Housekeeping* article stated that she was "a stubborn, obstructionist woman, deaf and blind to the evidence of overwhelming statistics," with a quick smile and a "lively spark" in her eyes.[3] This was controversial press for the 1962 winner of the Distinguished Federal Civilian Service Award for keeping the drug thalidomide, which caused dramatic birth defects, out of the American market.

Born in 1914 in British Columbia, Canada, Frances was the daughter of a retired British officer. She always wanted to be a scientist.[4] She received her bachelor and master of science degrees from McGill University and then moved to Chicago to attend graduate school on a fellowship. (She became a naturalized U.S. citizen in 1956.) Following her graduation from the University of Chicago, she joined the faculty in the Department of Pharmacology. There, Dr. E.M.K. Geiling, a leader in the development of the discipline of pharmacology, instilled in her the importance of pharmacology and the need for high standards of research.[5]

While at the University of Chicago, Frances met Dr. Fremont Ellis Kelsey and they were married in December 1943. Because of regulations concerning married couples, Frances resigned from the faculty and began medical school. She earned her M.D. degree in 1950 from the University of Chicago and while in school gave birth to two daughters. Following graduation from medical school, Dr. Kelsey worked as an editorial associate for the American Medical Association.

When her husband was offered a job at the University of South Dakota, Frances went with him. She worked part-time in the University of

South Dakota Department of Physiology and Pharmacology from 1954 to 1957, and from 1957 to 1960 she interned in South Dakota. At the same time she engaged in private practice, which sometimes required her to leave her family for weeks at a time while she traveled throughout South Dakota relieving doctors in rural communities.

In 1960, when Dr. F. Ellis Kelsey joined the Division of General Medical Sciences of the National Institutes of Health, Frances began her career with the Federal Drug Administration (FDA). Her responsibility as medical officer was to review license applications by drug companies that wished to market new drugs in the United States. New over-the-counter or prescription drugs had to be approved as safe by a panel of physicians, pharmacologists, and chemists.

When Frances joined the FDA, it was in the midst of a review by the Antitrust and Monopoly Subcommittee of the Senate Judiciary Committee, headed by Senator Estes Kefauver of Tennessee. **Dr. Barbara Moulton**, who preceded Frances in the New Drug Section, had resigned in disgust because of constant repeals of her decisions by her superiors. Moulton testified before the committee concerning the close relationship between the representatives of the drug industry and officials of the FDA. Frances attended the hearing. She and Moulton became close friends.[6]

The first request to cross Frances's desk was from the William S. Merrell Company of Cincinnati, Ohio. The routine application, submitted on September 8, 1960, was for the drug thalidomide, which would be marketed under the trade name Kevadon. The FDA was required to respond to the pharmaceutical firm within 60 days of application.[7]

The drug thalidomide had first been synthesized by a Swiss pharmaceutical house in 1954. It was developed in West Germany by the drug firm Chemie Grunenthal and had been marketed there since 1957, accounting for 46 percent of Grunenthal's gross profits in the early 1960s.[8] Thalidomide, first sold over the counter and later as a prescription medication, was present throughout the European market under various trade names and was touted mainly as a sleeping pill that had various other uses: for grippe, depression, neuralgia, and asthma; to calm nerves; in cough medications; for morning sickness; and for irritable children. It had the effect of giving a quick, deep, and restful sleep without the side effect of a hangover. It was also inexpensive and did not appear to be habit-forming. Additionally, there was not a danger of an accidental overdose or suicide.[9]

Dr. Kelsey found that the research included in the application did not meet her standards. It incorporated incomplete studies and clinical work that was not well documented. Before the review period was over, she asked for more data on how the drug worked. The application now could be considered for another 60 days.

In 1961, Kelsey read an article from the December 1960 *British Medical Journal* concerning patients who were taking thalidomide and had experienced a numbing of the arms and legs.[10] Additional reports confirmed damage to the nervous system. The symptoms usually stopped once the drug was discontinued, but not always.[11] Kelsey notified the drug company of the side effects and requested additional research.

The Merrell Company repeatedly contacted both Kelsey and her superiors to try to win approval of the drug, but she continued to delay and to request more information and research on the drug. She felt that the evidence of the drug's safety was "incomplete in many respects."[12]

Because of her resistance to approving the drug, Kelsey bore the brunt of much criticism and pressure. She was especially concerned by the fact that the drug did not make animals sleepy in the same way that it did humans. She also questioned the effect of the drug on special populations, such as those with liver or digestive problems, and she was concerned with the effect of the drug on the fetus.[13] The pharmaceutical company had no research to answer any of her questions. Kelsey recalled that

> if this drug were a cancer treatment or helpful in really dread diseases, I wouldn't have worried about minor side effects. But I could anticipate the results of an enthusiastic sales effort. Everybody— sick and well, old and young—would be taking it. That's why I felt I couldn't be too careful. I had to hold out against every pressure.[14]

Kelsey was supported in her stand against approval by her husband, also a pharmacologist, who spent many evenings with his wife trying to evaluate the research. Early in her career, she and her husband had conducted a study that showed the effect of quinine on pregnant rabbits. Thalidomide was much stronger than quinine, and she could not forget the results of that study. Again, Frances requested more data from the drug firm. Dr. Moulton also provided her advice and support.[15]

In the meantime, the Merrell Company had distributed tablets of thalidomide to some 1,267 physicians throughout the United States to be used on an experimental basis. The drug reached an estimated 20,000 patients, far larger than the usual distribution. At this time, it was normal practice for drugs to be distributed by pharmaceutical companies to physicians before approval by the FDA. The medications could be given to the patients, but not sold. Quite often, medications were distributed to patients without their being aware of the drugs' experimental status. The FDA was not usually aware that new drugs were being tested until the manufacturer made application for a license to sell them.[16]

In November 1961 a German scientist, Dr. Widukind Lenz, reported on a rise in cases of phocomelia, a deformation of limbs in newborn babies. This condition usually results in the lack of one arm, with rudimentary fingers that look like the flippers of a seal coming from the stub below the shoulder.[17] This was a condition that was seen rarely, if ever, by physicians. In eight Western German pediatric clinics there were no cases reported between 1954 and 1959, but there were 12 cases in 1959 and 83 in 1960. The number of cases in 1961 rose to 302. Children were born without both arms, both legs, or with no limbs at all. In some cases, external ears were missing. There were also malformations in the eyes, esophagus, and intestinal tract.[18] An article concerning the condition was published in the December 16 issue of *Lancet*. Dr. William McBride connected phocomelia with thalidomide in Australia. Soon reports were received from other parts of Europe, the Middle East, South America, Canada, and other parts of the world.[19]

A cablegram was sent from Europe to the Merrell Company on November 29, 1961, describing the abnormalities. Fourteen months after the company first submitted the application for approval, it was withdrawn. Frances urged the company to contact doctors who had been given the free samples for distribution. She saw to it that FDA inspectors retrieved the remaining stock of the drugs.[20]

Frances was nominated for the Distinguished Federal Civilian Service award not by someone in the FDA but by Senator Kefauver. In the nomination, he noted her "great courage and devotion to the public interest by preventing the sale in this country of the drug thalidomide."[21] On August 7, 1962, President John F. Kennedy presented the Distinguished Federal Civilian Service Award to Kelsey. The citation recognized that her refusal to approve thalidomide "prevented a major tragedy of birth deformities in the United States," and she was praised for her "high ability and steadfast confidence in her professional decision."[22] Kelsey invited to the ceremony those who had supported and helped her in her fight against Merrell.

As a result of this near-tragedy, President Kennedy called for legislation to strengthen the FDA's power over distribution of drugs for experimental use that had not received approval for market. Kelsey testified before a Senate subcommittee in August 1962 that the FDA needed to be able to control both the testing and the marketing of drugs.[23] She helped in writing the legislation, which was passed in October 1962 and signed by President Kennedy, with Kelsey present. The new law called for the formation of the investigational drug branch to oversee the testing of new drugs for safety. Kelsey was appointed to head this new branch in December 1962.[24]

Caution, reserve, integrity, and high standards remain the foundations of Dr. Frances Kelsey's work. Because of her questioning and insistence on answers, the United States was spared the tragedy of thalidomide that was suffered in other parts of the world. Now in her eighties, she is still active at the FDA in her pursuit of safe and effective drugs for the United States.

Notes

1. "Drug Market Guardian: Frances Oldham Kelsey," *New York Times* (Aug. 2, 1962).

2. W. Jonathan, "The Feminine Conscience of FDA: Dr. Frances Oldham Kelsey," *Saturday Review* 45 (1962): 43.

3. J.L. Block, "Dr. Kelsey's Stubborn Triumph," *Good Housekeeping* 155 (1962): 12, 17.

4. W. Hoffman and J. Shields, "Frances Oldham Kelsey and Thalidomide," in *Doctors on the New Frontier: Breaking through the Barriers of Modern Medicine* (New York: Macmillan Publishing, 1981), p. 138; M. Truman, "The Doctor Who Said No," in *Women of Courage* (New York: Morrow, 1976), p. 219.

5. Truman, "The Doctor Who Said No," p. 220.

6. *Ibid.*, p. 222.

7. Hoffman and Shields, "Frances Oldham Kelsey," p. 142; Truman, "The Doctor Who Said No," p. 223.

8. Truman, "The Doctor Who Said No," p. 224.

9. Block, "Dr. Kelsey's Stubborn Triumph," p. 18; M. Mintz, "Dr. Kelsey Said No," *Reader's Digest* 81 (1962): 86–88; Hoffman and Shields, "Frances Oldham Kelsey," p. 141.

10. Truman, "The Doctor Who Said No," p. 229.

11. *Ibid.*, p. 225.

12. Mintz, "Dr. Kelsey Said No," p. 87.

13. Truman, "The Doctor Who Said No," p. 226; Block, "Dr. Kelsey's Stubborn Triumph," p. 20.

14. Hoffman and Shields, "Frances Oldham Kelsey," p. 145.

15. Jonathan, "The Feminine Conscience of FDA," p. 43; Block, "Dr. Kelsey's Stubborn Triumph," p. 20; Hoffman and Shields, "Frances Oldham Kelsey," p. 148; Truman, "The Doctor Who Said No," pp. 232–33.

16. Hoffman and Shields, "Frances Oldham Kelsey," p. 140.

17. Mintz, "Dr. Kelsey Said No," p. 88.

18. *Ibid.*

19. Truman, "The Doctor Who Said No," p. 237.

20. Hoffman and Shields, "Frances Oldham Kelsey," p. 149; Truman, "The Doctor Who Said No," p. 237.

21. "Frances Oldham Kelsey," *Current Biography* (New York: Wilson, 1965), p. 219.

22. *Ibid*, pp. 218–19.

23. N. Lichtenstein, ed., *Political Profiles: The Kennedy Years* (New York: Facts on File, 1976), p. 264.

24. Truman, "The Doctor Who Said No," p. 239.

Bibliography

American Men and Women of Science, Physical and Biological Science, 14th ed. New York: Bowker, 1947.

Block, J.L. "Dr. Kelsey's Stubborn Triumph." *Good Housekeeping* 155 (1962): 12, 17–18, 20.

"Drug Market Guardian: Frances Oldham Kelsey." *New York Times* (Aug 2, 1962).

"Frances Oldham Kelsey." *Current Biography.* New York: Wilson, 1965.

Hoffman, W., and J. Shields. "Frances Oldham Kelsey and Thalidomide," in *Doctors on the New Frontier: Breaking through the Barriers of Modern Medicine.* New York: Macmillan, 1981.

Jonathan, W. "The Feminine Conscience of FDA: Dr. Frances Oldham Kelsey." *Saturday Review* 45 (1962): 41–43.

Kelsey, Frances O. "The Evolution of New Drug Legislation." *Boston Medical Quarterly* 17 (1966): 72–81.

———. "Thalidomide Update: Regulatory Aspects." *Teratology* 38 (1988): 221–26.

Lichtenstein, N., ed. *Political Profiles: The Kennedy Years.* New York: Facts on File, 1976.

Mintz, M. "Dr. Kelsey Said No." *Reader's Digest* 81 (1962): 86–89.

Truman, M. "The Doctor Who Said No," in *Women of Courage.* New York: Morrow, 1976.

Who's Who of American Women, 1993–94. 18th ed. Chicago: Marquis, 1993.

JUDY F. BURNHAM

CHUNG-HEE KIL

(1899–1990)

Physician

Birth	February 3, 1899
1919	Participated in Declaration of Korean Independence
1923	M.D., Tokyo Women's Medical College
1925	Married Dr. Tak-Won Kim
1928	Founded (with her husband and Dr. Rosetta Hall) the Chosen Women's Medical Training Institute
1959	Recognized by the Korean Minister of Public Health for her contribution to medical education in Korea
1960	Recognized by the City of Seoul as a dedicated public servant
1961	Acknowledged by Ewha Woman's University as a pioneer in the medical education of women
1964	Retired from the practice of medicine

Chung-Hee Kil. Photo reprinted by permission of Sangduk Kim.

1979	Emigrated from Korea to the United States
1980	Established a fellowship fund to promote scientific education at Korea University College of Medicine
1981	Published an autobiography
Death	1990

Chung-Hee Kil was born in Seoul, Korea, in 1899. She was one of the first women medical doctors in Korea and co-founded the first medical school in Korea devoted to training women doctors. During her career as a physician and educator, she struggled with the prejudices of Korean society toward women medical doctors, the antipathy of the male-dominated Korean medical establishment to female medical doctors, and the Japanese occupation of Korea.

She was the second daughter of three children of Mr. and Mrs. Hyun-Suk Kil. In the Korea of her youth, girls were by custom not educated but raised to serve as good wives and good mothers. Dr. Kil was raised by her paternal grandparents after her father passed away when she was in her early teens. Her grandfather, Mr. In-Soo Kil, was the Chung-3rd degree in the Lee Dynasty and was very progressive for his times.[1] He encouraged her to get a formal education.

In 1910, Japan annexed Korea. The Japanese harshly discriminated against Koreans even to the extent of prohibiting them from speaking the Korean language. Dr. Kil enrolled at the Tokyo Women's Medical College at age 19. She suffered from the prejudices of the Japanese toward Koreans while in Japan, always being watched by her classmates. Dr. Kil participated in the Declaration of Korean Independence on March 1, 1919, and pledged to fight for independence by swearing an oath in her own blood. While in Japan, she witnessed an incident that confirmed her belief in the nobility of her chosen profession. During her senior year at the Medical College, in the midst of the great Kwang-To earthquake near Tokyo, she witnessed a young nurse crushed by a collapsing wall while saving an infant. She graduated in 1923 as only the second Korean woman to receive a medical degree from the Tokyo Women's Medical College.

Dr. Kil returned to Korea after graduation and began an internship at the Chosen Government General Hospital (now Seoul National University Hospital). In 1925 she married Dr. Tak-Won Kim, a physician specializing in internal medicine/neuropsychiatry. Soon thereafter, in order to further their medical training, they separated for almost three years. She returned to Japan and he went to Beijing, China. While in Japan during her senior year at the Tokyo Women's Medical College, Dr. Kil had been visited by Dr. Rosetta Hall, a missionary physician from New York State and the president of the Tong-Dae-Moon Women's Hospital in Seoul. Dr. Hall had impressed on Dr. Kil the need for women doctors in Korea. Indeed, Korean women were hesitant to be seen by male doctors, and many died as a result of failing to seek medical care. After returning to Korea, Dr. Kil joined the Dong-Dae-Moon Women's Hospital and decided that only a medical school devoted to the training of women medical doctors would fill the need for women doctors in Korea.

In 1928, Dr. Hall, Dr. Kil, and her husband, Dr. Kim, co-founded the Chosen Women's Medical Training Institute. At the time, there was no medical school in Korea that would train women. Dr. Kil served as an associate dean and lecturer in obstetrics/gynecology and pediatrics. In 1933, Dr. Hall retired from the Institute. Dr. Kil and her husband continued to expand the curriculum of the Institute, so that within a short period of time the courses taught there were the equivalent of the courses at one-year premedical and four-year medical colleges. A hospital also was attached to the Institute.

However, the male-dominated medical establishment in Korea had not accepted the idea of women medical doctors; as a consequence, funding for the Institute was always a problem. Drs. Kil and Kim aggressively lobbied prominent members of Korean society on the pressing need for women medical doctors in Korea. At the time the nation was ruled by a Japanese governor-general who actively discouraged the development of the Chosen Women's Medical Institute and put pressure on the Institute. Japan's policy at that time was to suppress even pre-existing educational institutions for male Koreans. Despite these difficulties, the Chosen Women's Medical Training Institute (renamed the Seoul Women's Medical Training Institute) grew and had an enrollment of 64 students by 1938.[2]

In 1938 the Seoul Women's Medical Training Institute became a medical college, the Seoul Women's Medical College (now the Korea University College of Medicine). However, once the Institute had been elevated to the college level, the Japanese prohibited Drs. Kil and Kim from joining the faculty as punishment for their anti-Japanese activities in Korea. Dr. Tak-Won Kim had been arrested for leading anti-Japanese activities. The Japanese themselves took charge of operating the Seoul Women's Medical College, favoring Japanese medical students over their Korean counterparts.

In 1939, Dr. Kil's husband passed away. Dr. Kil never remarried; she continued to practice medicine in Seoul for the next 25 years. During her practice she continued to suffer under the prejudices of Korean society. She recalled that her patients and colleagues referred to her as a "nurse" or "midwife," until she successfully performed an operation that her male colleagues had refused to perform.

Dr. Kil served as president of the Korean Women's Medical Association and also as a chief physician for the Korean royal family. She was also an instructor at the Ewha Women's University. While practicing and teaching medicine, she raised two daughters, both medical doctors in the United States, and a son, a man of letters.

In 1959 the Korean minister of public health recognized Dr. Kil for her contribution to medical education in Korea. In 1960 the City of Seoul recognized her as a dedicated public servant, and in 1961 the Ewha Woman's University acknowledged her as a pioneer in the medical education of women.

In 1964, Dr. Kil retired from the practice of medicine and engaged in her hobby of calligraphy. In 1979 she emigrated to the United States to live with her daughter. In 1980 she established a fellowship to promote scientific education at Korea University College of Medicine. The fund provides graduates of Korea University College of Medicine the opportunity to study medical sciences in the United States. In 1981, Dr. Kil's autobiography was published. It has served as a source on the history of women's medical education in Korea and the history of the Korea University College of Medicine. Throughout her life, Dr. Kil was a devoted Roman Catholic. She passed away at age 92 in 1990 at Cheltenham, Pennsylvania.

Notes

1. In Korea during the Lee Dynasty, the titles of high government officials were divided into classifications of 1st to 8th grades of Chung and Chong.

2. In 1933 the Japanese government renamed the Chosen Women's Medical Training Institute as the Seoul Women's Medical Training Institute.

Bibliography

"History of Korea University College of Medicine (Interview with Sangduk Kim)." *Ho-Il-Lyong* (student magazine) 13 (1993): 138–59.

Kim, Sangduk. "Women's Medical Training Institute, 1928–1938." *Korean Journal of Medical History* (Seoul, Korea) 1 (1993): 80–84.

Kim, Sanghee, and Sangduk Kim. "History Should Be Flawlessly Recorded." *Korea University Medical College Newspaper* (Seoul, Korea) no. 117 (1988): 3.

SANGDUK KIM

HELEN DEAN KING

(1869–1955)

Zoologist

Birth	September 27, 1869
1892	A.B., Vassar College
1899	Ph.D., Bryn Mawr College
1899–1904	Assistant in Biology, Bryn Mawr College
1899–1907	Teacher of Science, Miss Florence Baldwin's School
1906	"Starred" in first edition of *American Men of Science*
1906–1908	Fellow in Biology, University of Pennsylvania
1908–13	Assistant in Anatomy, Wistar Institute of Anatomy and Biology
1913–27	Assistant Professor of Anatomy, Wistar Institute
1927–49	Professor of Embryology, Wistar Institute
1932	Shared Ellen Richards Prize, Association to Aid Scientific Research for Women
Death	March 9, 1955

Helen Dean King. Photo reprinted by permission of The Archives of The Wistar Institute, Philadelphia, Pennsylvania.

Helen Dean King's research on the breeding of rats earned her a star in the first edition of *American Men of Science*. The star indicated that she was among America's top 1,000 scientists, as voted by her peers.

Helen Dean King was born in 1869 in Owego, New York. She was the eldest of two daughters born to William and Lenora King. Both parents were from prosperous New York families. Her father continued in the family leather business as president of the King Harness Company. Like her father, Helen began her education near home at the Owego Free School. At age 18 she entered Vassar College. This was an excellent choice for the aspiring scientist, for at that time Vassar College led all other women's colleges in expenditures for scientific equipment and re- sources. King earned her baccalaureate degree in 1892. In 1895 she began her graduate studies at Bryn Mawr College. Her major course of study was in the area of morphology under the direction of Thomas Hunt Morgan. She also studied physiology and paleontology.

King was fortunate to have Morgan as a teacher and mentor. He was known not only for his own research but also for his success as a teacher. Many of his students from Bryn Mawr went on to be prominent scientific researchers. He aided his students by including them as co-authors of his published studies, and he encouraged them to pursue independent, original research. Many of his Bryn Mawr students published articles as sole authors while in graduate school. Morgan obviously recognized and respected the talent and intellect of his women students, a perspective not prevalent in the late nineteenth century.

King received her Ph.D. in 1899. In 1901 she published her dissertation on the embryonic development of the common toad, "The Maturation and Fertilization of the Egg of *Bufo Lentiginosus*." King remained at Bryn Mawr, continuing her research in association with Morgan and assisting in the biology laboratory. Her early work continued to focus on amphibians, reflecting Morgan's interest in regeneration and developmental anatomy. At this time King also taught science at Miss Florence Baldwin's School.

In 1906, King became a university fellow for research in zoology at the University of Pennsylvania. She remained there until 1908, when she accepted a teaching position at the Wistar Institute of Anatomy and Biology in Philadelphia. King remained associated with the Wistar Institute for over 40 years, until her retirement in 1949. She held several positions there and was promoted to full professor of embryology in 1927. She served on the Institute's advisory board for 24 years and was editor of its bibliographic service for 13 years. In addition, she served as editor of the *Journal of Morphology and Physiology* from 1924 to 1927.

While at the Wistar Institute, King changed the focus of her research from amphibians to rats. Her most outstanding contribution was her successful experimentation in breeding pure strains of rats for use as laboratory animals. She was most interested in the effects of close inbreeding, experimenting with brother-sister matings of albino rats. She published numerous studies analyzing the growth and activity of the inbred animals, comparing them with stock albino rats. King concluded that the inbred animals compared favorably with the stock albinos. She rejected the almost universal prejudice against inbreeding. This notion caught the attention of the popular press, which reported that King considered incest taboos unnecessary.

King was also interested in the process and effects of domesticating wild animals. She conducted a series of breeding experiments with the Norway rat, an animal found wild in the streets of Philadelphia. She started with six pairs of the wild rats and bred the animals through the twenty-eighth generation. In the course of this study she documented many mutations, including curly-haired rats, waltzing rats, and chocolate-colored rats. She concluded that captivity leads to diversity in a wild species, rather than producing a homogeneous population.

Numerous awards and affiliations attest to the respect King was shown by her scientific contemporaries. Her name was starred in several editions of *American Men of Science*. In 1932 she was awarded the Ellen Richards Research Prize of the Association to Aid Scientific Research for Women. She was a member of the American Society of Zoologists and served as its vice president in 1937. Other distinctions included election as a fellow of the New York Academy of Science and affiliation with the American Association for the Advancement of Science, the American Society of Naturalists, the American Society of Anatomists, the Society of Experimental Biology and Medicine, Phi Beta Kappa, and Sigma Xi.

After a long and successful career as a genetic researcher, Helen Dean King died in Philadelphia on March 9, 1955, at age 85.

Bibliography

Bailey, Brooke. *The Remarkable Lives of 100 Women Healers and Scientists*. Holbrook, Mass.: Bob Adams, 1994.

Bailey, Martha J. *American Women in Science: A Biographical Dictionary*. Santa Barbara, Calif.: ABC-CLIO, 1994.

Dictionary of Scientific Biography, Vol. 17, Supplement II, s.v. "Helen Dean King." New York: Charles Scribner's Sons, 1990.

King, Helen Dean. *Life Processes in Gray Norway Rats during Fourteen Years in Captivity*. Philadelphia: Wistar Institute of Anatomy and Biology, 1939.

———. "The Maturation and Fertilization of the Egg of *Bufo Lentiginosus*." *Journal of Morphology* 17, no. 2 (1901): 293–350.

———. *Studies in Inbreeding*. Philadelphia: Wistar Institute of Anatomy and Biology, 1919.

Ogilvie, Marilyn Bailey. *Women in Science, Antiquity through the Nineteenth Century: A Biographical Dictionary with Annotated Bibliography*. Cambridge, Mass.: MIT Press, 1986.

Oppenheimer, Jane M. "Thomas Hunt Morgan as an Embryologist: The View from Bryn Mawr." *American Zoologist* 23 (1983): 845–54.

ANN LINDELL

LOUISA BOYD YEOMANS KING
(1863–1948)
Horticulturist

| Birth | October 17, 1863 |
| 1890 | Married Francis King |

1912–15	President, Garden Club of Michigan
1921	Awarded George Robert White Medal for eminent service in horticulture by the trustees of the Massachusetts Horticultural Society
1923	Medal of Honor, Garden Clubs of America
1931	Distinguished Service Award, National Home Planting Bureau
Death	January 16, 1948

Louisa Boyd Yeomans King was instrumental in creating the Garden Club of America and popularizing gardening through her many books and articles. She promoted better knowledge of plants, flowers, fruits, and vegetables as well as the notion that modern life should be filled with nature's beauty.

Louisa Yeomans grew up in a home that was family-centered, fun-loving, religious, and intellectually stimulating. She was the third of five children and the elder of two daughters of a Presbyterian minister, Alfred Yeomans, of Washington, New Jersey. After being educated in private schools, she married Francis King of Chicago when she was 26 years old.

The first 12 years of her married life, during which three children were born, were spent in the home of her husband's mother, Mrs. Henry. W. King, an able and devoted gardener. The elder Mrs. King's old-fashioned, formal garden in Elmhurst, Illinois, with its 210 varieties of herbs, provided Louisa with an apprenticeship in soil management, pruning, spraying, and general horticultural practices. Her mother-in-law was both knowledgeable and scholarly. Although she hired gardeners to do the hard work, she planned and supervised every aspect of her garden. She was accompanied on her daily rounds by her daughter-in-law, Louisa. When the work in the garden was done, on rainy days, or on winter days when the garden was dormant, they could retire to an extensive library of horticultural books and magazines. Mrs. Henry King purchased new books on horticultural subjects as soon as they were published and subscribed to a variety of garden magazines. Louisa was able to read about a revolution in the garden world through the writings of Gertrude Jekyll and William Robinson. This exposure to theory and practice inspired in her a love of gardening and an interest in developing a garden of her own.

When in 1902 she moved to her own home, called Orchard House, in Alma, Michigan, Louisa had an opportunity to do so. She also began, at this time, her correspondence with the writers whose books she had read in her mother-in-law's library. As her expertise and interest grew, Louisa began writing about her gardening enthusiasm. Her first contribution was to *Garden Magazine,* and later she published in *House Beautiful* and *Country Life.* Her first book, *The Well-Considered Garden,* was published in 1915. It chronicles her beginner's attempts at planning and establishing a garden. The book was followed by nine more over the next 15 years. Gertrude Jekyll had become a valued friend and wrote introductions or forewords to two of her books. In these books, Louisa King expanded on the ideas of Robinson and Jekyll and introduced these concepts to her American readers. These writers led the revolution that resulted in more naturalistic plantings and subtler uses of color than had been seen in Victorian gardens.

King despised packets of mixed seed, encouraging nurseries to package their seed by individual color. In her own garden she often mixed just two varieties together, each variety of a single color, carefully chosen to complement one another. In *The Well-Considered Garden* she confessed to "a faint prejudice against stripes, flakes, or eyes in phloxes, principally because, as a rule, the best effects in color groupings are obtained by the use of flowers of clear, solid tones."[1] King's books emphasized the careful use of color as well as planning and design. Readers were expected to turn to other sources for practical advice regarding soil management, tool care, and general cultivation.

As Louisa King developed her horticultural knowledge and wrote her immensely popular books, she cultivated other interests as well. There were at the time only a few garden clubs in the eastern states and none elsewhere in the country. At a luncheon meeting held in Philadelphia in 1913, the Garden Club of America was established, with Louisa King one of four vice presidents. As a result of the work done by this organization, with Louisa King playing a vital role, garden clubs were established all over the country. King had definite ideas regarding the purposes of a garden club. In *Chronicles of the Garden* she describes advice she had been given for her own garden by a fellow gardener, saying, "The best gardeners always speak out boldly to each other; they share the serious interest in the garden, and know that truthful comment is the tonic thing." She then advised that the reader turn to the counsel of a garden club, which exists to give "mutual pleasure through mutual help." She suggested that all members of a garden club provide cooperation to the program committee and offer suggestions for programming. Then,

> every member of a garden club should be up and doing in some capacity. Some can govern, lead; some can suggest, inspire; some can write, speak; some can sketch, paint, photograph gardens; some will have a gift for design; others the happy ability in organization

necessary for the shows; and the inference is that all are able to garden. . . . I repeat, there is no member of any garden club who should be considered in good standing therein, who does not, will not, help.[2]

Besides mutual help to members, King asserted that another important purpose of a garden club is education. She advocated expanding flower shows to include fruits and vegetables, by way of educating viewers and members as to their cultivation, uses, and benefits. Garden competitions or contests also were suggested as means for education. Such competitions, King suggested, could be judged on artistic effect of planting, best use of available space, and artistic merit of the photographs submitted. Further suggestions for education included a meeting devoted to the study of nursery catalogs. Finally, she suggested that each garden club develop a collection of a specific plant and develop expertise in this plant. She expected that successful garden clubs would become arbiters of outdoor beauty in their communities. King admonished gardeners to study botany and to always use botanical names, thus doing away with imprecision and confusion when discussing plants. In addition, she claimed, botanical nomenclature is just as easy to learn as local names.

Not content with her work in the Garden Club of America, Louisa King was instrumental in 1914 in the formation of the Women's National Farm and Garden Association. She served as its first president. This organization, established by university women investigating business and professional opportunities for women in horticultural and agricultural fields, helped isolated farm women come together over matters of mutual interest. The membership of the Association grew to several thousand, with its own magazine and international affiliations. With the outbreak of the World War I in 1918, women in agriculture assumed new importance, as many women managed farms while the men were away fighting.

After the death of King's husband in 1927, Orchard House was sold. The trauma associated with leaving the garden she had built and beginning anew at Kingstree in New York is described in *From a New Garden*. Although her style was always direct and personal, this is the most personal of her books; she describes finding the house, discovering what was already planted there, and establishing her own vegetable garden, trial gardens, and various garden "rooms." Her love and appreciation of glorious trees is expressed as follows:

Long ago one with an eye to fine planting saw to it that this house should be flanked at one end by a grove of locust and maple trees and at the other by one single elm—and what an elm! The talk of the country-side! . . . a great tree . . . its girth near the ground more than fourteen feet. This miracle among trees, amber with leaf-buds against the spring blue, sets the seal of age upon the quiet manor-house below it and stands to show that good houses and noble trees are companions inevitable.[3]

Her personal style is again evidenced as follows:

[I]f you have sloping ground and . . . do not terrace it and get the beauty of variety that comes from skillful terracing, the loveliness of well-laid steps . . . delicious backgrounds for the planting of shrubs and flowers . . . shall I say it?—you do not deserve a garden.[4]

Louisa King's contributions to horticulture were recognized in both England and the United States. In England she was made a vice president of the Garden Club of London and a fellow of the Royal Horticultural Society. In 1921 she was the first woman recipient of the George Robert White Medal of the Massachusetts Horticultural Society, the highest gardening distinction in America at that time. In 1923 she received the Medal of Honor of the Garden Club of America. She was awarded the Distinguished Service Award by the National Home Planting Bureau in 1931. An extensive planting of dogwoods in the National Arboretum was established in her memory by the Women's National Farm and Garden Association.

Louisa King was not just a theorist, writer, lecturer, and organizer. She was also a woman who genuinely loved her garden, her privacy, and her plants. She did not merely plan and direct other gardeners but wore old clothes herself and dug, planted, and weeded. Her contributions to American horticulture include her books, which are still valuable for their advice on planning, layout, and the use of color; the impetus she provided for the establishment of garden clubs throughout the country; and the inspiration she provided to gardeners and women through her writings, her lectures, and her visits to garden clubs and gardeners. She died in 1948 at age 84.

Notes

1. Louisa King, *The Well-Considered Garden*, Preface by Gertrude Jekyll (New York: Scribner's, 1915), p. 5.

2. Louisa King, *Chronicles of the Garden* (New York: Scribner's, 1925), pp. 192, 194.

3. Louisa King, *From a New Garden* (New York: A.A. Knopf, 1930), p. 5.

4. *Ibid.*, p. 18.

Bibliography

Hollingsworth, Buckner. "Mrs. Francis King, 1863–1948: Gardener's Guide and Friend," in *Her Garden Was Her Delight*. New York: Macmillan, 1963.

King, Louisa. *From a New Garden*, Introduction by A.P. Saunders. New York: A.A. Knopf, 1930.

———. *The Well-Considered Garden*, Preface by Gertrude Jekyll. New York: C. Scribner's, 1915.

Notable American Women, 1607–1950. Cambridge and London: Belknap Press, 1970.

Saunders, Louise S.B. "Dean of American Gardening." *House and Garden* (Mar. 1940): 42–43ff.

Who Was Who in America with World Notables. Chicago: Marquis Who's Who, 1973.

CAROL W. CUBBERLEY

MIMI A.R. KOEHL

(1948–)

Biologist, Biomechanic

Birth	October 1, 1948
1969	Phi Beta Kappa; Columbia University Teachers College Book Prize
1970	B.A., biology, *magna cum laude*, Gettysburg College; Henry W.A. Hanson Scholarship Award
1976	Ph.D., zoology, Duke University
1976–77	Postdoctoral Fellow, Friday Harbor Laboratories, University of Washington
1977–78	Postdoctoral Fellow, University of York, England
1978–79	Assistant Professor, Div. of Biology and Medicine, Brown University
1979–	Faculty Member, Dept. of Zoology (now Dept. of Integrative Biology), University of California, Berkeley (Assistant Professor, 1979–83; Associate Professor, 1983–87; Professor, 1987–)
1983	Presidential Young Investigator Award
1985	Young Alumni Achievement Award, Gettysburg College

Mimi A.R. Koehl. Photo courtesy of Mimi A.R. Koehl.

1986	Visiting Scholar, Centre for Mathematical Biology, Oxford University
1987	Helen Homans Gilbert Lectureship, Harvard University
1988	John Simon Guggenheim Memorial Foundation Fellowship
1990	MacArthur Foundation Fellowship Award
1993	Fellow, California Academy of Science; Visiting Professor, Zoologisches Institut der Universität Basel

Dr. Mimi A.R. Koehl is a woman whose curiosity has taken her far. As an undergraduate art student she had to take a biology course and liked it enough to switch to biology as a major. Koehl said she was "fascinated by natural form, and aside from art, biology was another way to pursue that interest in form."[1] Now, after years of study, she has become well known in biomechanics, a field that relates physical form to fluid and solid mechanics. Koehl has been recognized as an excellent researcher whose creativity and desire for knowledge fuel her quest to understand how nature works. She has changed the way students view the natural world around them and has shed light on how organisms survive in their environments.

Koehl met with few stumbling blocks in her educational path. After graduating *magna cum laude* in biology from Gettysburg College she went to Duke, where she studied zoology. There she explored sea anemones and how their body structure withstood or altered the forces they experienced in moving water. "There were not many women around, but that didn't bother me too much," said Koehl. "I had role models and mentors who were men. At the time, there were low numbers of women going through the programs, but things were beginning to change. When I started out as an undergraduate, however, I was not taken seriously; but because I loved the questions, organisms, and learning, I persisted." Koehl has been working hard to prove that she is a worthy scientist, not just a woman scientist who has been given some breaks. Her perseverance resulted in her Ph.D. degree and two postdoctoral appointments before becoming an assistant professor at Brown for a year. Koehl then took a position at the University of California, Berkeley, where she is now a professor of integrative biology.

Koehl's main research focuses on "how organisms interact with the physical environment—the consequences of their morphology to how well they function. I use engineering techniques as a bag of tools, to allow me to look at organisms quantitatively." Some of Koehl's topics address the questions of how marine animals cope with wave action; how organisms feed in water; how embryos develop; and why insect wings evolved as they did. Her research has been published in journals representing a variety of fields including mathematics, evolution, marine biology, and fluid mechanics.

Observant of nature, Koehl finds many of her research questions by noticing patterns in the forms of organisms. She says:

> along wave-swept rocky shores, the biggest organisms are flexible, so how does this help in dealing with wave action? Or, many organisms in different phyla have hairs on their appendages for swimming, flying, and catching food. How do these hairs work in air and water? How does it effect performance? What are the basic physical rules about how organisms interact with the air and water around them?

These questions are complex and can be addressed in ecological or evolutionary terms. Once a question is proposed, Koehl believes in a strong mix of field research and laboratory tests. "In the field you get to know the organisms and can measure important aspects of their environment and of their behavior there. In the lab you can do more controlled experiments to get the details piece by piece." Research is, however, not always clear-cut.

> I love surprises—when people think they understand something and it actually works another way. . . . At one point I was studying appendages on zooplankton that looked like combs, and were thought to be used for filter feeding. During my study, I found that water wouldn't go through the gaps because they were so tiny and the water was too sticky. So these appendages were not filters at all, as had been previously assumed.

A lot remains to be learned from biomechanics, as it is still a relatively small field. Koehl says that the biomechanics of organisms other than humans appeals to her partly because "the community is friendly, not cut-throat. You are judged simply on the merits of the work you have done. The price I pay is that biomechanics is not mainstream, it is inter-disciplinary—you can do new types of work, but you have no niche." Yet because of her unique work, Koehl has had her research presented in public forums in addition to research journals, books, and meetings. Her research has been highlighted in *Science News, Science, Scientific American, Discover,* and *Newton* (a Japanese publication), and on television science programs in the United States, Canada, and Australia.

Additional interest in biomechanics also comes from the side benefits of research. Koehl notes that

> our work on how crabs run in water and air will help in the design of walking robots for uses such as cleaning toxic wastes. Drag forces pulling on marine organisms that foul oil platforms or docks can increase the forces these structures have to withstand; but by knowing how big the forces are on the organisms, one can figure out how much stronger to build the platforms.

Koehl sees good uses from her work in such cases, but her main drive remains to explain the basic physical principles of how organisms work in nature.

A disadvantage of studying the biomechanics of organisms other than humans is that not being an easily recognizable field, funding is often difficult to find. Koehl has done well, however, in receiving both competitive grants and special awards. Her awards include a Presidential Young Investigator Award (in the first year these awards were presented), a Guggenheim Fellowship, and a MacArthur Foundation Fellowship. She is also a fellow of the California Academy of Science. Koehl's MacArthur Fellowship, which is given to creative scholars in a variety of fields, has allowed her to "pursue really new and different ideas, and given headway for future funding."

In sharing her knowledge of biomechanics, Koehl is not just a researcher but a teacher as well. "Academics is hard; it takes a lot of time and energy," she says. She finds being an educator at Berkeley "frustrating because I can't get to know undergraduate students or see their progress because the university is so big. I love graduate students because I can follow their successes. It's extremely disappointing to teach at a large-scale university with insufficient funding."

Along with her hard work as a scientist, Koehl takes some time off for jogging, backpacking, and bicycling. She says, "Fortunately, my husband is also a scientist—he's an oceanographer—so he understands my drive to do research whenever I can get some free time away from my duties at the university."

Koehl enjoys spreading the word "that biomechanics is not just boring equations and physics." She has given many invitational lectures across the United States and internationally. For future research, Koehl plans to focus more on the physics of how the forms of organisms come to be during development and growth. In describing her goals as a teacher, Koehl says, "I want to make a positive difference, to improve students' understanding, respect, and appreciation of the natural world."

Note

1. Quotations for this entry have been taken from conversations between Mimi Koehl and the contributor.

Bibliography

Emerson, S.B., and M.A.R. Koehl. "The Interaction of Behavior and Morphology in the Evolution of a Novel Locomotor Type: 'Flying Frogs.' " *Evolution* 44 (1990): 1931–46.

Franklin, D. "The Shape of Life." *Discover* 12 (1991): 10–15.

ould, S.J. "Not Necessarily a Wing: Which Came First, the Function or the Form?" *Natural History* 94 (1985): 12–25.

Kingsolver, J.G., and M.A.R. Koehl. "Aerodynamics, Thermoregulation, and the Evolution of Insect Wings: Differential Scaling and Evolutionary Change." *Evolution* 39 (1985): 488–504.

Koehl, M.A.R. "Effects of Sea Anemones on the Flow Forces They Encounter." *Journal of Experimental Biology* 69 (1977): 87–105.

———. "Hairy Little Legs: Feeding, Smelling, and Swimming at Low Reynolds Number." *Contemporary Mathematics* 141 (1993): 33–64.

———. "The Interaction of Moving Water and Marine Organisms." *Scientific American* 247 (1982): 124–34.

Koehl, M.A.R., and R.S. Alberte. "Flow, Flapping, and Photosynthesis of Macroalgae: Functional Consequences of Undulate Blade Morphology." *Marine Biology* 99 (1988): 435–44.

Lewin, R. "Flights of Conjecture." *Scientific American* 254 (1986): 66B.
————. "On the Origin of Insect Wings." *Science* 230 (1985): 428–29.
Morell, V. "The Origin of Flight." *Equinox* 32 (1987): 13–16.

JOY SCHABER

MARIE-LOUISE LACHAPELLE
(1769–1821)
Midwife, Teacher

Birth	January 1, 1769
1795–97	Chief Midwife, Hôtel Dieu, Paris
1797–1821	Director and Instructor, Maison d'Accouchements, Hôtel Dieu
Death	October 4, 1821

By the end of the eighteenth century, France had an enviable tradition of capable and well-trained midwives. The Hôtel Dieu in Paris, originally a hospital for the poor connected to the Cathedral of Notre Dame, became the foremost maternity hospital in Europe. The first school for midwives had been set up at the Hôtel Dieu during the second half of the sixteenth century, and some of the most distinguished midwives/obstetricians taught or studied there well into the nineteenth century.

Marie-Louise Lachapelle was one of the most famous products of that environment. She was born in Paris in 1769 to Marie Jonet Dugès, who was descended from a long line of midwives, and Louis Dugès, a health officer in Paris. Marie Dugès (1730–1797) had been a midwife at Châtelet; after her appointment as midwife-in-chief at the Hôtel Dieu in 1775, she oversaw the maternity ward there. Marie-Louise assisted her mother from an early age onward; when she was only 15 years old, she acted as midwife at a very difficult birth and saved both mother and child.

In 1792 she married a surgeon at the Hôtel St. Louis, but when her husband died after only three years she had to support herself and her daughter. She worked again as a midwife. The maternity wards at the larger French hospitals had developed training schools modeled on the apprentice system, whereby the midwife-in-chief taught the practical aspects of delivery supplemented by lectures from professors from various medical faculties. When Mme. Dugès had started working at the Hôtel Dieu, she reorganized the maternity ward and Marie-Louise joined her mother as associate chief midwife. After her mother's death in 1795 she succeeded her as head of the maternity department.

Like most hospitals of that period, the Hôtel Dieu was a charitable institution founded to serve the poor, and only the poor went to the hospitals. Rooms were crowded, patients shared beds, hygiene was practically nonexistent, and infections spread almost unchecked. Childbed fever, in particular, posed a great danger to mothers and babies, often reaching epidemic proportions in the hospitals. Periodically attempts were made to improve conditions, especially after the French Revolution when the government took over most of the charity hospitals.

In 1797, Marie-Louise Lachapelle was asked to help organize a new maternity department to be part of the old Hôtel Dieu but located in a former religious institution at Port Royal de Paris. From its very beginnings, this new Hospice de la Maternité (later named Maison d'Accouchements, although the old name was still sometimes used) was set up as a teaching hospital by its main organizer, Jean-Louis Baudelocque (1746–1810). He was the foremost French obstetrician of his time and the surgeon-in-chief and director of the Hôtel Dieu's maternity ward. He had long felt the need for a systematically organized school that would combine practical and theoretical instruction for midwives, who in turn would train midwives in the provinces.

Marie-Louise Lachapelle spent some time studying at the University of Heidelberg in Germany with the famous obstetrician Franz Carl Naegele (1778–1852), who also headed the maternity clinic in Heidelberg. She then became resident director and first instructor of the Maison d'Accouchements at Port Royal under Dr. Baudelocque. Lachapelle and Baudelocque worked together extremely well. Baudelocque in particular had the greatest respect for her practical knowledge and her manual dexterity when delivering babies.

The course of study they developed for training midwives lasted one year. The students lived at the hospital and attended lectures at the Ecole de Médecine as well as a special course in midwifery. In addition, they were instructed daily by Lachapelle. They assisted with the management of labor and delivery, and they joined the attending physician in his daily visit. They were also required to attend the postmortems of all women who died in the hospital. At the end of this rigorous training, the students had to pass an examination and then received a diploma from the Ecole de Médecine. The school soon attracted students from the whole of France, then from all over Europe, and served as a model for similar institutions in other countries.

Marie-Louise Lachapelle seems to have been an inspired teacher as well as an exceptionally skillful midwife. She used a manikin to explain the mechanics of the birth process and showed a special talent in managing problem births. She made notes of all the difficult deliveries she attended, collecting about 40,000 case histories, which formed the basis for her book *Pratique des Accouchements; ou Mémoires et Observations Choisies, sur les Points les Plus Importants de l'Art*. Unfortunately, she was only able to work on the first volume before she died of stomach cancer in Paris on October 4, 1821. Her nephew, Dr. Antoine Dugès, professor of medicine at the University of Montpellier, edited her notes and published them in three volumes in 1825. The work was republished repeatedly and translated into German. It became an important reference work for nineteenth-century obstetricians and midwives. Lachapelle also contributed five case histories to the journal *L'Annaire Médico-Chirugical* for its first volume in 1819.

In her book Mme. Lachapelle covered all aspects of midwifery, including case histories of difficult deliveries, operations, and autopsy findings. Dr. Baudelocque had listed 94 possible birth positions for the fetus in his writings. Lachapelle disagreed and pointed out that in her 30 years of practice she had observed only 22 distinct positions (presentations). Her approach to midwifery focused on the practical aspects of the birth process; she was opposed to unnecessary and potentially dangerous interventions by an attending doctor or midwife. She insisted that instruments such as forceps be used as little as possible and never just to shorten labor. According to her statistics, she used forceps only 93 times. She taught her students a method of inserting and handling the instrument with only one hand, so that trauma to mother and especially the baby would be minimal. Remarkably, she performed only one Caesarian section in her 40,000 recorded cases. And in spite of the fact that the mechanism of transmission of childbed fever was not yet properly understood, Mme. Lachapelle's methods ensured that the incidence of puerperal fever in her maternity ward remained low. She also tried to limit the number of people attending a birth because she thought that the crowd around the mother's bed might cause the dreaded fever.

The clear, well-organized, and systematic presentation of material in Lachapelle's book reflects the high level of instruction at the Maison d'Accouchements. Under the leadership of Professor Baudelocque and Marie-Louise Lachapelle, this institution became the foremost school for midwives during the first half of the nineteenth century. Mme. Lachapelle and her students, among them the famous **Marie Boivin** (1773–1847), contributed immensely to raise the position of midwives as well-trained, competent professionals.

Bibliography

Alic, Margaret. *Hypatia's Heritage: A History of Women in Science from Antiquity through the Nineteenth Century*. Boston: Beacon Press, 1986.

Biographisches Lexikon der hervorragenden Ärtzte aller Zeiten und Völker, 2nd ed. Berlin: Urban & Schwarzenberg, 1931.

Cutter, Irving S., and Henry R. Viets. *A Short History of Midwifery*. Philadelphia: W.B. Saunders, 1964.

Delacoux, Alexis. *Biographie des Sage-Femmes Célèbres, Anciennes, Modernes et Contemporaines*. Paris: Trinquart, 1834.

Herzenberg, Caroline L. *Women Scientists from Antiquity to the Present: An Index*. West Cornwall, Conn.: Locust Hill Press, 1986.

Hurd-Mead, Kate Campbell. *A History of Women in Medicine from the Earliest Times to the Beginning of the Nineteenth Century*. Haddam, Conn.: Haddam Press, 1938. (Reprinted: New York: AMS Press, 1977.)

Lachapelle, Marie-Louise. *Pratique des Accouchements; ou Mémoires et Observations Choisies, sur les Points les Plus Importants de l'Art*. Paris: J.B. Bailliere, 1821–1825.

Lovejoy, Esther Pohl. *Women Doctors of the World*. New York: Macmillan, 1957.

Marks, Geoffrey, and William K. Beatty. *Women in White*. New York: Scribner's, 1972.

Michaud, J. Fr. *Biographie Universelle Ancienne et Moderne*. Graz, Austria: Akademische Verlagsanstalt, 1966–1968. (First printed in Paris, 1854.)

Mozans, H.J. (Zahm, John Augustine). *Woman in Science*. Notre Dame, Ind.: University of Notre Dame Press, 1991. (First printed in New York: D. Appleton, 1913.)

Ogilvie, Marilyn Bailey. *Women in Science: Antiquity through the Nineteenth Century*. Cambridge, Mass.: MIT Press, 1986.

IRMGARD WOLFE

REBECCA CRAIGHILL LANCEFIELD

(1895–1981)

Bacteriologist

Birth	1895
1916	B.A., zoology, Wellesley College
1918	M.A., bacteriology, Columbia University
1922	Began lifelong career at Rockefeller Institute
1925	Ph.D., bacteriology, Columbia University
1960	T. Duckett Jones Memorial Award
1964	American Heart Association Achievement Award
1965	Presented the T. Duckett Jones Memorial Lecture and the Armine T. Wilson Memorial Oration

Rebecca Craighill Lancefield. Photo courtesy of Dr. Vincent A. Fischetti, Professor, The Rockefeller University, NY.

1970	Member, National Academy of Sciences
1973	New York Academy of Medicine Medal; Research Achievement Award from the journal *Medicine*; Research Achievement Award, Wellesley College; D.Sc., *honoris causa*, The Rockefeller University
1976	D.Sc., *honoris causa*, Wellesley College; Honorary fellowship in the Royal College of Pathologists
Death	March 3, 1981

Many people are familiar with the ubiquitous "strep" bacteria. Few, however, know anything about the pioneering woman bacteriologist who studied streptococci for almost 60 years. Rebecca Craighill Lancefield began her scientific career as a research assistant in 1918. In a time when few women were encouraged to pursue their interest in scientific research, Lancefield made significant contributions to her field.

Born Rebecca Price Craighill in 1895 on Staten Island, she was the daughter of West Point graduate William Edward Craighill and Mary Wortley Montague Byram. After a nomadic childhood, Rebecca Craighill attended Wellesley College in 1912. Originally considering a French and English major, she changed to zoology after her interest was piqued by her roommate's zoology text. She then immersed herself in science, taking as many biology courses as possible and also developing a strong background in chemistry.

After her graduation in 1916 she followed the prescribed path for women of that time. She accepted a teaching position at a girl's boarding school, even though the science she taught was physical geography. Lancefield then was offered a graduate scholarship at the Teacher's College at Columbia University.

Accepting the scholarship, she decided to study bacteriology or genetics. However, finding no such course at the Teacher's College, she elected to pursue her degree at Columbia. After spending much of her next year in the laboratory, she received her master's degree in 1918. During her time at Columbia she met fellow graduate student Donald Lancefield. They married and shared a love of science for the rest of their time together.

During the same time O.T. Avery and A.R. Dochez, two well-known researchers at the Rockefeller Institute, returned from a trip with 120 cultures of streptococci. Lancefield, having completed her graduate work, had applied to work at the Institute. She was assigned as a technician on the streptococcus study. During the course of her work she identified at least four immunological types among the cultures.

After funding was discontinued for the streptococcus classification study, Lancefield spent one summer at Woods Hole at the Marine Biological Laboratory. Upon her return to New York, she went to work as a research assistant in the fruit fly laboratory of Dr. C.W. Metz. This work resulted in three publications, including cytological and genetic studies performed by Lancefield. Donald then finished his requirements for the Ph.D. in 1921 and was offered a position in his home state at the University of Oregon.

After a year in Oregon, the Lancefields returned to New York and Columbia University. Rebecca was to complete her doctoral studies in the laboratory of Hans Zinsser, who suggested she take a position that was available at the Rockefeller Institute. She accepted the position to study rheumatic fever with Dr. Homer Swift. Lancefield remained at the Rockefeller Institute for the rest of her career.

The next three years proved frustrating as Lancefield completed her Ph.D. work on *Streptococcus viridans*. The medical community had incorrectly linked the "green" bacterium with rheumatic fever, a conclusion Lancefield documented in the publications from that work.[1] After receiving her doctorate, Lancefield returned to her previous work on the hemolytic bacteria, expanding on the work done with Avery and Dochez. She found a polysaccharide common to all streptococci studied at that time, which were from severe human infections. Later she found that the polysaccharide was typical of the serological group A.

In 1933 she found that the immune sera could be reacted with extracts of streptococcus, resulting in the formation of a precipitate. This precipitin test allowed her to divide the strep into five groups. Designating the strains she studied earlier as group A, she assigned later groups successive letters of the alphabet. The publication detailing this work, "A Serological Differentiation of Human and Other Groups of Hemolytic Streptococci," was to be one of the most significant of her career.[2] The methods she described allowed her to classify more than 60 distinct strains and resulted in a classification system that is still in use.

Over the next several years Lancefield continued her work on the streptococcal antigens, which included transatlantic collaborations. Although Lancefield knew the classification work was important, she considered it to be the means through which to study the biology of the bacteria.[3] She collaborated with Swift in presenting a paper in 1936 at the Second International Conference for Microbiology in London. In 1940, while an associate of the Rockefeller Institute for Medical Research, she delivered the eighth and last Harvey Society lecture at the New York Academy of Medicine. The title of her speech was "Specific Relationship of Cell Composition to Biological Activity of Hemolytic Streptococci."

Dr. Lancefield was truly a bench scientist, always preferring research to giving lectures or attending to administrative work. During World War II she was a member of the Office of Scientific Research and Development and a consultant to the Armed Forces Epidemiology on the Commission on Streptococcal and Staphylococcal Diseases. Although that commission no longer exists, it informally continued its meetings. In 1977 the name "The Lancefield Society" was adopted.[4] This is just one of many testaments to the importance and impact of Lancefield's work.

The Lancefield laboratory continued to be the world source for identification of strep strains and for antisera. After Dr. Swift retired, Maclyn McCarty joined Lancefield's lab to continue the study of rheumatic fever. Lancefield continued her own study of the streptococcal antigens.

Although Lancefield tried to remain primarily in her laboratory, the larger scientific community called on her for service. In the 1940s she was elected president of the Society of American Bacteriologists (later the American Society for Microbiology). The American Association of Immunologists elected her its first woman president in 1961. Many honors were bestowed on her as well. The Rockefeller University awarded her an honorary degree in 1973, as did her alma mater, Wellesley, in 1976.

As a mentor, Lancefield nurtured many junior scientists and participated fully in laboratory social affairs. She was well known as an attentive hostess, and she liked socializing. Enjoying time with her husband and daughter, the family summered at Cape Cod to avoid the New York heat and to continue their research at the Marine Biological Laboratory.

Dr. Lancefield retired in the mid-1960s but continued doing research and publishing papers. Elected a member of the National Academy of Sciences in 1970, she was only the tenth woman to earn that honor. Lancefield preferred to downplay the importance of being the first woman to receive recognition and awards. She did agree that it was harder for women to get ahead in science, but she thought it was "partly women's own fault" because they usually would "consider home careers ahead of other careers and often take advantage of that."[5]

Regardless of her reluctance to be in the public eye, Rebecca C. Lancefield became a legendary figure among her fellow scientists. She loved science and had good relations with her colleagues and students. Her work on classifying streptococcal strains and the detailed experimental analysis she performed have become the basis for our understanding of these disease-causing organisms. Fellow researcher Lewis W. Wannamaker wrote her obituary for the American Society for Microbiology; commenting on her status as a microbiologist legend, he said, "She was a patient, hard-working scientist and a warm, soft-spoken person who made her contribution without fanfare and without feeling the need to compete with anyone except, possibly, herself."[6] Rebecca Craighill Lancefield died on March 3, 1981.

Notes

1. For an exhaustive biography of Lancefield, see Elizabeth Moot O'Hern, *Profiles of Pioneer Women Scientists* (Washington, D.C.: Acropolis Books, 1985), Chap. 5.

2. See Elizabeth M. O'Hern, "Rebecca Craighill Lancefield, Pioneer Microbiologist," *ASM News* 41, no. 12 (1975): 805–10.

3. For the most recent biography of Lancefield, see Judith N. Schwartz, "Mrs. L.," *The Rockefeller University Research Profiles* (Summer 1990): 1–6.

4. See Lewis W. Wannamaker, "Obituary: Rebecca Craighill Lancefield," *ASM News* 47, no. 12 (1981): 555–59. This moving tribute contains unique stories about Lancefield and her work.

5. *Ibid.*, p. 558.

6. *Ibid.*, pp. 558–59.

Bibliography

Corner, George W. *A History of the Rockefeller Institute: 1901–1953. Origins and Growth.* New York: Rockefeller Institute Press, 1964.

Foster, W.D. *A History of Medical Bacteriology and Immunology.* London, Fakenham, and Reading: William Heinemann Medical Books, 1970.

O'Hern, Elizabeth Moot. *Profiles of Pioneer Women Scientists.* Washington, D.C.: Acropolis Books, 1985.

————. "Rebecca Craighill Lancefield: Pioneer Microbiologist." *American Society for Microbiology News* 41 (1985): 805–10.

Schwartz, Judith N. "Mrs. L." New York: The Rockefeller University, 1990.

Wannamaker, Lewis W. "Obituary: Rebecca Craighill Lancefield." *American Society for Microbiology News* 47 (1981): 555–59.

BARBARA I. BOND

ELISE DEPEW STRANG L'ESPERANCE

(ca. 1879–1959)

Pathologist

Birth	ca. 1879
1896–99	Woman's Medical College of the New York Infirmary for Women and Children
1900–1902	Residency at Babies Hospital, N.Y.
1900–1908	Private practice in pediatrics, New York City
1910–12	Assistant in Pathology, Cornell Medical College
1910–44	Pathologist and Director of Laboratories, New York Infirmary for Women and Children
1912–20	Instructor, Dept. of Pathology, Cornell Medical College
1919–32	Instructor in Surgical Pathology, Bellevue Hospital
1920–32	Assistant Professor, Dept. of Pathology, Cornell Medical College
1932–59	Director, Strang Tumor Clinic
1940–50	Director, Strang Cancer Prevention Clinics
1944–50	Assistant Professor, Dept. of Preventative Medicine, Cornell Medical College
1946	Friendship Award for Eminent Achievement, American Women's Association
1948	Clement Cleveland Medal, New York City Cancer Committee
1950	Elizabeth Blackwell Citation, New York City Cancer Committee

1950–59	Professor, Dept. of Preventative Medicine, Cornell Medical College
1951	Lasker Award, American Public Health Association
Death	January 21, 1959

Elise Strang was the second daughter of Albert Strang, a physician, and Elise (Depew) Strang. She entered medical school at age 16 and married David L'Esperance, a lawyer, while she was a student. Her medical residency was in pediatrics at Babies Hospital in New York City. She entered into private practice in pediatrics. However, she soon became interested in tuberculosis and subsequently chose to do further study in pathology, the branch of medicine that studies the microscopic changes in tissues and organs to determine the cause of diseases. The pathologist and cancer specialist with whom she wished to study, Dr. James Ewing, had previously refused to employ a woman, but he accepted Dr. L'Esperance as his assistant. This was a menial position for a person with a medical degree. She worked as his assistant for two years and "earned" an academic appointment as an instructor.

After her mother died of cancer in 1930, Dr. L'Esperance and her unmarried sister, May, established the Strang Tumor Clinic, using an inheritance from her maternal uncle, Chauncey Depew, a deceased financier and U.S. senator. She subsequently founded the Kate Depew Strang Cancer Prevention Clinic in 1937 in the New York Infirmary in honor of her mother. Dr. L'Esperance believed that cancer could be terminated if detected early. In fact, in the clinic's first year of operation, 71 women who thought they were in good health had examinations by Dr. L'Esperance and three were found to have cancer.

The clinic was set up to detect cancer in women and children. It became so popular that a second clinic was opened in 1940 at Memorial Center for Cancer and Allied Diseases. The clinics were subsequently expanded to include men and adolescents. By 1950 the clinics had examined 35,000 patients, including 1,300 children and 8,500 men. Cancer was detected in between 1 and 2 percent, according to age and sex; but precancerous conditions were found in 18 percent. Patients under age 45 returned annually, and patients over 45 were examined every six months.

Dr. L'Esperance's goal was not detection of cancer but its prevention. She believed that if physicians knew what to look for and how to look for precancerous conditions, the prevention of fully malignant cancers could be achieved. To meet this goal, medical students from Cornell University's Department of Preventative Medicine began taking a mandatory session at the Strang Cancer Prevention Clinic, beginning in 1946. In 1950 the clinic took its first interns. Dr. L'Esperance's clinics became the prototypes for cancer detection and prevention clinics throughout the United States and in other countries.

Dr. L'Esperance published approximately 30 peer-reviewed papers in medical journals. These included descriptions of Hodgkin's disease, carcinoma of the cervix, carcinoma of the ovary, and articles on the prevention of cancer as practiced at the Strang clinics. She was also at one time an editor of *Medical Women's Journal* and the *Journal of the American Medical Association*. She was a fellow of the New York Academy of Medicine and a member of a number of learned societies and associations, including the American Medical Association, the American Cancer Society, the Harvey Society, the Woman's American Medical Association, the American Association of Pathology and Bacteriology, the American Association of Immunologists, and the American Radiologists Society.

There is no documentation that Dr. L'Esperance had children of her own. She was once described as a "tall, fast-moving, strongly built woman." She enjoyed wearing hats—even in the office while seeing patients—and attributed this habit to time she spent in X-ray. "There never was any place to hang the thing. So I kept it on. Got in the habit. Now I'd feel headless without it," she said.[1]

Dr. L'Esperance was also known as a breeder of show horses. She owned a stable, and for many years she drove her own harness ponies in the National Horse Show at New York's Madison Square Garden. During her later years she lived with her sister in suburban Westchester County, New York, where they enjoyed keeping horses, cats, and other animals.

Note

1. *Country Life* 1 (May 1948): 41.

Bibliography

Current Biography (1950): 340–41.
Current Biography (1959): 256.
"Dr. L'Esperance, Specialist, Dead." *New York Times* (Jan. 22, 1959): 31.
"Medicine. Prevention Is Her Aim." *Time* 55 (Apr. 3, 1950): 78–79.
Notable American Women: The Modern Period. Cambridge, Mass.: Harvard University Press, 1980.

KAREN JAMES

RITA LEVI-MONTALCINI

(1909–)

Embryologist

Birth	April 22, 1909
1936	M.D., *summa cum laude*, University of Turin
1945–47	Research Assistant to Professor Giuseppe Levi (no relation), University of Turin
1947–51	Research Associate with Viktor Hamburger, Washington University
1951–58	Associate Professor of Zoology, Washington University
1958–77	Professor of Neurobiology, Washington University
1961	Established (with Pietro Angeletti) the Laboratory of Cellular Biology, Rome
1969–79	Director, Institute of Cell Biology, Italian National Council of Research in Rome
1977	Sc.D., University of Uppsala, Sweden
1977–	Professor Emerita, Washington University
1982	Ph.D., Washington University School of Medicine
1986	Nobel Prize in Physiology or Medicine (with Stanley Cohen); Albert Lasker Basic Medical Research Award (with Stanley Cohen)
1987	National Medal of Science, USA; Honorary degrees, University of London; Biophysical Institute, University of Brazil
1989	Honorary degree, Harvard University
1990	Honorary degree, University of Urbino, Italy

Rita Levi-Montalcini has enjoyed a long career in basic biology research, primarily in the United States and Italy. She has said, "The moment you stop working, you are dead."[1] She has had many difficult and frustrating moments, including persecution during World War II, nonrecognition of the importance of her discovery of nerve growth factor in the 1960s, and recent accusations that the Italian pharmaceutical company Fidia "bought" the Nobel Prize for her in 1986.

Rita and her twin sister Paola, an artist, were the youngest children of Adele Montalcini, a painter, and Adamo Levi, an electrical engineer. Her mother was a model Victorian woman in a home filled with affection. However, her father controlled all the details of their lives—even forbidding them to wear the straw hats with flowers and ribbons that the girls loved but that he thought were in terrible taste. Rita had to beg for her father's permission to attend medical school. She had been afraid of him in life but especially mourned his death, because she came to realize too late how she had inherited from him "a remarkable tenacity in following the path I believe to be right, and a way of underestimating the obstacles standing between me and what I want to accomplish." All four of the children grew up with "a tendency to look on others with sympathy and without animosity and to see things and people in a favorable light."[2]

The lack of serious studies in her girls' high school most affected Rita, because her sisters were interested in the arts and literature and would not require an advanced degree. Her timidity, social isolation, and discomfort with the subordinate role she saw for women convinced her to tell her parents early in life that she would not marry and have children. Within eight months of tutoring in Greek, Latin, and mathematics, she and her cousin Eugenia Lustig passed their entrance examinations for medical school.

Salvador Luria and Renato Dulbecco (who also would eventually win Nobel Prizes) attended medical school with Rita and Eugenia. All of them joined others in the Institute of Anatomy to work with Professor Giuseppe Levi, a master in histology, in the preparation of sections of body tissues for microscopic study. Later Rita would "confess that I hated histology" but that she had a passionate and superb teacher.[3] Luria and Dulbecco remained her friends and were supportive when she was in the United States and her research seemed to be stalled.

The fascist regime in Italy and its anti-Semitic laws kept Rita and Professor Levi from holding official positions at the university. In fact, the laws kept Rita from practicing medicine at all. However, they continued working under the most primitive and dangerous conditions during the early war years in her home, using fertilized chicken eggs that she begged for her nonexistent children. Afterwards, her family ate omelets. Possibly she was able to watch patiently, away from a hectic university setting, the dynamic character of the nervous system development. Rita published an important paper in Belgium on nerve growth control in chick embryos that countered the views of Viktor Hamburger, a well-known embryologist. She finally had to stop working when her family went south in the face of the Nazi invasion.

Soon after her resumption of research with Professor Levi, the invitation came to visit Professor Hamburger, the chair of zoology at Washington University in St. Louis. Rita's observations—that (1) in some places neurons that began development normally died, (2) neurons that would normally innervate a limb died when the limb was amputated, and (3) neurons migrated on predictable pathways—elicited a reversal of Hamburger's earlier thesis on the mechanism of control that the limb tissues had on the growth of neuron cell fibers, the extensions of the neurons that normally associate with the cells they innervate.[4]

Rita then stopped grafting normal tissues onto chick embryos and repeated work done by a former student of Hamburger's, Elmer Bueker. He had found enhanced growth of neuron cell fibers when he transplanted a certain mouse tumor, S. 180, near the embryo's posterior limb. Early experiments demonstrated (1) greatly increased fiber growth, particularly from sympathetic ganglia (collections of autonomic neuron cell bodies), and (2) less from sensory ganglia (sensory cell bodies) when S. 180 and a related type mouse tumor, S. 37, were grafted onto the sides of three-day-old chick embryos. The fibers of the nerve cells that grew out in response to something in the tumors did not actually associate with embryo or tumor cells, but seemed to be randomly located within the organs. "The penetration of the nerve fibers into the veins, furthermore, suggested to me that this still-unknown humoral substance [a type of secretory factor] might also be exerting a neurotropic effect, or what is also known as a chemotactic directing force, one that causes nerve fibers to grow in a particular direction."[5]

Rita next tested to see whether there was a humoral factor by transplanting pieces of tumor onto the noninnervated vascular chorioallantoic membrane of the chick embryo that would only allow communication with its embryo through the blood vessels. As predicted, she found, upon sectioning the embryos, that the experimental ganglia were enlarged and that some fibers had grown into the organs and even into the embryonic veins. The result confirmed that there was a humoral factor released from the tumor and acting through the blood on parts of the nervous system.

Amid this work, there were relaxed times as well. Rita traveled to Rome in 1952 for a visit and then went on to the Institute of Biophysics in Rio de Janeiro for work with Hertha Meyer, a tissue culture expert. Rita carried two tumor-bearing mice munching on apple in a box in her coat pocket. "The discomfort they suffered from their confined quarters was compensated by their feeling safe inside the little box."[6] After several failed attempts and letters of despair to Washington University, she was able to establish the tumor-embryo system in culture and was even able to show that merely an extract from the embryos with tumors would enhance the growth of the ganglion cells in tissue culture. She had found the agent that grows nerves.

From 1953 to 1959 Rita collaborated with Stanley Cohen, a biochemist, to name the agent "nerve growth factor," or NGF. They purified it in 1954 and then further characterized its effects and found other sources. During the testing of various tissues, Cohen made the suggestion that

the large male mouse submandibular salivary glands might be a prob-
able source, a homologous gland to that of a snake's venomous glands
that already were known to contain NGF. "The following morning we
marveled, with a mixture of happiness and incredulity, at results far
surpassing our greatest expectations: fibrillar halos had grown around
the ganglia cultured in a medium in which a couple of adult male mouse
glands had been diluted 10^4 times."[7]

The large amount and inexpensiveness of this new source allowed
Levi-Montalcini to study the effects of NGF on newborn mice and rats,
and allowed Cohen to purify enough to make antiserum. When the antiserum to NGF was injected into newborn rodents, the affected neurons
were almost totally atrophied. This discovery marked the beginning of
the work that would demonstrate the dependence on NGF for full differentiation of several types of nerve cells and one type of immune cell.
During a talk in 1975 before a group celebrating the birthday of a fellow
neurobiologist, Rita spoke about this period with Cohen: "In a biological
science, perhaps to a larger extent than in any other experimental science,
chance and good luck play a notoriously great role."[8] The entire field of
the dependence of the development of various embryonic systems on
growth factors was born on June 11, 1959, with the in vivo embryo work
and reached new heights with the Nobel Prize on December 10, 1986.

Levi-Montalcini and Pietro Angeletti alternated their collaborative efforts between St. Louis and Rome as they started a laboratory closer to
her beloved family. For two years she commuted to Turin to be with her
mother and sister, Paola; but after her mother's death in 1963, her sister
moved to Rome. They continue to share the closeness they had experienced as children.

The Institute of Cell Biology in Rome was created in 1969. Levi-
Montalcini served as its director from 1969 to 1979. Molecular biology
research sometimes seemed to overshadow the neurobiology efforts.
Also, "the desire to work independently in the scientific realm—a desire
much stronger in Italy than in the United States—causes a withering of
productive forces, and scatters them into a thousand separate rivulets."[9]

In 1971 a new purification technique for NGF allowed the group to
sequence the amino acids of the protein; this allowed two teams in the
United States to locate the gene for NGF in 1983. Knowledge of the three-
dimensional structure of the dimer (two identical copies of the protein)
will be useful for studying how NGF works in the body. Pure human
NGF has been synthesized by molecular engineering. Levi-Montalcini
especially credits Pietro Calissano and Luigi Aloe for much work on the
mechanism of action of NGF. One of her later papers, published in the
United States, discussed the effects of aggressive behavior in male mice
and the levels of NGF messenger RNA and protein in the brain. A new
and exciting field of research involves the hypothesis that there may

be a disturbance in the level of NGF in several nervous system dysfunctions, including Alzheimer's. Levi-Montalcini recalls that she was reading an Agatha Christie thriller when the phone call came from Stockholm in 1986, so she wrote a quick note and date in the page margin as she quickly returned to find out about the criminal.

Currently, Levi-Montalcini is president of the Italian Multiple Sclerosis Association. She and her sister have begun a joint venture to provide mentoring and financial help for young people interested in the arts and sciences. Prospective students talk with her in her laboratory at the Institute in Rome. She has started new research on NGF in the immune and endocrine systems. "So even now I am doing something entirely different. Just in the same spirit as when I was a young person. And this is very pleasing to me. I mean, at my old age, I could have no more capacity. And I believe I still have plenty."[10]

Notes

1. Rita Levi-Montalcini, "Interview," *Omni* 10 (Mar. 1988): 70–104.

2. Rita Levi-Montalcini, *In Praise of Imperfection*, trans. Luigi Attardi (New York: Basic Books, 1988), p. 4.

3. Rita Levi-Montalcini, "NGF: An Uncharted Route," in *The Neurosciences: Paths of Discovery*, eds. Frederic G. Worden, Judith P. Swazey, and George Adelman (Cambridge, Mass.: MIT Press, 1975), p. 246.

4. Viktor Hamburger and Rita Levi-Montalcini, "Proliferation, Differentiation and Degeneration in the Spinal Ganglia of the Chick Embryo under Normal and Experimental Conditions," *Journal of Experimental Zoology* 111 (1949): 457–502.

5. Levi-Montalcini, *In Praise of Imperfection*, p. 148.

6. *Ibid.*, p. 153.

7. *Ibid.*, p. 165.

8. Levi-Montalcini, "NGF: An Uncharted Route," pp. 252–53.

9. Levi-Montalcini, *In Praise of Imperfection*, p. 197.

10. Marguerite Holloway, "Finding the Good in the Bad," *Scientific American* 254 (Jan. 1993): 36.

Bibliography

Cohen, S., R. Levi-Montalcini, and V. Hamburger. "A Nerve Growth-Stimulating Factor Isolated from Sarcomas 37 and 180." *Proceedings of the National Academy of Sciences USA* 40 (1954): 1014–18.

Hamburger, Viktor, and Rita Levi-Montalcini. "Proliferation, Differentiation and Degeneration in the Spinal Ganglia of the Chick Embryo under Normal and Experimental Conditions." *Journal of Experimental Zoology* 111 (1949): 457–502.

Heady, Judith E. "Rita Levi-Montalcini: 1986," in *The Nobel Prize Winners: Physiology or Medicine*, ed. Frank N. Magill, pp. 1507–15. Pasadena: Salem Press, 1991.

Holloway, Marguerite. "Finding the Good in the Bad." *Scientific American* 254 (Jan. 1993): 32, 36.

Kass-Simon, G. "Biology Is Destiny, Rita Levi-Montalcini: Nerve Growth Factor," in *Women of Science Righting the Record*, eds. G. Kass-Simon and Patricia Famcs, pp. 247–49. Bloomington: Indiana University Press, 1990.

Leon, A., A. Buriani, R. Dal Toso, M. Fabris, S. Romanello, L. Aloe, and R. Levi-Montalcini. "Mast Cells Synthesize, Store, and Release Nerve Growth Factor." *Proceedings of the National Academy of Sciences USA* 91 (Apr. 1994): 3739–43.

Levi-Montalcini, Rita. *In Praise of Imperfection*, trans. Luigi Attardi. New York: Basic Books, 1988.

———. "Interview." *Omni* 10 (Mar. 1988): 70–74ff.

———. "NGF: An Uncharted Route," in *The Neurosciences: Paths of Discovery*, eds. Frederic G. Worden, Judith P. Swazey, and George Adelman, pp. 244–65. Cambridge, Mass.: MIT Press, 1975.

Levi-Montalcini, Rita, and Pietro Calissano. "The Nerve Growth Factor." *Scientific American* 240 (June 1979): 68–77.

Marx, Jean L. "Nerve Growth Factor Acts in Brain." *Science* 232 (1986): 1341–42.

Randall, Frederika. "The Heart and Mind of a Genius." *Vogue* 177 (Mar. 1987): 480–81ff.

JUDITH E. HEADY

MARGARET ADALINE REED LEWIS

(1881–1970)

Anatomist, Physiologist

Birth	November 9, 1881
1901	A.B., Goucher College
1904–1907	Lecturer in Physiology, New York Medical College for Women
1905–1906	Lecturer in Zoology, Barnard College
1906–1907	Preparator in Zoology and Assistant to T.H. Morgan, Columbia University
1907–1909	Instructor in Biology, Columbia University; Instructor in Biology, Barnard College
1910	Married Warren Harmon Lewis on May 23
1915–1926	Collaborator, Dept. of Embryology, Carnegie Institute of Washington (in Baltimore)

1927–1946	Research Associate, Carnegie Institute of Washington
1938	William Wood Gerhard Gold Medal, Pathological Society of Philadelphia (joint recipient with her husband); honorary LL.D., Goucher College
1940–1946	Guest Investigator, Wistar Institute of Anatomy and Biology, Philadelphia
1946–1964	Member (Emerita Member after 1958), Wistar Institute of Anatomy and Biology
Death	July 20, 1970

How much biological and medical knowledge do we owe to Margaret Adaline Reed Lewis? It is probably impossible to estimate. Although she was identified primarily as an anatomist and physiologist, Margaret Lewis had a long career that encompassed the fields of embryology, cytology, bacteriology, virology, and cancer research. She often did seminal work, especially in tissue culture, that has been expanded by many others since.

Margaret was born in Kittaning, Pennsylvania, in 1881, the daughter of Joseph Cable and Martha Adaline (Walker) Reed. She graduated with a bachelor's degree from Goucher College in Baltimore in 1901 and did graduate work in several institutions, including Bryn Mawr College and the Universities of Zurich, Paris, and Berlin, although this did not result in a graduate degree. During much of this time, Margaret also worked as a teacher and researcher at a number of schools and laboratories: Columbia University, Barnard College, New York Medical College for Women, Miss Chapin's School, Johns Hopkins Training School for Nurses, Woods Hole Marine Biological Laboratory in Massachusetts, and others.

In 1915, Margaret joined the Carnegie Institute of Washington (located in Baltimore) as a collaborator in the Department of Embryology. She became a research associate there in 1927. Although she remained at the Carnegie Institute until 1946, she became affiliated in 1940 (as a guest investigator) with the Wistar Institute of Anatomy and Biology, located in Philadelphia. In 1946, Margaret became a member of the Wistar Institute, staying until 1964; her last six years there were as an emerita member. She often spent her summers doing research in other laboratories: Woods Hole; the Harpswell Laboratory and the Mt. Desert Island Biological Laboratory, both in Maine; and Stanford University.

Margaret's early research involved investigations into regeneration in crayfish and embryology of amphibians. The first successful culturing of cells in vitro was accomplished by Ross G. Harrison in 1907, using nerve cells from frogs and frog lymph as a growth medium. In 1908, Margaret was in Berlin working with another researcher, Rhode Erdmann, who was culturing amoebae on agar supplemented with physiological salt solution. This led Margaret to try culturing bone marrow and spleen cells from a guinea pig, with good results. In fact, she is credited with achieving the first successful mammalian tissue culture.

After her marriage in 1910 to Warren Harmon Lewis, the two collaborated on mammalian tissue culture research, determining what kinds of cells could be cultured and formulating different culture media for varying purposes. In 1911 and 1912 they published a series of papers on this work. Because the Lewises were interested in investigating cell structure, they developed a clear nutritive medium, which became known as the Locke-Lewis solution. Their method of culture—suspending the cells in a drop of medium hanging from the underside of a glass slip, which was then placed over a hollow on a microscope slide—became known as the Lewis culture. The cells would grow on the underside of the glass slip and could be observed easily with the microscope. This technique allowed Margaret and her husband to describe many of the cell's organelles, such as the nucleus and mitochondria, as well as a number of physiological processes, including the locomotion of white blood cells and the contraction of smooth muscle cells.

As the importance of pH (acid and base balance) in living systems became increasingly apparent to biologists, Margaret and a co-worker, Lloyd D. Felton, began to investigate the effects of different pH values in the tissue culture media. In 1921 they published a paper in *Science* entitled "The Hydrogen-Ion Concentration of Cultures of Connective Tissue from Chick Embryos." A second collaborative paper followed in the next year, entitled "The Hydrogen-Ion Concentration of Tissue Growth *in vitro*." This information allowed tissue culture researchers to optimize growing conditions so that the cells remained alive longer. The research, along with additional work done by Warren Lewis, helped the Lewises to determine that under certain circumstances the white blood cells called monocytes became transformed into macrophages, or "scavenging" cells, that are capable of ingesting and destroying fairly large microorganisms and foreign particles. This disproved an existing hypothesis that monocytes and macrophages were of two distinct cell types.

Later, Margaret turned her efforts to the study of immunity and cancer, often still collaborating with her husband. In fact, she has been referred to as a "world-renowned authority on tumors" and part of the "colony of talented women scientists . . . [that] formed around the Johns Hopkins University in the 1920s and 1930s."[1] During the 1930s her interest included the behavior of chromosomes in malignant cells. The Lewises observed that the Walker rat sarcoma cells they studied, which contained the normal number of chromosomes, were just as malignant as those containing abnormal numbers. This work led the Lewises to suggest that abnormal chromosome number alone did not cause malignancy. Subsequently, Margaret and her co-workers investigated the tumor-inhibiting effects of various chemotherapeutic agents, among them dyes, on cancer cells. This was a direct result of work she had done with Philip P. Goland, using orally administered dyes to stain tumors in living organisms. They observed during these experiments that some of the dyes actually retarded the growth of the tumors.

During her long and distinguished career, Margaret Lewis authored or co-authored nearly 150 papers. Because of their prominence in the field, the Lewises were asked to provide a chapter entitled "Behavior of Cells in Tissue Culture" for the book *General Cytology: A Textbook of Cellular Structure and Function for Students of Biology and Medicine* (1924), which was widely used. The Lewises also made motion pictures of cellular events, which were employed as teaching aids for many years.

Margaret Lewis received several honors in recognition of her work. She was a member of Phi Beta Kappa, Sigma Xi (the honorary research society), and an honorary life member of the Tissue Culture Society. In 1938 her alma mater, Goucher College, awarded her an honorary Doctor of Laws (LL.D.) degree. In the same year Margaret and her husband were joint recipients of the William Wood Gerhard Gold Medal of the Pathological Society of Philadelphia.

In 1964, the year Margaret retired, she and her husband were honored by the Wistar Institute with a symposium at which the leaders in the field of tissue culture presented papers. In the introduction to the symposium in tribute to their work on the development of tissue culture media, Louis B. Flexner said: "Perhaps it would be fair to say that many of the more recent far-reaching advances which have been made in the field of tissue culture go back to early media of this kind, described in part by the Lewises as early as '11."[2]

Margaret also received the highly coveted star next to her name in the sixth edition (1938) of Cattell's *American Men of Science*. The star originally indicated that the awardee was among the top 1,000 scientists in the nation. (This number eventually expanded to considerably more than 1,000 but was still a distinct honor.) In that year, only 10 of the 251 names added to the list were those of women; and in all seven editions, only 52 women ever received stars. Margaret Rossiter has noted that most women scientists were unmarried at the time, and that most married women scientists were unemployed; she concluded that "by 1938 unemployment had become largely a married woman's problem."[3] It is clear, however, that many women managed successfully, even in those days, to combine marriage and a career in science. A number of these women were married to scientists. Some, like Margaret Lewis, collaborated with their husbands to varying degrees.

In addition to her illustrious career, Margaret had three children: Margaret Nast Lewis, Warren Reed Lewis, and Jessica Helen Lewis Myers.

She enjoyed music and mountain climbing.[4] Indeed, she was as active in her life outside the laboratory as she was in her scholarly pursuits. Margaret Lewis died on July 20, 1970, but her legacy to biology and medical science lives on.

Notes

1. Margaret W. Rossiter, *Women Scientists in America: Struggles and Strategies to 1940* (Baltimore: Johns Hopkins University Press, 1982), pp. 207–8.

2. Louis B. Flexner, Introduction to *Retention of Functional Differentiation in Cultured Cells*, Wistar Institute Symposium Monograph No. 1, ed. Vittorio Defendi (Philadelphia: Wistar Institute Press, 1964), p. 1.

3. Rossiter, *Women Scientists*, p. 142.

4. See Durwood Howes, ed., *American Women, 1935–1940: A Composite Biographical Dictionary*, Vol. 1, A–L (Detroit: Gale Research, 1981).

Bibliography

Cattell, James McKeen. *American Men of Science*, 6th ed. New York: Science Press, 1938.

Cowdry, Edmund V., ed. *General Cytology: A Textbook of Cellular Structure and Function for Students of Biology and Medicine*. Chicago: University of Chicago Press, 1924.

Farquhar, S. Edgar, ed. *The Progress of Science: A Review of 1940*. New York: Grolier Society, 1941.

Flexner, Louis B. Introduction to *Retention of Functional Differentiation in Cultured Cells*, Wistar Institute Symposium Monograph No. 1, ed. Vittorio Defendi. Philadelphia: Wistar Institute Press, 1964.

Herman, Kali. *Women in Particular: An Index to American Women*. Phoenix: Oryx Press, 1984.

Howes, Durwood, ed. *American Women, 1935–1940: A Composite Biographical Dictionary*. Detroit: Gale Research Co., 1981.

The National Cyclopedia of American Biography. Clifton, N.J.: James T. White and Co., 1979.

Rossiter, Margaret W. *Women Scientists in America: Struggles and Strategies to 1940*. Baltimore: Johns Hopkins University Press, 1982.

Siegel, Patricia Joan, and Kay Thomas Finley. *Women in the Scientific Search: An American Bio-Bibliography, 1724–1979*. Metuchen, N.J.: Scarecrow Press, 1985.

KIMERLY J. WILCOX

MARTHA DANIELL LOGAN

(1704–1779)

Horticulturist

Birth	December 29, 1704
Death	June 28, 1779

Martha Daniell Logan was born to a prominent family in St. Thomas Parish, South Carolina, in 1704. Her father, Robert Daniell, the deputy-governor, owned land, ships, and slaves. She was the second of four children born to him and his second wife, Martha. Nothing is known of the younger Martha's education, but she was taught to read and write and to do needlework. Apparently she also learned something about horticulture and gardening. At age 14, following her father's death, she married George Logan, Jr. The couple made their home 10 miles from Charleston on a plantation, which Martha had inherited from her father. During the following 17 years eight children were born, six of whom lived.

Despite Martha's heritage and her inheritance, the Logans apparently encountered financial difficulties. Martha placed an advertisement in the *South Carolina Gazette* in 1742 in which she offered to board children, teach them to read and write, and do plain work, embroidery, tent and cut work. In subsequent years she offered to teach reading and drawing, offered her estate for sale, and announced the opening of a school in Charleston. By 1753 her son, Robert, was advertising imported seeds, flower roots, and fruit stones at his mother's house. After that time Martha Logan earned an income from a nursery business. She is said to be the author of the "Gardener's Kalendar," first published in John Tobler's *South Carolina Gazette* in 1751. This calendar, or variations of it, appeared frequently in South Carolina almanacs into the 1780s.

However, Logan is best known because of her association and correspondence with the noted botanist John Bartram. Because some of her letters to Bartram were saved, it is through them that we can best know her. Bartram traveled widely in the New World collecting and studying native plants. He sent specimens to collectors in England and grew them in his own gardens in Pennsylvania, which were visited by George Washington, Thomas Jefferson, and Benjamin Franklin. He was appointed botanist to His Majesty King George III and wrote extensively about the flora he saw and collected during his travels. Bartram met Martha Logan very briefly on one of his trips to Charleston. As a result of this meeting, they began a correspondence and an exchange of seeds, roots, and plants. In a letter to Peter Collinson, a merchant and botanist in London to whom he sent specimens, Bartram wrote:

> I received a lovely parcel (of seeds and cuttings) in the Spring from Mistress Logan, my facinated widow. Her garden is her delight and she has a fine one: I was with her about 4 minutes in her company yet we contracted such a mutual correspondence that one silk bag of seed hath repast several times.[1]

In her letters to Bartram, Martha Logan often referred to the little bag as well as to tubs in which plants were sent by boat. She mentioned several times a Captain North, who delivered plants between the two. Also mentioned in the letters is Dr. Alexander Garden, a Charleston physician, botanist, and naturalist. Garden gave his name to the gardenia and was the author of *Flora Caroliniana*. In a postscript to one letter, Logan wrote:

> When you Send or Wright to me againe, be pleased to Direct for me to the Caire of John Logan, Merchant in Charles Town, for Dr. Garden has to much business he has not time to Think of me. Wheirefore your Letters have Some times layen a good While and I never known of them.[2]

Captain North apparently was more conscientious, for of him she said, "Capt. North here is the Person brings your box. Will be a Safe hand when you favious me with the roots and seeds you mentioned."[3] With each letter, Logan included a list of the seeds, roots, and plants she was sending. These included olive tree, English honeysuckle, passion flower, white crocust, and grape hyacinth (blue). Referring to the little bag, she wrote:

> I have Last weak Recived boath your faveours with the Seeds theirin mentioned, for which am much Obliged, and wish you Woulde bein so kinde to Lett me known what we have that woulde bein most Exceptable to you; but as you did not, have sent inClosed the Littel Bagg which Containe Some Varity but fue of a kinde, (as you Requested).[4]

Another time she wrote:

I am Sorry the Holley Did not Prove what you Wanted but Shall make Inquirey after the trees in Mrs. Wraggs and by Mr. Rapers garden, and Send you Some of theire Berryes as Soon Ripe. As Likewise all the Other things you Mention in theire Season for Removing.[5]

In 1761, Bartram's wife responded to Martha Logan in her husband's absence. Logan replied:

I Recived your favour by Capt. North and am much Obligd for taking the trouble of Answering Mine in Mr. Bartrams Absence. . . . I shoulde be Very glade he would tell mr Ratlive what the Andromed on the road he mentioned to me is, and I will Most Sertinly get it and Send at a proper Season. . . . I hearwith Send him Some roots of the Indian, or Worm Pink, as the Seeds weire all fallen before I had yrs about them. In the Same Tub is Some Slips of Mrs Bees Littel flower, Least the Seeds Shoulde faile. The Berryes on the Tree are not yett Ripe Enough but if I live, yr Spouse may Sertinly Expect them with the Other things.[6]

Although she was clearly interested in horticulture, Martha Logan's advice in a "Gardener's Kalender" shows her to be knowledgeable about vegetable and herb culture as well. She passed on traditional lore and knowledge gained from her personal experience. "Early onions and carrots may be sowed the full of the moon." In November, she advised, "Trim your monthly roses, and the full of the moon open their rootes and dung them."[7] Martha Logan carried on the traditions of English women who gardened, collected, and exchanged knowledge, plants, and seeds. These traditions were continued in the colonies and the new republic, and they are alive today.

Notes

1. Buckner Hollingsworth, "Martha Logan, 1704–1779. John Bartram's 'Facinated Widow'—and a Notable Early American Florist," in *Her Garden Was Her Delight* (New York: Macmillan, 1962), p. 18.
2. "Letters of Martha Logan to John Bartram, 1760–1763," ed. Mary Barbot Prior, *South Carolina Historical Magazine* 1 (1958): 40.
3. *Ibid.*
4. *Ibid.*
5. *Ibid.*, p. 42.
6. *Ibid.*, p. 43.
7. *A Gardener's Kalendar Done by a Colonial Lady*, ed. Alice Logan White (Charleston: National Society of the Colonial Dames of America in the State of South Carolina, 1976).

Bibliography

A Gardener's Kalendar Done by a Colonial Lady, ed. Alice Logan White. Charleston: National Society of the Colonial Dames of America in the State of South Carolina, 1976.

Hollingsworth, Buckner. "Martha Logan, 1704–1779. John Bartram's 'Facinated Widow'—and a Notable Early American Florist," in *Her Garden Was Her Delight.* New York: Macmillan, 1962.

"Letters of Martha Logan to John Bartram, 1760–1763," ed. Mary Barbot Prior. *South Carolina Historical Magazine* 1 (1958): 38–46.

Notable American Women, 1607–1950. Cambridge and London: Belknap Press, 1971.

Spruill, Julia C. *Women's Life and Work in the Southern Colonies.* Chapel Hill: University of North Carolina Press, 1938.

CAROL W. CUBBERLEY

JANE LUBCHENCO

(1947–)

Marine Ecologist

Birth	December 4, 1947
1969	B.A., biology, Colorado College
1971	M.S., ecology, University of Washington
1975	Ph.D., ecology, Harvard University
1975–77	Assistant Professor of Ecology, Harvard University
1976–	Principal Investigator, National Science Foundation
1977	Visiting Assistant Professor, Discovery Bay Marine Lab
1977–83	Field research in Panama
1978–84	Science Advisor, Ocean Trust Fund
1978–88	Associate Professor, Oregon State University
1978–	Research Associate, Smithsonian Institution
1979	Mercer Award
1981–88	Science Advisor, West Quoddy Marine Station
1982–84	Council Member, Ecological Society of America
1983–86	Awards Committee Chair, Ecological Society of America

1985	Visiting Associate Professor, Universidad de Antofagasta, Chile
1987	Visiting Associate Professor, Institute of Oceanology, Qingdao, China
1987–89	National Lecturer, Phycological Society of America
1988–	Professor of Zoology, Oregon State University
1989–92	Chair, Dept. of Zoology, Oregon State University
1991	Sustainable Biosphere Initiative, Ecological Society of America
1992	Pew Scholar in Conservation and the Environment
1992–93	President, Ecological Society of America
1993	MacArthur Foundation Fellowship Award
1995	President-Elect, American Association for the Advancement of Science

Jane Lubchenco is an important ecologist who has made an impact on research policy and been recognized for her contributions by a Mac-Arthur Foundation Fellowship. She specializes in experimental marine ecology. Her work has involved marine plant-herbivore interactions, chemical ecology, predator-prey interactions, algal ecology, and life histories. She is also interested in biodiversity and sustainable ecological systems. She has been involved with groups that are trying to alter research to concentrate more on studying how the earth can survive the pollution humans have created.

Jane Lubchenco was born in 1947 in Denver, Colorado. She had five sisters, and her mother was a pediatrician who practiced part-time when her daughters were young so that she could be with them. One of Jane's mother's sayings was, "Where there's a will, there's a way!"[1] This attitude must have rubbed off on Jane, because she certainly found a way to accomplish a great deal.

Jane earned her bachelor's degree from Colorado College in 1969 with a major in biology, and her master's degree in ecology in 1971 from the University of Washington. She married fellow ecologist Bruce Menge at about the same time and kept right on with her work. In 1975, Jane received her Ph.D. in ecology from Harvard University. She then became an assistant professor of ecology at Harvard for two years. In 1976, Jane was a principal investigator with the National Science Foundation and in 1977 a visiting assistant professor at the Discovery Bay Marine Lab.

At this time, Jane and Bruce had two full-time jobs, the ideal situation for many two-career couples. However, they were not completely happy. They loved teaching and research but wanted time for family. They began to seek an institution that would be willing to try a fairly unique arrangement—they wanted to split a position. They were very particular. They did not want part-time positions that would take them off their career tracks. On the contrary, they wanted tenure-track positions involving both teaching and research, but ones that allowed them time for family.[2] Luckily, they found an institution willing to try this arrangement: Oregon State University.

Although Jane's position was only half-time, she put in a lot of effort in research. In 1977 she was a member of an advisory panel for a long-term ecological research program. From 1977 to 1983 she conducted field research in Panama. From 1978 to 1984 she was a science advisor for the Ocean Trust Fund.

The year 1978 was a very good one for Jane Lubchenco. She was promoted to associate professor while still half-time. She also became a research associate at the Smithsonian Institution. In addition, she published a paper that won her the Ecological Society of America's Mercer Award in 1979 for the best ecological paper published in 1978.

Lubchenco's first son was born around this time and the second in 1982, but her professional involvement remained strong. She was a science advisor from 1981 to 1988 at the West Quoddy Marine Station and a council member of the Ecological Society of America from 1982 to 1984. From 1983 to 1986 she was the Ecological Society of America awards committee chair. In 1985 she was visiting associate professor at Universidad de Antofagasta, Chile, and in 1987 she was visiting associate professor at the Institute of Oceanology in Qingdao, China.

In 1987, Lubchenco needed less time at home and the university increased her position to three-quarters time instead of half-time. From 1987 to 1989 she was national lecturer for the Phycological Society of America, and in 1988 she became full professor of zoology at Oregon State University. In 1989 her position became full-time and she became department chair, a position she held until 1992.

In 1991, Lubchenco was part of the group that wrote the Sustainable Biosphere Initiative. This was an Ecological Society of America research agenda with priorities on global change, biological diversity, and sustainable ecological systems. Also in 1991 an adaptation panel of the National Academy of Sciences issued a report stating that greenhouse warming will have ecological and economic effects, but that society's rate of change is fast enough to keep up with those effects. Lubchenco was a strong critic of the panel's belief that change will be gradual with few effects.[3]

In 1992, Lubchenco became Pew Scholar in Conservation and the Environment. From 1992 to 1993 she was president of the Ecological Society of America. In 1993 she and her husband wrote an article on their split positions in the hopes that others could learn from their beneficial arrangement. They are advocates for women in science and believe that split positions can be very useful to women trying to balance career and family.

Notes

1. Jane Lubchenco and Bruce A. Menge, "Split Positions Can Provide a Sane Career Track—A Personal Account," *Bioscience* 43, no. 4 (1993): 243.
2. *Ibid.*
3. Leslie Roberts, "Academy Panel Split on Greenhouse Adaptation," *Science* 253 (Sept. 6, 1991): 1206.

Bibliography

American Men and Women of Science. New York: Bowker, 1992/1993.

Lubchenco, Jane, and Bruce A. Menge. "Community Development and Persistence in a Low Rocky Intertidal Zone." *Ecological Monographs* 48, no. 1 (1978): 67–94.

———. "Split Positions Can Provide a Sane Career Track—A Personal Account." *Bioscience* 43, no. 4 (1993): 243–48.

Lubchenco, Jane, and Steven D. Gaines. "A Unified Approach to Marine Plant-Herbivore Interactions. I. Populations and Communities." *Annual Review of Ecology and Systematics* 12 (1981): 405–38.

Lubchenco, Jane, et al. "The Sustainable Biosphere Initiative: An Ecological Research Agenda." *Ecology* 72, no. 2 (1991): 371–412.

Roberts, Leslie. "Academy Panel Split on Greenhouse Adaptation." *Science* 253 (Sept. 6, 1991): 1206.

Who's Who in Technology Today. Highland Park, Ill.: J. Dick, 1986.

JILL HOLMAN

MADGE THURLOW MACKLIN

(1893–1962)

Geneticist

Birth	February 6, 1893
1914	A.B., Goucher College
1919	M.D., Johns Hopkins Medical School
1938	Honorary LL.D., Goucher College
1945–59	Research Associate, National Research Council

1957	Elizabeth Blackwell Award
1959	President, American Society of Human Genetics
Death	March 14, 1962

Madge Thurlow Macklin was a remarkable geneticist. Her contributions to medical genetic research were varied and great. Through her active campaign, she demonstrated the need for genetics to be part of medical education. She convinced colleagues of the importance of family history, and her careful experiments contributed to sound statistical methodology in human genetics. Later in her career, she explored the connection of heredity to specific types of cancers.

Madge Thurlow was born in 1893 in Baltimore, Maryland. Her grade school and high school years were spent in the Baltimore schools. She attended Goucher College, receiving her A.B. degree in 1914. Then for one year she conducted physiological research at the Johns Hopkins Medical School. While attending college and medical school, she was awarded several scholarships and fellowships. She was a brilliant student. In 1919 she completed her M.D. at Johns Hopkins University. In the same year she married Dr. Charles C. Macklin, and they subsequently had three daughters. During 1920 she worked as instructor of physiology at the School of Hygiene of Johns Hopkins University. In the following year she and her husband moved to the University of Western Ontario.

In 1922, Dr. Macklin was appointed part-time instructor in histology and embryology at the University of Western Ontario. She was appointed to this post as her husband was appointed professor of histology and embryology. After eight years she was promoted to part-time assistant professor. Dr. Macklin remained at that rank until her dismissal in 1945. During her 23 years at the university she taught embryology to first-year students and assisted her husband in the histology course. They collaborated on research in histology. She was a gifted, well-liked teacher and prolific author. Dr. Macklin published nearly 150 research papers in the area of medical genetics. By 1961 her publications had grown to 200. Her writings were clear, logical, and always supported by solid data and evidence. She worked tirelessly to introduce human genetics as a subject in medical school curricula, asserting in 1933 that

> medical genetics should be taught in the medical school in the final year of medicine, by a medically trained person conversant with the broader aspects of science of genetics. It should not be given by a geneticist who is not medically trained, inasmuch as a person unacquainted with the phenomena of disease, and their diagnostic significance. It should not be given before the final year in medicine

because the student is not yet familiar with the signs and symptoms of disease, and hence cannot appreciate even the terms used in the discussion.[1]

Dr. Macklin was meticulous in her experiments. The contributions she made to data analysis, statistical techniques, and controlled experiments were significant to the field of genetics. She investigated hereditary and environmental factors and their direct influence to stomach and breast cancer. In 1960 she reported that

gastric cancer was found significantly more often in the relatives of index cases with gastric cancer than in the general population, but cancer of the large intestine was not. . . . These facts indicate that the basis for increased frequency probably lies in the genetic simi-larity between the members of the affected pairs, rather than in the environmental similarity.[2]

Previously she had found that "grandmothers, aunts and sisters of women with breast cancer have had breast cancer with a frequency which is significantly greater than that of women in a similar age range either in the general population or in two sets of two selected control samples. This excess cannot be entirely environmental in origin."[3]

Dr. Macklin's human experiments complemented the mice experiments conducted by **Maud Slye**. Although approximately 50 of Macklin's publications were on cancer, she was convinced early on that inheritance, defined by genetic studies, was the key to understanding some cancers. Thus the eugenics movement, at its height during the 1920s and 1930s, found a great supporter in Dr. Macklin; she wrote in 1933 that "it is my firm conviction that the Eugenics Societies of any country can make no greater contribution to the progress of their cause than by the repeated agitation to put on a course in Medical Genetics in every medical school." She went on to suggest that state examination boards include questions on "the inheritance of disease in Man" in their examinations.[4]

Her numerous publications led to invitations to give speeches through-out North America and abroad. In 1938, Goucher College, her alma ma-ter, recognized her with an honorary LL.D. degree. She was awarded the Elizabeth Blackwell Award in 1957. Two years later she was elected president of the American Society of Human Genetics. Dr. Macklin also belonged to many American and Canadian biological societies.

Dr. Macklin's scientific contributions and exemplary teaching were not rewarded by the University of Western Ontario. She was never given a full-time professorship, and she was promoted only once to the rank of part-time assistant professor. Her appointments were sessional, she was grossly underpaid, and during the Depression years she was not paid at all. Her years at Western Ontario were difficult. The university did not openly accept husband and wife professors in the same faculty or department.

The two-career family was not yet fashionable. Dr. Macklin was an aggressive person whose views and strong will were thought to have offended people. Many found her views on eugenics controversial. Ironically, it was her strong personality that directed her to become one of the world's pioneering geneticists.

The National Research Council in the United States offered Dr. Macklin a research associateship in 1946. She conducted her cancer research for the Council at Ohio State University in Columbus, Ohio. She was granted appointments in the zoology department and the medical school. While at Ohio State University, she worked with Dr. Lawrence Snyder and taught a course in human genetics. Snyder was an advocate of including genetics in the medical curriculum.

In 1959, Dr. Macklin retired from Ohio State University. She returned to London, Ontario, Canada, where her husband had remained to do research and teach. He died in December of that year. After his death she moved to Toronto, Canada, to be near her three married daughters. In 1962, at age 69, she died of coronary thrombosis. She was remembered at the 1962 annual meeting of the American Society of Human Genetics:

It's been a great year for genetics, saddened by the death of a former President of our Society, Madge Macklin, who played a pioneer role in introducing genetics to the medical fraternity, and whom we all loved for her energy, tenacity of purpose, humour, compassion and the whimsical tirades with which she could castigate her male colleagues. She did a fine job and had a good time; her memory remains as an inspiration for all of us.[5]

Dr. Macklin faced considerable professional and personal discrimination. The medical community did not accept the significance of her research, two-career families were uncommon, and women academics in the sciences were not readily accepted into the fraternity. Her outspoken personality led to clashes with her academic superiors and colleagues. Yet despite these difficulties, she persevered and established herself as a pioneer in the field of medical genetics.

Notes

1. M.T. Macklin, "Medical Genetics: An Essential Part of the Medical Curriculum from the Standpoint of Prevention," *Journal of the Association of American Medical Colleges* 8 (1933): 291.

2. M.T. Macklin, "Inheritance of Cancer of the Stomach and Large Intestine in Man," *Journal of the National Cancer Institute* 24 (1960): 551.

3. M.T. Macklin, "Comparison of Number of Breast Cancer Deaths Observed in Relatives of Breast Cancer Patients, and in the Number of Expected on the Basis of Mortality Rates," *Journal of the National Cancer Institute* 22 (1959): 927.

4. M.T. Macklin, "The Need of a Course in Medical Genetics in the Medical Curriculum: A Pivotal Point in the Eugenic Programme," *Edinburgh Medical Journal* 40 (1933): 20.

5. F. Clarke, presidential address—extract, American Society of Human Genetics, Annual Meeting, Corvallis, Oregon, August 30, 1962.

Bibliography

Macklin, Madge Thurlow. *The Role of Inheritance in Disease.* Baltimore: William and Wilkins Co., 1935.
Notable American Women: The Modern Period. Cambridge and London: Belknap Press, 1980.
Siegel, Patricia Joan, and Kay Thomas Findey. *Woman in the Scientific Search.* Metuchen, N.J., and London: Scarecrow Press, 1985.
Soltan, Hubert C. "Madge Macklin—Pioneer in Medical Genetics." *University of Western Ontario Medical Journal* 38 (Oct. 1962): 6–11.

MARGARET A. IRWIN

BARBARA MCCLINTOCK

(1902–1992)

Geneticist

Birth	June 16, 1902
1923	B.S., Cornell University
1925	M.A., Cornell University
1927–31	Instructor in Botany, Cornell University
1928	Ph.D., Cornell University
1931–33	Fellow, National Research Council
1933–34	Fellow, Guggenheim Foundation
1934–36	Research Associate, Cornell University
1936–41	Assistant Professor, University of Missouri
1942–67	Staff Member, Carnegie Institution of Washington, Cold Spring Harbor, NY

Barbara McClintock. Photo reprinted by permission of the Cold Spring Harbor Laboratory, Cold Spring Harbor, NY.

1947	Achievement Award, American Association of University Women; Honorary Sc.D., University of Rochester
1949	Honorary Sc.D., Western College for Women
1954	Visiting Professor, California Institute of Technology
1957	Merit Award, Botanical Society of America; Honorary Sc.D., Smith College
1963–69	Consultant, Agricultural Science Program, Rockefeller Foundation
1965–74	Andrew D. White Professor-at-Large, Cornell University
1967	Kimber Genetics Award, National Academy of Sciences

1967–92	Distinguished Service Member, Carnegie Institution of Washington, Cold Spring Harbor, NY
1968	Honorary Sc.D., University of Missouri
1970	National Medal of Science
1972	Honorary Sc.D., Williams College
1978	Lewis S. Rosensteil Award for Distinguished Work in Basic Medical Research; Louis and Bert Freedman Foundation Award for Research in Biochemistry
1979	Honorary Sc.D., The Rockefeller University; Harvard University
1980	Salute from the Genetics Society of America
1981	Thomas Hunt Morgan Medal, Genetics Society of America; Honorary Member, Society for Developmental Biology; Wolf Prize in Medicine; Albert Lasker Basic Medical Research Award; MacArthur Prize Fellow Laureate; Honorary D.H.L., Georgetown University
1982	Honorary Member, Genetical Society, Great Britain; Louisa Gross Horwitz Prize for Biology or Biochemistry; Charles Leopold Mayer Prize, Académie des Sciences, Institut de France; Honorary Sc.D., Yale University; University of Cambridge
1983	Nobel Prize for Medicine; Honorary Sc.D., Bard College; State University of New York; New York University
1984	Albert A. Michelson Award; Honorary Member, American Medical Women's Association; Regents Medal of Excellence, State University of New York; Honorary Sc.D., Rutgers University
1985	Honorary Sc.D., La Faculté des Sciences Agronomiques de l'Etat à Gambloux, Belgium; Honorary Member, Medical Women's International Association; Honorary Member, American Society of Naturalists; Honorary Member, New York Academy of Sciences; Honorary Vice President for the 16th International Congress of Genetics (Ontario, Canada, 1988)
1986	National Women's Hall of Fame Award
1989	Foreign Member, Royal Society, London
1990	125th Anniversary Medal, University of California, San Francisco
1991	Honorary Fellow, Indian Society of Genetics and Plant Breeding
Death	September 2, 1992

Barbara McClintock is among the best-known and best-loved women scientists. This is in part because, like James Watson and the double helix, she has a great story. Unlike Watson's story, however, McClintock's is not one of undiluted success. It is more like a tragic play. During the first two acts, McClintock matured as a scientist and made a name in science. In the third act, intellectual tragedy befell her. But in the final act she was exonerated and given the honor she was due. Like all myths, that of Barbara McClintock serves a purpose, has a moral, an explanatory power. Yet like other myths it is a mixture of truth and fiction, a story created to explain something we don't understand.

What is the myth of Barbara McClintock? The tiny lady with sparkling eyes, born in Hartford, Connecticut, in 1902, was the first person to discover that genes can move around on the chromosomes. She discovered transposable elements, or "jumping genes." These small genetic elements, McClintock found, could move from one position to another on a chromosome or even between chromosomes—as though a word, such as "gene," spontaneously moved from this essay to the one on Rachel Carson [or someone else in this book], or to the next book on the shelf. McClintock first published this discovery in 1950, but the idea's real test came in the next year. She presented a paper at the Cold Spring Harbor Symposium, a meeting of many of the brightest, most famous, and most powerful biologists. No one understood her revelations. She felt that the reactions ranged from disinterest to hostility. She tried again five years later, and the reaction was even colder. She stopped publishing, gave up on her colleagues, and continued her work alone. Finally, in 1983, 35 years later, her work was recognized and she was awarded the Nobel Prize.

The McClintock myth is used as an example of injustice toward women in science; of the lack of acceptance of an intuitive, "feminine" style of science; and of the ultimate triumph of truth over systematic ignorance. McClintock herself, however, hated the myth; ironically, her very refusal to confront it publicly allowed it to survive.

What's wrong with this myth? First, McClintock was never ignored. From her graduate student days in the 1920s she was recognized as one of the brightest young maize geneticists in the country. During the intervening 20 years, she published a steady stream of new data and new concepts. In the early 1930s she found chromosomes that formed rings. Later she discovered that the ring chromosomes were a special case of broken chromosomes, which could be produced by X-raying corn. This sometimes produced a pattern of instability in the chromosomes that is now recognized as being related to the genetic damage seen in certain cancers. McClintock predicted structures, which she called telomeres, on the ends of normal chromosomes, which maintained the chromosome's stability and integrity and which were lost when the chromosomes were broken. Telomere research today is a rapidly growing area of biology, with implications for both cancer and aging. In the 1940s she discovered a source of this genetic instability: transposable elements. These specialized gene fragments seemed to cause specific chromosome breakage, a genetically controlled example of the phenomenon she had been observing for years. By the time she made this discovery (the one that flopped at the 1951 Cold Spring Harbor Symposium), she was among the most famous geneticists in the country. She was elected president of the Genetics Society of America in 1944. In the next year she was elected to the National Academy of Sciences. Later she was given other honors, including the Kimber award and a lifetime MacArthur "genius" award.

It is true that McClintock had a less-than-enthusiastic reception at the 1951 Cold Spring Harbor Symposium. There are several reasons for this. First, McClintock studied maize, or Indian corn. Maize was by 1951 out of fashion among biologists. Bacteria and viruses were the vogue. Worse, many of the virologists were physicists and chemists and had little biological training. They had a hard time following McClintock's intricate series of genetic crosses. Second, McClintock made a challenging situation worse by the fact that she tended not to give her audience a lot of explanation; she just presented the data (the appropriate conclusions no doubt seemed obvious to her). One of the most famous and brilliant geneticists of the day, Alfred Sturtevant, remarked of her talk, "I didn't understand a word of it, but if Barbara says it is so, it must be so!" Finally, maize is a peculiar organism, genetically. Those scientists who did understand her work thought genes might jump around in corn but not necessarily in bacteria, fruit flies, or humans.

It is also true that it took 35 years for McClintock to win the Nobel Prize (the average is 10 to 15). She won the Nobel because she profoundly shook up the field of genetics. The science of genetics in the late 1940s was rooted in the idea that genes were fixed. The previous era of genetics had been dominated by a group of scientists who used the fruit fly to "map" genes, or pinpoint their location on the chromosomes. If genes moved around, how could they be mapped? To scientists of the day, it must have seemed as though the brilliant woman were telling them that a small area in Southern California is sometimes on the West Coast, sometimes on the East Coast!

McClintock gave her audience a second shock as well: transposable elements, she argued, were a way for the genes to regulate themselves. Jumping genes could act as switches, turning genes on and off, for example. In McClintock's view, the genes were not simply a static blueprint for an organism but a dynamic system—almost an organism in itself— that interacted with the environment. She emphasized that she felt she had discovered not just a new kind of gene but something altogether different from genes. She called them "controlling elements" to connote that they regulated the genes.

Put yourself in the shoes of one of McClintock's colleagues. A woman whose intellect and science you greatly admire shows you an enormous number of complicated experiments you don't understand. Based on spots in corn kernels and stripes in leaves, she tells you that these little pieces of chromosomes are moving around spontaneously. But she is not part of the mainstream. Her organism, corn, is rather old-fashioned— too complex, as compared with bacteria and viruses—to tell you much about what genes are made of. What little conclusion she gives you from this exhausting presentation shakes the very foundation of the science to which you've dedicated 10, 20, maybe 30 years. Yet you've never seen anything like this in bacteria or fruit flies. It probably doesn't apply. If she's right, maybe all you and your colleagues have done is meaningless. Little wonder she had a cool reception.

Transposable elements and genome regulation were eventually accepted into the fold of mainstream science. But we still haven't answered the question of why it took so long. The answers are many, but the simplest is that scientists are skeptics. McClintock, after all, never saw a transposable element. She reasoned that they must exist on the basis of her observations of the pigmentation in corn kernels—much the same way as no one has ever actually seen a black hole, but we know they are there. No one saw a transposable element until the late 1970s. By that time, the science of molecular biology had developed to the point at which scientists could pull a jumping gene out of a chromosome and put it in a test tube. When McClintock studied transposable elements, no one knew for certain what a gene was made of. Beginning in the late 1960s, molecular biologists began to find transposable elements not only in corn but in fruit flies, bacteria, yeast, and humans. Then, when the importance of McClintock's work became clear, she began winning the prestigious medals and awards in science that lead to the Nobel Prize.

Another part of the McClintock myth is that she was a recluse, an asocial genius working day after day, decade after decade, solitary and silent. The grain of truth here is that she carried out nearly all of her experiments alone. She rarely took students; and after her early papers, she rarely published collaborative studies. But she was far from a recluse. She had scores of friends. When she wasn't in the field or analyzing corn, she often had company. She never married, so she never retreated from her friends into family life. One had to be careful when one visited her— a 15-minute chat often turned into a two-hour discussion. She loved young people. In her later years, many of her friends were much younger than her, in their twenties and thirties. She loved to hear about their science, to ask them questions. Other friends were groundskeepers and gardeners. An avid amateur botanist, she could talk to a gardener just as long as to a scientist.

McClintock hated publicity and hated to have a fuss made over her. She said that was the worst thing about winning the Nobel. She often said, even until she died in 1992, that she didn't think her story was typical, so it wouldn't be of interest to anyone. This may be part of why the McClintock myth grew—she never corrected it. The worst thing about myths is that they oversimplify. Barbara McClintock was a complex person, not a caricature. She dressed boyishly, in men's shirts and slacks, and wore her hair short. Thus some might have thought she was not feminine and was uninterested in men. The truth is, however, that she loved to watch men, even when she was in her eighties. This cool, detached scientist had a wicked sense of humor and a warm heart.

Myths serve a purpose, but eventually they need to be replaced. The real lessons to be gleaned from McClintock's story are about the intuitive nature of genius more than sexism, the essential conservatism of science more than narrow-mindedness. The true story of Barbara McClintock is perhaps even more instructive and inspirational than the myth.

Bibliography

Fedoroff, Nina, and David Botstein, eds. *The Dynamic Genomes: Barbara McClintock's Ideas in the Century of Genetics.* Plainview, N.Y.: Cold Spring Harbor Laboratory Press, 1992.

Keller, Evelyn Fox. *A Feeling for the Organism: The Life and Work of Barbara McClintock.* San Francisco: W.H. Freeman, 1983.

McClintock, Barbara. *The Discovery and Characterization of Transposable Elements: The Collected Papers of Barbara McClintock.* New York: Garland Press, 1987.

McGrayne, Sharon Bertsch. *Nobel Prize Women in Science.* New York: Carol Publishing Group, 1993.

Opfell, Olga S. *The Lady Laureates,* 2nd ed. Metuchen, N.J.: Scarecrow Press, 1986.

Shiels, Barbara. *Women and the Nobel Prize.* Minneapolis: Dillon Press, 1985.

NATHANIEL COMFORT

MARIA SIBYLLA MERIAN

(1647–1717)

Entomologist, Naturalist

Birth	April 2, 1647
1665	Married artist Andreas Graff
1685	Joined the Labadists
1699	Departed for Surinam

1701	Returned to Amsterdam
Death	1717

Maria Sibylla Merian was born in 1647 in Frankfurt on the Main. Her father, Matthaus Merian, was a publisher and engraver. He reportedly had a vision that his daughter would grow to be a woman of extraordinary talent. He died, however, when Sibylla was 3 years old. It was Sibylla Merian's stepfather, artist Jacob Marell, who started her on her way as an artist. He taught her drawing, painting, and engraving.

Sibylla's ability soon exceeded that of her stepfather. When it did, she studied with Marell's first pupil, Abraham Mignon. Sibylla lived in a world of artists, including her brother, Matthew Merian, a well-known portrait painter whose works include "Artemisia," a portrait of Pierre Serini.

Sibylla's journals begin in 1660. In that year, she reported acute observations of the life cycle of the silkworm: the transformation of caterpillar to moth. Her work antedated by nine years that of Marcello Malpighi, the Italian physician who was given credit for describing the metamorphosis of the silk moth.

In 1665, Sibylla married another artist, Andreas Graff. Her first daughter, Johanna Helena, was born in 1668. In 1670 the Graff family moved to Nuremburg, where Sibylla became a very popular painter of flowers. She painted flowers on tablecloths and embroidery pieces. She also made copper plate engravings. So many people wanted her work that she had many pupils who helped her and apprenticed to her.

By 1680 it seemed no one in Europe could get enough flower paintings. Art critics called the phenomenon "tulipomania." The critics scorned people who paid great sums of money for the paintings. Not surprisingly, the artists and buyers ignored the critics. However, Merian had departed significantly from other flower painters. By 1669 she had begun to add insects to her flowers. A long-time informal student of the six-legged creatures, Merian knew a lot about insects. She began systematically collecting and depicting caterpillars in 1674, putting the immature insects in containers, feeding them, and then observing the adults that unfolded from pupal (resting) stages. She painted what she saw. In fact, Merian's second daughter, Dorothea Maria Henrietta, born in 1678, became a great and accomplished painter of flowers and insects in her own right. In 1681, Merian went to Frankfurt to help her mother. Soon thereafter, Maria's marriage to Graff ended in divorce. There are hints about what contributed to the demise of the marriage: "[His] affairs became so much involved, and his conduct in other respects so censurable, that he was obliged for a time to leave the country."[1]

Sibylla Merian joined the Labadist community in the Schloss Walta-

State in 1685. The community members were followers of John Labadie, a native of Bourg in Guinne who had renounced the doctrines of the Church of Rome. The Labadists resembled in part the Quietists. But Labadie, who was purported to have had a charismatic effect on women in particular, added many of his own ideas. Merian used her time at the Labadist community to study Latin. She also worked on a series of flower and butterfly paintings. In what was perhaps the most important juncture in her life, she saw a collection of butterflies from the West Indies made by Cornelis van Sommelsdijk. That experience kindled her growing interest in the insects of South America. When the community of Labadists broke up in 1688, Maria eventually made her way to Amsterdam.

On April 23, 1699, at age 52, Maria Sibylla Merian and her daughter Johanna Helena left Europe for South America. For 21 months they lived in Surinam. There they painted and wrote about the insect and plant associations they observed. They also studied reptiles, crabs, and snails.

Sugar, citrus fruits, coffee, bananas, minerals, and timber are some of what attracted most seventeenth-century Europeans to South America. A curious woman, Sibylla Merian was attracted by something else. She wanted to study the natural world, particularly insects, in this tropical country. During the seventeenth century, Dutch explorers and colonists often took animal and plant specimens from Surinam back to Europe. Merian had long been intrigued by what she saw.

Only determination got Merian to Surinam. She had no money for such a trip, yet she had the courage to ask the city leaders of Amsterdam for travel funds. It was the city of Amsterdam that made it possible for Maria and her daughter to travel to Surinam.[2] Merian's art had been well known in the Amsterdam of 1699. She had lived in the city since 1691. But she was going to Surinam as more than a gifted artist. She was going as a student of nature and would use her art to advance science in a profound way.

In Surinam, the efforts of Merian and her daughter puzzled the colonists from The Netherlands. The Dutch colonists saw the natural world as something to be conquered. They could not understand why the two women sat in the intense heat and humidity to observe and to draw. In fact, they lent them no assistance. Merian and her daughter built a camaraderie with the Indian and African residents of Surinam who had been enslaved by the Dutch.

Despite the complexities of the situation, Merian and her daughter worked diligently. Although they had planned to remain for a long time, the climate forced them to leave in less than two years. Unfortunately, no records remain of Merian's journeys across the Atlantic. When Merian and her daughter returned to Amsterdam on September 23, 1701, she had enough material to keep her busy until her death in 1717. Daughter

Dorothea had to complete the painting for the work on insects of Surinam. There are unverified reports that Merian sent a daughter to Surinam a second time.

Merian recorded for the first time the associations between caterpillars and adults of many species of insects. Although she was extraordinarily fond of butterflies, she studied all the insects she could: ants, beetles, and flies, to name a few. She captured them in situ, showing their activities and lending clues to their ecology. Other invertebrates, amphibians, and some birds, particularly those interacting with insects, also captured her attention. Even the "white potato" caught Merian's eye well before it was taken back to Europe, where it became a staple. As it happened, she was looking for caterpillars that feed on the potato, which, she noted, people did not eat.

In the eighteenth century, Carolus Linnaeus tapped much of Sibylla Merian's work. Linnaeus developed the system of assigning scientific names to living things (binomial nomenclature). Indeed, Merian's work provided a foundation for many scientists and all modern natural science. Yet the credit due Sibylla Merian is largely absent for most of the nineteenth and twentieth centuries. Her work was eagerly and easily appropriated and copied, but she was cited only rarely.

There are at least two explanations for the oversight of Sibylla Merian's achievements. First, she was a woman working in the sphere of men— those producing high-fidelity renderings of the natural world—at a time when women did not compete with men in any domain. Second, lamentably, the efforts to copy many of her plates became so frenzied in the eighteenth century that the copies were made without attention to appropriate color. Some of the copies became absolutely fanciful. Although Sibylla Merian was not responsible for those copies, she was often held accountable for them by men working in the disciplines of science and art.

Merian has been called "one of the most remarkable women of the 17th century."[3] She was a talented artist, a careful observer of the natural world, and an industrious person. For centuries, her contributions to science have been known to all too few. Every caterpillar and butterfly should be a reminder to tell a friend about Maria Sibylla Merian.

Notes

1. James Duncan, *The Natural History of British Moths, Sphinxes, etc.*, vol. 4 of *The Naturalist's Library* (Edinburgh: W.H. Lizars, 1836), p. 20.

2. Sibylla Merian records Johanna Helena as the daughter who traveled with her. However, some secondary references refer to Dorothea as the traveling companion.

3. Dennis Landis, "#40, Maria Sibylla Merian, Dissertatio de Generatione et Metamorphosibus Insectorum Surinamensium," in *The Literature of Encounter: A*

Selection of Books from European Americana, pp. 66–67 (Providence: John Carter Brown Library, 1991), p. 66.

Bibliography

Duncan, James. *The Natural History of British Moths, Sphinxes, etc.* Vol. 4, *The Naturalist's Library.* Edinburgh: W.H. Lizars, 1836.

Landis, Dennis. "#40, Maria Sibylla Merian, Dissertatio de Generatione et Metamorphosibus Insectorum Surinamensium," in *The Literature of Encounter: A Selection of Books from European Americana,* pp. 66–67. Providence: John Carter Brown Library, 1991.

Merian, Maria Sibylla. *Flowers, Butterflies and Insects; All 154 Engravings from Erucarum Ortus.* New York: Dover Pictorial Archive Series, 1991.

Pohlmann, Olga. *Maria Sibylla Merian.* Berlin: Wolg. Kruger Verlag, 1935.

Stedman, John Gabriel. *Narrative of a Five Years' Expedition against the Revolted Negroes of Surinam.* Transcribed from the 1790 original edition, edited and with an introduction and notes by Richard Price and Sally Price. Baltimore and London: Johns Hopkins University Press, 1988.

Valiant, Sharon. 1993. "Maria Sibylla Merian: Recovering an Eighteenth-Century Legend." *Eighteenth Century Studies* 26, no. 3 (1993): 467–79.

DIANE M. CALABRESE

YNES MEXIA

(1870–1938)

Botanist

Birth	May 24, 1870
1885	Attended Saint Joseph's Academy, Emmitsburg, MD
1921	Enrolled in natural science classes at the University of California, Berkeley
1925	Took a course in flowering plants at Hopkins Marine Station, Pacific Grove, CA
1925–26	Expeditions to Western Mexico
1928	Expedition to Mount McKinley, AK
1929	Expedition to Brazil
1934	Expedition to Ecuador
1935	Expedition from Peru south to the Straits of Magellan

| 1937 | Expedition to southwestern Mexico |
| Death | July 12, 1938 |

At age 55, Ynes Mexia happily exchanged the comforts of San Francisco for the dirt trails of western Mexico and the jungles of South America. To collect plants she battled insects, scaled cliffs, suffered broken bones, and slept in banana groves and on beaches tracked by jaguar. From 1925 to 1938, Mexia obtained 137,600 plant specimens, more than any other woman botanical collector of her era.

From childhood, Mexia's life was checkered with unhappiness and financial reversals. Ynes was born in 1870 in Georgetown, Washington, D.C., to Enrique Antonio Mexia, an agent for the Mexican government, and Sarah R. Wilmer, a divorcee with six children from a previous marriage. Her parents separated when Ynes was 3 years old. Sarah and the children moved to Texas. At age 15, Ynes returned east to attend Saint Joseph's Academy in Emmitsburg, Maryland. She spent the next ten years living on her father's Mexico City hacienda, which she inherited after his death.

Married in 1897 and widowed seven years later, Mexia started a successful poultry and pet stock-raising business at the hacienda. Her second marriage to a man 16 years younger was disastrous. When the 39-year-old Mexia traveled to San Francisco for medical reasons in 1909, her husband nearly bankrupted the business. Mexia remained in San Francisco, sold her business, divorced her husband, and did part-time social work. She was depressed; her life lacked purpose.

Over the next five years, a succession of events brought happiness to Mexia's life: trips with the local Sierra Club; enrollment in natural science classes at the University of California, Berkeley, through which she discovered an interest in botany; and, in the fall of 1925, a collecting trip to western Mexico with Stanford botanist Roxanna Stinchfield Ferris.

Mexia's familiarity with the language and customs of Mexico, her interest in plants, and her love of adventure made botanical collecting an ideal career. Her trip to western Mexico was the first of seven botanical expeditions. Cut short by a fall from a cliff that injured her hand and fractured several ribs, the two-month expedition yielded 500 botanical species. During her recuperation in San Francisco, Mexia decided to finance future botanical trips by collecting and selling plant samples to various institutions.

In 1926, Mexia returned to western Mexico. From Mazatlan, she explored the coastline, mountains, and jungles in the states of Nayarit, Sinaloa, and Jalisco. She fought biting gnats and swarms of dot-sized ticks, scaled a cliff, and slept in a banana grove where a week earlier a panther had killed two of her host's dogs. Her modes of transportation varied: she rode horseback out of Tepic; paddled a dugout canoe on the Rio San Pedro; sailed on a steamer to Puerto Vallarta; and accompanied a packtrain to the crest of the Sierra Madre. She returned home with 33,000 undamaged specimens including 50 new species and a new genus, *Mexianthus mexicanus*, named in her honor.

Mexia spent the summer of 1928 collecting plants around the slopes of Alaska's Mt. McKinley. She added 6,100 specimens to her burgeoning collection from this expedition. Then in October 1929 she traveled to Brazil. Arriving in Rio de Janeiro, she went to the highlands in Minas Gerais and remained there for a year and a half. Back in Rio, Mexia set out to cross the South American continent, an adventure she vividly described in her article, "Three Thousand Miles up the Amazon," which was published in the February 1933 issue of the *Sierra Club Bulletin:*

> On August 28, 1931, with a truckload of equipment, I boarded the river-steamer "Victoria" and started up the famous river. Surely there was no roughing it on the steamer. Screened cabins, electric fans, ice plant aboard, as well as fresh meat "on the hoof."[1]

Whenever the *Victoria* stopped for supplies or wood, Mexia would "jump into boots and khaki (much to the amusement of the passengers) and walk the plank to investigate the forest."[2]

> Beautiful as is the forest seen from the river, it is repelling to enter. The canopy is so dense it cuts off all sunlight, prohibiting under-growth. There are no trails; it is dark and dank, with crowding tree-trunks, tangling *lianas* rotting logs everywhere, and oozy, treacherous soil. No flowers are to be seen; such trees as are in bloom keep their color and fragrance for the forest roof where the real life of the forest displays itself.[3]

On the twenty-fourth day out of Rio, the *Victoria* ended its 2500-mile journey at Iquitos, Peru. Mexia repacked her equipment, laid up sup-plies, and hired three men and a launch named *Alberto.* Seven days later the crew abandoned the *Alberto* for two canoes and four Indian paddlers. Their westward journey continued on the Rio Maranon, a tributary of the Amazon. Mexia was enthralled as she sat "amidships under a little palm-leaf shelter." At night they slept on sandy beaches "often tracked by jaguar and tapir." Inching upstream "past monstrous stranded trees," they set up camp at the mouth of the Rio Santiago, a tributary of the Rio Maranon, because "the eastern-flung chain of the mighty Andes" ended their journey westward. For three months Mexia collected botanical spec-imens. She and her party bartered trade goods with the local Indians and consumed "toucans, monkeys, and parrots." By January, a home-made raft of balsa logs was loaded with her "precious collection of plants and birds and insects." Sporting a palm-leaf thatch over the platform, a chicken coop, and a fireplace at the rear, the raft was guided downstream. During the two-week journey, Mexia wrote notes and prepared her botanical specimens. This was, in her words, "the most delightful mode of transportation that I have encountered."[4]

At Iquitos, Peru, Mexia dispatched her collection to California while she fulfilled her wish to cross the continent.

I took a hydroplane up the Amazon and the Ucayalli rivers to Massesea, an airplane across country to San Ramon in the lower Andes, mule and automobile to chilly Tarma at 10,000 feet in a valley of the Andes, then the Transandean Railroad, which, after crossing a pass at nearly 16,000 feet, drops down a steep incline to Lima and Callao and the Pacific.[5]

At last, Mexia was able to say: "I had fulfilled my wish to cross South America at its greatest breadth."[6] Almost two and a half years had passed since her arrival in Brazil, but she returned home with 65,000 specimens to her credit.

Over the next two years she enjoyed outings with the Sierra Club and botanical trips in the southwestern United States, but she did not abandon her love of travel. After months of negotiating with the U.S. Department of Agriculture, Mexia obtained permission to study a rare wax palm in Ecuador. This particular species grew at higher altitudes and cooler temperatures than other palms and, consequently, might thrive in California. In September 1934 she left for "the Land of the Equator!"— her description of Ecuador in the February 1937 *Sierra Club Bulletin* article entitled "Camping on the Equator."[7]

In the mountains between Ecuador and Colombia, she found a native who agreed to guide her to a wax palm "half a day distant by rough trail." To describe the trail as "rough" was an understatement. She braved choked forests, precipitous slopes, and hot sun until she "saw the beautiful palm, its slender trunk white against the dark forest and its spreading crown waving above the forest canopy."[8] And then:

We reached it—a photograph—and with a pang (unfelt by my companions) I gave the order. The ax bit in, and the great tree crashed to earth.

Unluckily, it was not in fruit so no seeds were available, but at least it was in flower. I photographed the great spathe and flower-cluster, so heavy the two men could hardly lift it; made measurements and notes; and took portions of the great arching fronds.[9]

Back in Tulcan, Mexia spent three days pressing, drying, and caring for her 5,000 specimens before joining a botanical expedition organized by T. Harper Goodspeed of the botany department, University of California. From October 1935 to January 1937 she traveled from Peru south to the Straits of Magellan and collected 15,000 specimens. In his book, *Plant Hunters in the Andes*, Goodspeed described Mexia as "the true explorer type and happiest when independent and far from civilization."[10] At age 67, Mexia returned to Mexico. In spite of chest pain, she collected 13,000 botanical specimens before returning home in late May 1938. On July 12, 1938, Ynes Mexia died of lung cancer. Goodspeed expressed the sentiments of many when he wrote, "Her death was a distinct loss to botanists, who had learned much that was new about the vegetation of the territories in which she collected."[11]

Notes

1. Ynes Mexia, "Three Thousand Miles up the Amazon," *Sierra Club Bulletin* (Feb. 1933): 88.
2. *Ibid.*, p. 90.
3. *Ibid.*, pp. 90–91.
4. *Ibid.*, pp. 92–96.
5. *Ibid.*, p. 96.
6. *Ibid.*
7. Ynes Mexia, "Camping on the Equator," *Sierra Club Bulletin* (Feb. 1937): 85.
8. *Ibid.*, p. 89.
9. *Ibid.*, p. 90.
10. T. Harper Goodspeed, *Plant Hunters in the Andes* (Berkeley and Los Angeles: University of California Press, 1961), p. 12.
11. *Ibid.*

Bibliography

Bonta, Marcia Myers. *Women in the Field: America's Pioneering Women Naturalists.* College Station: Texas A&M University Press, 1991.

Goodspeed, T. Harper. *Plant Hunters in the Andes.* Berkeley and Los Angeles: University of California Press, 1961.

Mexia, Ynes. "Camping on the Equator." *Sierra Club Bulletin* (Feb. 1937): 85–91.
———. "Three Thousand Miles up the Amazon." *Sierra Club Bulletin* (Feb. 1933): 88–96.

Notable American Women, 1607–1950. Cambridge and London: Belknap Press, 1971.

REBECCA LOWE WARREN

AGNES MARY CLAYPOLE MOODY

(1870–1954)

Zoologist

Birth	January 1, 1870
1892	Ph.B., Buchtel College, Akron, OH
1894	M.S., Cornell University
1896	Ph.D., University of Chicago
1896–98	Instructor in Zoology, Wellesley College
1898–1900	Assistant in Histology and Embryology, Cornell University
1900–1903	Instructor, Throop Polytechnic Institute (now California Institute of Technology, Pasadena)
1903	Married Dr. Robert Moody
1906	"Starred" in first edition of *American Men of Science*
1918–23	Lecturer, Mills College, Oakland, CA
Death	1954

Agnes Claypole Moody belonged to the first group of women scientists who took advantage of the academic openings made possible by the determined pioneers in the years immediately before and after the Civil War. She had a successful career as a teacher and influenced many students. But at the same time, she faced the discrimination and hurdles that made it very hard for women to make careers for themselves as researchers and teachers at the university level.

Agnes Mary Claypole and her twin sister, **Edith Jane Claypole**, were born in 1870 in Bristol, England, to Edward Waller Claypole and Jane (Trotter) Claypole. An older brother had died in early childhood and the twins' mother died in 1870, a few weeks after their birth. Edward Claypole (1835–1901) had started as a teacher of classics and mathematics in England, but his interest in the sciences motivated him to matriculate at the University of London, where he received his degrees. When his liberal views on evolution caused him difficulties in finding teaching jobs, he moved to the United States in 1872, leaving his young daughters in England. He started teaching natural sciences at various American colleges in 1873 and actively participated in research in geology and paleontology. As one of the founders of the journal *American Geologist* and an editor of that publication until his death, he played an important part in the scientific life of his time.

In 1879, Edward married Katherine B. Trotter of Toronto, Canada, a second cousin of his first wife, and in the same year his daughters joined them in America. Professor Claypole has been described as a born teacher who in his long career taught a great variety of subjects, from the classics to geology. He and his second wife educated their daughters at home. The girls grew up in a family where science and research were discussed as a matter of course. His example and his encouragement influenced his daughters deeply, and they soon developed an abiding interest in science. Both attended Buchtel College in Akron, Ohio, where their father taught natural sciences for 15 years. The twins took the same science courses together and graduated with bachelor's degrees in 1892.

Since its beginnings in the 1830s and 1850s, higher education for women had made tremendous strides, especially after the establishment of independent women's colleges between 1860 and 1880. Many state universities also started permitting women to enroll as undergraduates, and toward the end of the century admitted them to master programs, albeit grudgingly. Cornell University in upstate New York in particular became a focus for women students seeking advanced study. The Claypole sisters enrolled in its graduate program in the fall of 1892.

While attending Cornell University, the sisters took advantage of a unique opportunity for furthering their scientific education when they participated repeatedly in the seaside summer school at Woods Hole, Massachusetts. The Marine Biological Laboratory at Woods Hole had been set up in 1888 as a permanent research opportunity and school for teachers and researchers. In its early stages, it had included influential women among its trustees.

Although the sisters took many classes together, their aspirations and careers began to diverge at Cornell University. Agnes Claypole received her M.S. degree in 1894 and wrote her thesis on the digestive tract of the Cayuga Lake Lamprey, an eel-like fish. The thesis was also published as "Prize Paper in Animal Histology" in the *Proceedings of the American Microscopical Society* of 1894 as her first publication. She decided to pursue a doctorate in zoology after graduating from Cornell, and she went to the University of Chicago. Founded in 1891, the University of Chicago had, from the very beginning, accepted women on the same footing as men in its undergraduate as well as its graduate programs. Women interested in higher education immediately gravitated to the University of Chicago, especially because so many other universities made doctoral study almost impossible for them. Agnes Claypole received her degree from Chicago in 1896. Her dissertation, entitled "The Embryology and Oogenesis of Anurida Maritima," researching and illustrating the early development and egg production of an order of small insects, was published in the *Journal of Morphology* in 1898 and later reprinted as a monograph.

But now Dr. Claypole faced the next hurdle for women with advanced degrees, namely, finding a job that would allow her to use her training. Possibilities for employment were offered by the independent women's colleges. As a rule, these colleges were not administered by a denomination and thus drew talented students from a variety of backgrounds. They offered the same curricula as men's schools and also offered courses in the sciences. By the 1890s these institutions were trying to upgrade their faculty and started hiring women with doctorates. Agnes Claypole taught at Wellesley College in Massachusetts from 1896 to 1898. Because her sister Edith acted as head of the zoology department during those two years, it seems a reasonable assumption that she had recruited her twin.

In the fall of 1898, Agnes Claypole started to teach histology and embryology at Cornell University. She was the first woman to teach laboratory classes that were required courses for all students. The move to a coeducational university meant, however, facing the opposition these institutions aimed at women who wanted to teach there. Although most universities had admitted females as undergraduates and (much more slowly) as graduate students for almost two decades by now, faculty appointments for women, especially in the sciences, called forth intense resistance. If women were hired at all, they were often not given professorial rank but instead were employed as lecturers or instructors, and hence had low status and were paid accordingly. Cornell University had only started hiring women late in the 1890s, so Agnes Claypole belonged to this early group. The university would not permit a higher rank than "assistant" for her, even though she held a doctorate in zoology and was a fully qualified professional.

In 1898, Agnes's father accepted a teaching job in geology and biology at the Throop Polytechnic Institute in Pasadena, California (now the California Institute of Technology), and in 1900 she joined him there as instructor. Edward Claypole had moved to California because of his wife's health, so family reasons may have been partly responsible for Agnes Claypole's decision to leave; but she may also have become discouraged at Cornell University. After her father's sudden death in August 1901, she succeeded him in his appointment at Throop Institute. In 1903 she quit teaching and married Robert Orton Moody, a professor of anatomy at the University of California, Berkeley. (Robert Moody had received a doctorate in medicine at Yale University in 1894, taught at Yale from 1891 to 1900 and at Cornell from 1900 to 1901, and then moved to California. He was the son of another distinguished scientist and medical doctor, Mary Blair Moody, who in 1894 had been the first woman elected to join the Association of American Anatomists.)

Agnes Claypole Moody did not teach again until 1918, when she joined the faculty of Mills College in Oakland, California, as a lecturer in sociology. She taught there until 1923. Even though she would have liked to teach in her field, she might not have found a suitable position because (1) married women were discouraged from holding faculty appointments, and (2) most universities would not hire spouses of faculty members already employed at their institutions. Agnes took an active part in communal politics in Berkeley, where she served on the School Board and the Berkeley Commission of Public Charities. In 1923 she was elected to the Berkeley City Council. Although she still pursued her research interests, she did not publish any more in her field.

Agnes Moody was recognized as an outstanding zoologist at a time when few women worked in the demanding speciality of animal histology. She was included among the 150 American zoologists listed in the first edition of the biographical directory *American Men of Science*. She was awarded a star in the first through the seventh editions, which meant she was recognized as one of the top 1,000 scientists in America. She died in Berkeley, California, in 1954.

Bibliography

American Men of Science: A Biographical Dictionary, eds. 1–11. Lancaster, PA.: Science Press, 1906–1970.

Bailey, Martha J. *American Women in Science: A Biographical Dictionary*. Santa Barbara, Calif.: ABC-CLIO, 1994.

Claypole, Agnes Mary. *The Embryology and Oogenesis of Anurida maritima (Guér)*. Boston: Ginn and Co., 1898.

———. "The Enteron of the Cayuga Lake Lamprey." *Proceedings of the American Microscopical Society* 16 (1895): 125–64.

Herzenberg, Caroline L. *Women Scientists from Antiquity to the Present: An Index*. West Cornwall, Conn.: Locust Hill Press, 1986.

Moody, Agnes Claypole. "The True Story of May First, Nineteen Twenty-Three." *Mills Quarterly* 6, no. 2 (July 1923): 61–64.

National Cyclopedia of American Biography. New York: James T. White, 1906.

Ogilvie, Marilyn Bailey. *Women in Science: Antiquity through the Nineteenth Century*. Cambridge, Mass.: MIT Press, 1986.

Rossiter, Margaret W. *Women Scientists in America: Struggles and Strategies to 1940*. Baltimore: Johns Hopkins University Press, 1982.

Siegel, Patricia Joan, and Kay Thomas Finley. *Women in the Scientific Search: An American Bio-Bibliography, 1724–1979*. Metuchen, N.J.: Scarecrow Press, 1985.

Visher, Stephen Sargent. *Scientists Starred, 1903–1943, in "American Men of Science."* Baltimore: Johns Hopkins University Press, 1947.

Williamson, Mrs. Burton M. "Some American Women in Science." *The Chautauquan* 28 (Nov.–Mar. 1898–1899): 361–68.

Woman's Who's Who of America, ed. John William Leonard. New York: American Commonwealth, 1914. (Reprinted: Detroit: Gale Research, 1976).

IRMGARD H. WOLFE

ANN HAVEN MORGAN
(1882–1966)
Zoologist, Ecologist

Birth	May 6, 1882
1912	Ph.D., Cornell University
1916–47	Chair, Dept. of Biology, Mount Holyoke College
1918	Taught at Marine Biological Laboratory, Woods Hole, MA
1920	Visiting Fellow, Harvard University
1921	Visiting Fellow, Yale University
1926	Went to Tropical Laboratory, Kartabo, British Guiana
1930	Published *Field Book of Ponds and Streams*
1947	Retired from Mount Holyoke College
1955	Published *Kinships of Animals and Man*
Death	June 5, 1966

Ann Haven Morgan (born Anna), the eldest of three children of Stanley Griswold and Julia Douglass Morgan, began her career as a noted zoologist and ecologist by exploring the woods and streams near her home in Waterford, Connecticut. From these early experiences, she became a famous teacher and popularizer of the cause of conservation.

Ann entered Wellesley College in 1902 but, finding the atmosphere somewhat restrictive, chose to continue her education at Cornell University.[1] She received her undergraduate degree in zoology in 1906. After her graduation she worked as an assistant and instructor at Mount Holyoke College from 1906 to 1909. Morgan then returned to Cornell to pursue her advanced degree. There she studied under James G. Needham of the Limnological Laboratory. Needham, she noted, was a great inspiration to her and "helped her to see things in the water."[2] He apparently thought highly of her work because even before she became his

Ann Haven Morgan. Photo courtesy of the Mount Holyoke
College Library Archives.

student, Needham proposed her name to the Entomological Society in
1908.[3]

Her area of interest was the aquatic biology of insects—in particular
the biology of mayflies, the subject of her dissertation. Her enthusiasm
for this topic earned her the nickname "Mayfly" Morgan from her students. She was awarded her Ph.D. in 1912 and served as assistant and
instructor at Cornell between 1909 and 1911. Around this time she
changed her name from Anna to Ann.[4] Morgan returned to Mount Holyoke in 1912 and served as instructor and, in 1914, as associate professor
of zoology. She became a full professor in 1918. By 1916 she had become
chair of the department, a position she held until her retirement in 1947.

In addition to her teaching duties at Mount Holyoke, Morgan spent
summers at the Marine Biological Laboratory at Woods Hole, Massachusetts, doing research and teaching courses on echinoderms. During the
summer of 1926, Morgan worked in British Guiana at the Tropical Lab-

oratory in Kartabo doing research on William Beebe's preserve. Most of her research, however, was done in the northeastern United States. She taught water biology courses and winter biology courses that included field trips to do research.

Morgan considered herself a "general zoologist," and this was reflected in her published works. Her book *Field Book of Ponds and Streams* (1930), for which Needham wrote the foreword, covered a variety of stream and pond life and introduced important techniques in the gathering and study, preservation, and mounting of specimens. It was to be her most popular work. She herself referred to it as an "angler's favorite." In the introduction, she wrote that she hoped the book would serve as a guide to the "vividness and variety" of the ways of water plants and animals and that she wanted the book to "help toward a wider enjoyment and further acquaintance in the field of water biology." The book emphasized the concept of the ecological niche of aquatic insects and their place in the ecosystem.

Morgan continued to expand her interests beyond the ecology and respiration of aquatic insects to include other freshwater biology and the habits and conditions of hibernating animals. Her publication *Field Book of Animals in Winter* (1939), a study of animals' efforts to survive in winter, reflected her growing interest not just in other animal life but also for the environment, ecology, and conservation—topics on which she would focus throughout her career. This was an interest most evident in her last book, *Kinship of Animals and Man* (1955), a study of the behavior patterns of animals and the competition and cooperation between different species. It was meant as a textbook for introductory zoology courses. Morgan herself drew the pictures and took the photographs for all her books, which are considered classics in the field. They helped to make the topics of ecology and conservation understandable to a wide audience. *Animals in Winter* was used by the *Encyclopedia Britannica* in 1949 to make the educational film "Animals in Winter."[5]

Morgan was also known for her efforts to reform education, in particular the science curriculum. She was inspired by a fellow zoology professor and trustee of the Woods Hole Marine Biological Laboratory, **Cornelia Clapp**, a strong advocate of educational reform and a mentor and friend to Morgan. Morgan wrote a tribute to Clapp in 1935 in the *Mount Holyoke Alumnae Quarterly*. Likewise, Morgan inspired and mentored other women in the field, such as Elizabeth Adams, another zoology professor at Mount Holyoke, to whom she dedicated *Field Book of Ponds and Streams*. She and Adams shared the same concerns about conservation and, after Morgan's retirement, traveled throughout the western United States as members of the National Commission on Policies in Conservation Education to study conservation projects there and in Canada. This trip inspired Morgan to become more active in conservation projects in the Connecticut River Valley area.

During the 1940s and 1950s, Morgan dedicated herself to further research as well as to the reformation of the science curriculum.[6] She hoped to see conservation and ecology become a part of the science curriculum in schools, and she taught summer workshops to teachers on these subjects. She also served on the National Committee on Policies in Conservation Education. Morgan apparently enjoyed teaching people about the outdoors. She herself, however, said that she was happiest wading in "some particularly oozy mudhole" looking for specimens. She encouraged her students to do the same.[7]

Through her writing and teaching, Morgan brought the need for conservation to the attention of many people. As she stated in the concluding chapter of *Kinships of Animals and Man,* "humanity is facing two very old problems, living with itself and living with its natural surroundings. Conservation is one way of working out these problems, an appreciation and intelligent care of living things and their environment. It is applied Ecology."[8] Morgan continued to support the cause of conservation and the need for humans to cooperate with and preserve the environment after her retirement in 1947.

Throughout her career, Morgan received recognition for her work. She won a star as one of the distinguished American scientists listed in *American Men of Science* (5th edition). She also received several grants and fellowships, including those from Cornell, Harvard, Yale, the American Association for the Advancement of Science, the Bache Fund of the National Academy of Sciences, the National Research Council, and the Rockefeller Foundation. She was a member of a number of national organizations.[9]

Ann Morgan died on June 5, 1966, of stomach cancer in South Hadley, Massachusetts. She is buried in Cedar Grove Cemetery, New London, Connecticut. She is remembered by her colleagues as a "true pioneer in the taxonomy and biology of mayflies" and one of the "outstanding teachers [who] attracted many students to the field of zoology."[10]

Notes

1. Marcia M. Bonta, *Women in the Field: America's Pioneering Women Naturalists* (College Station: Texas A&M University Press, 1991), p. 245.

2. Ann Haven Morgan, *Kinships of Animals and Man: A Textbook of Animal Biology* (New York: McGraw-Hill, 1955), p. vii.

3. Bonta, *Women in the Field*, p. 246.

4. *Ibid.*

5. Muriel Blaisdell, "Morgan, Ann Haven," in *Notable American Women: The Modern Period: A Biographical Dictionary,* eds. Barbara Sicherman et al. (Cambridge, Mass.: Belknap Press of Harvard University Press, 1980), p. 498.

6. Patricia Joan Siegel and Kay Thomas Finley, *Women in the Scientific Search: An American Bio-Bibliography, 1724–1979.* (Metuchen, N.J.: Scarecrow Press, 1985), p. 363.

7. Bonta, *Women in the Field,* p. 248.

8. Morgan, *Kinships,* p. 792.

9. Margaret W. Rossiter, *Women Scientists in America: Struggles and Strategies to 1940* (Baltimore: John Hopkins University Press, 1982), p. 174.

10. Charles P. Alexander, "Ann Haven Morgan, 1882–1966," *Eatonia* 8 (Feb. 15, 1967): 1.

Bibliography

Alexander, Charles P. "Ann Haven Morgan, 1882–1966." *Eatonia* 8 (Feb. 15, 1967): 1–3.

Blaisdell, Muriel. "Morgan, Ann Haven," in *Notable American Women: The Modern Period: A Biographical Dictionary,* eds. Barbara Sicherman et al., pp. 497–98. Cambridge, Mass.: Belknap Press of Harvard University Press, 1980.

"Deaths, Dr. Ann Haven Morgan, Prominent Conservationist, Dies at 84." *Holyoke* [Massachusetts] *Transcript Telegram* (June 6, 1966): 14.

Herzenberg, Caroline L. *Women Scientists from Antiquity to the Present.* West Cornwall, Conn.: Locust Hill Press, 1986.

Morgan, Ann Haven. "Cornelia Maria Clapp, March 17, 1894–December 31, 1934: An Adventure in Teaching." *Mount Holyoke Alumnae Quarterly* 19 (May 1935): 1–3.

———. *Field Book of Animals in Winter.* New York: G.P. Putnam Sons, 1939.

———. *Field Book of Ponds and Streams: An Introduction to the Life of Fresh Water.* New York: G.P. Putnam Sons, 1930.

———. *Kinships of Animals and Man: A Textbook of Animal Biology.* New York: McGraw-Hill, 1955.

"Prof. Ann Morgan, Taught Zoology at Mt. Holyoke," *New York Times* 115 (June 6, 1966): 41.

Rossiter, Margaret W. *Women Scientists in America: Struggles and Strategies to 1940.* Baltimore: John Hopkins University Press, 1982.

Siegel, Patricia Joan, and Kay Thomas Finley. *Women in the Scientific Search: An American Bio-Bibliography, 1724–1979.* Metuchen, N.J.: Scarecrow Press, 1985.

STEFANIE BUCK

BARBARA MOULTON

(1915–   )

Bacteriologist

Birth	1915
1937	A.B., University of Chicago
1940	M.A., George Washington University
1944	M.D., George Washington University
1947–48	Taught anatomy at George Washington University
1953	Instructor of Antibiotic Medicine, University of Illinois; Assistant Medical Director, Chicago Municipal Contagious Disease Hospital
1955–60	Director, Division of New Drugs, U.S. Food and Drug Administration
1960–61	Testified before the Kefauver congressional subcommittee investigating the drug industry
1962	Married E. Wayne Browne, Jr.
1961–79	Medical Officer, Division of Scientific Opinions, Bureau of Deceptive Practices, Federal Trade Commission
1967	Woman's Award for work in consumer protection

Barbara Moulton is perhaps best remembered for her key testimony about the relationship between government officials and drug company representatives in the approval of new drugs. Born in Chicago in 1915, Barbara was the daughter of Brookings Institute president Harold Moulton. She grew up with the first group of women who had legal rights to the vote. She was a student at both Smith College and the University of Vienna before receiving her bachelor's degree from the University of Chicago in 1937. She did postgraduate work there for the next two years in bacteriology and infectious diseases, developing a lifelong interest in these topics. Antibiotics were not widely used in clinical medicine before World War II, and she established a reputation among those who did pharmacologic research.

In 1940 she received her master's degree from George Washington University and in 1944, her M.D. Moulton was a surgical resident at St. Luke's in Chicago and at Suburban Hospital in Bethesda from 1945 to 1947. After finishing her medical training, Moulton taught anatomy at

George Washington from 1947 to 1948. She went into general practice with the Group Health Association for the next two years.

Moulton returned to her hometown for a period, serving as assistant director of the Student Health Service at the Illinois State Normal University. She then became assistant medical director of the Chicago Municipal Contagious Disease Hospital and instructor in antibiotic medicine at the University of Illinois in 1953. During this period she performed more antibiotics research, some of which was sponsored by drug companies.

Moulton returned to the Washington, D.C., area in 1954, where she went into general practice in Bethesda, Maryland. As a respected bacteriologist, she was invited to become the director of the Division of New Drugs of the U.S. Food and Drug Administration (FDA) in 1955. The FDA was then a branch of the Department of Health, Education, and Welfare.

Moulton was surprised by the pressure from drug company representatives, who expected her to act as a rubber stamp in approving their drugs without question. There was even more pressure from her colleagues and superiors in the FDA to accept the coziness between the regulators and the industry they oversaw. She began to have doubts about the process of drug approval. She said:

> A medical officer in the New Drug Branch has the power to release any new drug on his own initiative, without review by any of his colleagues. To refuse to release a new drug, however, he must have the unanimous support of the chief of the new drug branch, the director of the bureau of medicine, the Commissioner [of the FDA], and usually also the director of the bureau of enforcement and the general counsel.[1]

In 1960, Dr. Moulton resigned her post at the FDA. Fresh from her experience with pharmaceutical companies, she began preparing testimony for the Kefauver subcommittee, the U.S. Senate subcommittee on antitrust and monopoly investigating the drug industry. This subcommittee was chiefly interested in the case of Dr. Henry Welch, chief of the FDA antibiotics division. Welch was under fire for accepting a quarter of a million dollars over a seven-year period from several drugmakers, and specifically under investigation for his work marketing the drug Sigmamycin for Pfizer.

Moulton testified that the system was ripe for FDA medical officers to peddle their influence to drugmakers. She spoke of the shock felt by attendees at pharmaceutical conferences at the obvious commercialism of the FDA. Perhaps worst of all was her superiors' response to her concerns about Welch and the drug Sigmamycin: "The only answer I received was that Dr. Welch had a great reputation and added luster to the name of the Food and Drug Administration, and that nothing could or should be done against him."[2]

Moulton's philosophy on drug approval was far-sighted. She claimed quite simply that

> no physician, no one who has ever been responsible for the welfare of individual patients, will accept the idea that safety can be judged in the absence of a decision about efficacy. No drug is "safe" if it fails to cure a serious disease for which it is available. No drug is too dangerous to use if it will cure a fatal disease for which no other cure is available.[3]

Moulton also found the practice of drug company representatives spending many days a week looking over the shoulders of government chemists and pharmacologists as they performed their work "an almost insurmountable handicap." She thought it was "important that we . . . make some attempt to dispel the myth that we are protecting the consumer completely."[4]

The subcommittee did do something against the tarnished relationship between regulators and the pharmaceutical industry by proposing the bill that became the Kefauver-Harris amendments to the Food, Drug, and Cosmetic Act. These revisions led to better protection of consumers. Moulton's brave act did not ruin her career as a government official. The Federal Trade Commission appointed her as medical officer to the division of scientific opinions at the Bureau of Deceptive Practices in 1961.

On March 30, 1962, Moulton married E. Wayne Browne, Jr. She spent the rest of her career protecting the health and interests of the American people as a consumer advocate, as known and respected in this role as she had been in her role as a bacteriologist and physician. In 1967 she won the Woman's Award for her work to protect consumers from unsafe drugs and to prevent deceptive practices.

Moulton is a member of the American Association for the Advancement of Science, the American Public Health Association, the American Medical Women's Association, the American Society of Hematology, and the American Society of Microbiology. Her interest in raising cattle is also keen; she is a member of the West Virginia Farm Bureau, the Holstein-Friesian Association of America, and the West Virginia Holstein-Friesian Association. Moulton has also protected American land by serving on the board of directors for the Potomac Basin Federation and the West Virginia Eastern Panhandle Land and River Association.

Notes

1. John Lear, "Drugmakers and the Govt.—Who Makes the Decisions?" *Saturday Review* (July 2, 1960): 42.
2. *Ibid.*, p. 40.
3. *Ibid.*
4. *Ibid.*, pp. 41–42.

Bibliography

Bateman, Jeanne C., Barbara Moulton, and Nancy J. Larsen. "Control of Neoplastic Effusion by Phosphoramide Chemotherapy." *AMA Archives of Internal Medicine* 95 (May 1955): 713–19.

Lear, John. "Drugmakers and the Govt.—Who Makes the Decisions?" *Saturday Review* (July 2, 1960): 37–42.

Levin, Beatrice. *Women in Medicine.* Lincoln, Neb.: Media Publishing, 1988.

Moulton, Barbara. "Antibiotics in the Treatment of Viral Diseases." *Antibiotics Annual* (1955–1956): 719–26.

Who's Who of American Women 1974–1975, 8th ed. Chicago: Marquis Who's Who, 1975.

KELLY HENSLEY

ELIZABETH FONDAL NEUFELD
(1928–)
Molecular Biologist

Birth	September 27, 1928
1948	B.S., Queens College
1956	Ph.D., comparative biochemistry, University of California, Berkeley
1963–73	Research Biochemist, National Institute of Arthritis, Metabolism, and Digestive Diseases, National Institutes of Health (NIH)
1973–79	Chief, NIH Section on Human Biochemical Genetics
1977	Elected to the National Academy of Sciences and the American Academy of Arts and Sciences
1979–84	Chief, National Institute of Arthritis, Diabetes, and Digestive and Kidney Diseases (NIADDK), Genetics and Biochemistry Branch

1981–83	Deputy Director, NIADDK Division of Intramural Research
1982	Albert Lasker Clinical Medicine Research Award
1984–	Chair, Dept. of Biological Chemistry, School of Medicine, University of California, Los Angeles
1988	Wolf Prize in Medicine; Fellow, American Association for Advancement in Science
1992–93	President, American Society for Biochemistry and Molecular Biology
1994	National Medal of Science

Dr. Elizabeth Fondal Neufeld is a leading international authority on human genetic diseases. She is most renowned for her work involving Hurler and Sanfilippo syndromes, which are inherited disorders of the connective tissues. Hurler and Sanfilippo are in the group called mucopolysaccharidoses, or lysosomal storage diseases, which cause fatal neurological deterioration in patients whose cells lack certain enzymes needed to process complex sugars. The accumulation of sugars causes the cells to grow and put internal pressure on nerve tissues, which can die from too much pressure. Patients suffer from severe mental and motor deterioration, have vision and hearing problems, and die prematurely, usually before puberty. The better-known Tay-Sachs and I-cell diseases are related to the mucopolysaccharidoses. Dr. Neufeld's research has led to successful prenatal diagnosis and has contributed to the availability of genetic counseling for parents. There is also hope for future treatments such as gene replacement therapy and bone marrow transplants as a result of the work by Neufeld and others.

Neufeld was born in Paris, France, to Russian emigrant parents. At the beginning of World War II the family emigrated again, this time to New York. Neufeld was strongly encouraged by her parents to obtain an education, and her high school interest in biology led her to pursue an advanced degree in biochemistry. After completing her doctoral studies at the University of California, Berkeley, she began her career as a plant biologist. She studied cell division in sea urchins and later examined the biosynthesis of plant cell wall polymers; her work in these areas was critical to her later investigations into mucopolysaccharidoses (MPS).

In 1963, Dr. Neufeld began an impressive, 20-year career at the National Institutes of Health (NIH), where she achieved several high administrative positions. It was as a research biochemist at NIH's National Institute of Arthritis, Metabolism, and Digestive Diseases that Neufeld became interested in studying MPS disorders. As she recalls, "it was a propitious time to enter the field."[1] Investigations at the University of

Louvain, France, had recently proved the importance of lysosomal enzymes to cell metabolism. In the late 1960s and early 1970s, scientists at NIH were conducting hundreds of clinical trials using enzymes in the production of drugs and in the prevention of disease.[2] It was during this time of great discovery that someone described to her a patient with a genetic disorder that caught her interest.

Neufeld's research is important because, along with the discoveries of other scientists, it helped to provide the key that unlocked the door to treatment of MPS patients. Researchers were operating under the premise that cells were manufacturing more complex sugars than the patients could metabolize. Neufeld thought she could figure out what was wrong with the metabolism on the basis of her research with plant cell walls. "As it turned out, I was totally wrong," she says, "but by the time I had proved that, I was totally hooked."[3] Neufeld and Joseph Fratantoni were mixing normal cells with cells containing various forms of MPS disorders including Hunter syndrome. When Fratantoni mistakenly mixed cells from a Hunter patient with cells from a Hurler patient, they made an exciting discovery. The two cultures had "cross cured" each other, resulting in nearly normal cells. She and Fratantoni isolated the real problem: a defective gene that was causing the sugars to break down at an abnormally slow rate. Through further study, the corrective factors were identified as a series of enzymes that were lacking in Hunter-Hurler patients. With this knowledge, scientists were able to develop tests for successful prenatal diagnosis of MPS and related disorders.

Dr. Neufeld has gained international recognition for her life's work. She has received numerous awards and prizes for her research; chief among them is the prestigious Lasker Award, which she won jointly with Roscoe O. Brady. In 1984 she became the first woman department head at the School of Medicine at the University of California, Los Angeles. In 1994 she was one of eight researchers to be awarded the National Medal of Science, the highest award granted by the U.S. government for scientific achievement. She was presented the award by President Bill Clinton in October 1994 at a White House ceremony. Neufeld's important discoveries in biochemistry, her professional involvement in the scientific community, and her teaching have made her a role model for young scientists.

Notes

1. Elizabeth F. Neufeld, "Lessons from Genetic Disorders of Lysosomes," *The Harvey Lectures* 75 (1981): 41–60.

2. Joan Arehart-Treichel, "Enzymes: Medicine's New Gold Mine," *Science News* 114 (July 22, 1978): 58.

3. "Biologist at UCLA Wins Medal," *Los Angeles Times* (Sept. 9, 1994): B3.

Bibliography

Arehart-Treichel, Joan. "Enzymes: Medicine's New Gold Mine." *Science News* 114
    (July 22, 1978): 58.
"Biologist at UCLA Wins Medal." *Los Angeles Times* (Sept. 9, 1994): Sec. B.
Neufeld, Elizabeth F. "Lessons from Genetic Disorders of Lysosomes." *Harvey
    Lecture Series*, no. 75 (1981): 41–60.
————. "The Mucopolysaccharide Storage Diseases," in *The Metabolic Basis of
    Inherited Disease*, 6th ed. New York: McGraw-Hill, 1989.
                                    LESLIE O'BRIEN AND PATRICIA MURPHY

MARGARET MORSE NICE

(1883–1974)

Ornithologist

Birth	December 6, 1883
1906	A.B., French, Mount Holyoke College
1909	Married Leonard Blaine Nice
1913	Moved with her husband and children to Norman, OK; studied children's language development
1915	M.A., psychology, Clark University
1924	Co-authored (with husband) a book about state bird populations, *The Birds of Oklahoma*
1927	Moved to Columbus, OH; began to study territoriality in song sparrows
1937	First major song sparrow publication in English; elected fellow of American Ornithologists' Union; moved to Chicago
1938	First woman to be President of Wilson Ornithological Club
1942	Brewster Medal, American Ornithologists' Union
1955	Honorary D.Sc., Mount Holyoke College
Death	June 26, 1974

Margaret Morse Nice was one of the world's outstanding ornithologists.
Her best-known work focused on the behavior, or ethology, of a common

Margaret Morse Nice. Photo courtesy of the Mount Holyoke College Library Archives.

bird, the song sparrow. In fact, Nobel Prize–winning ethologist Konrad Lorenz once credited her with founding the science of ethology. A dedicated researcher and prolific writer, Margaret Nice succeeded despite an interrupted graduate education, lack of institutional support, and the demands of rearing five children.

A career as a "perfect housekeeper and homemaker" was expected for Margaret Morse at her birth in 1883 in Amherst, Massachusetts.[1] Her father, a history professor at Amherst College, encouraged his seven children to enjoy learning and nature. Professor Morse gave each child a garden plot, and Margaret's mother led the family on Sunday nature walks. Margaret began keeping a nature journal, focusing on birds, when she was 9 years old. She loved exploring the Massachusetts countryside as a child but grew increasingly depressed as a teenager. Her parents discouraged their daughters from pursuing professions. Margaret later wrote, "We three girls all wished we had been boys, since boys had far more freedom than girls did to explore the world and to choose exciting careers."[2]

The Morse family considered a college education valuable for home-makers, and Margaret enrolled at Mount Holyoke College in 1901. There she discovered a gift for languages, studying French, German, and Italian. Biology courses satisfied her less. Contemporary zoology emphasized animal classification and anatomy, and laboratory classwork emphasized dissection. Margaret later complained, "I could see very little connection between the courses in college and the wild things I loved."[3] She watched birds only on holidays, taking country walks, trail rides, and canoe trips. When her worried parents urged her to be more cautious, the headstrong Margaret bought a revolver for self-defense instead. She dutifully returned home after finishing college in 1906, but a lecture in Amherst by Dr. Clifton Hodge of Clark University gave her new direction. Hodge studied the economic impacts of animals on humans, particularly farmers. Margaret saw a chance to study live birds in graduate school and convinced her parents to let her enroll at Clark University in 1907. For a master's degree in biology, she began studying the food habits of bobwhite quail. Investigating a research question thrilled Margaret, and she planned to earn a Ph.D. at Clark. But in 1909 she married a doctoral student, Leonard Blaine Nice. To her parents' delight, she decided to abandon hopes for a doctorate and keep house while her husband finished his degree.

As the wife of a college professor, Margaret found herself moving with her husband to each new academic post. After two years at Harvard, Dr. Nice became a professor of physiology and pharmacology at the University of Oklahoma. Margaret set up an efficient household, minimizing housework to keep time free for "activities of more lasting value."[4] Research remained a favorite activity, and she published results from her bobwhite work. New research interests focused on her young daughters' language development. She published an average of one paper a year for 13 years. Clark awarded her a master's degree in psychology for the studies, but the articles attracted little attention. By 1918 she had four young children and was deeply frustrated by her limited opportunities for research. She later admitted, "I resented the implication that my husband and the children had brains, and I had none. He taught; they studied; I did housework."[5]

During the next summer a newspaper article revived her interest in bird research. The *Daily Oklahoman* reported that hunters wanted to open the mourning dove season in late summer when, they asserted, nesting was completed. Nice suspected that their facts were wrong and initiated a study of doves' nesting behavior. Her findings proved that doves nest into October and helped keep the season closed.

Fascination with doves soon generalized into a desire to learn about all of Oklahoma's birds. The Nice family piled into the car to search for birds, guided by **Florence Bailey's** *Handbook of Birds of the Western United*

States. Nice's articles on doves, cowbirds, nuthatches, and other local birds began appearing in ornithological journals. Family camping trips became research expeditions for gathering bird data from Oklahoma's diverse habitats. In 1924, Blaine and Margaret Nice published the region's first complete bird survey, *The Birds of Oklahoma.* To Nice, the work signaled a return to her childhood dream "of studying nature and trying to protect the wild things of the earth."[6]

Dr. Nice staunchly supported his wife's research by assisting in fieldwork, helping with childcare, and paying study expenses. However, his next academic appointment in Columbus, Ohio, tore Nice away from her Oklahoma studies. Her distress at the move deepened when one of their five daughters died during that winter. She found comfort in walks through a 60-acre weed patch between her house and the Olentangy River. What others viewed as "a tangled waste," Nice saw as an outdoor laboratory for bird research. In the spring of 1928 she witnessed a battle between two male song sparrows near her house. Ornithologists at the time theorized that male birds defend territories around their nests against males of the same species, but little evidence yet supported the theory. The song sparrow battle so intrigued Nice that she vowed to decipher the bird's territorial and reproductive behavior. Little was known about the abundant birds. "It was an unknown world," Nice said, "and each day I made fresh discoveries."[7]

Many discoveries were made possible by her pioneering techniques of bird banding. Other researchers marked birds with numbered aluminum leg bands to study the migration patterns of populations. But Nice made colored celluloid bands that enabled her to identify individual birds in the field. She could then study nest-building, mating, chick-raising, and other behaviors of individuals, season after season. She regularly rose before dawn and sat for hours taking careful notes at a sparrow nest. How large is a territory? Do both parents incubate eggs? Why do males sing? The more questions she answered, the more she wanted to learn.

Already known for her Oklahoma research, Nice was applauded by ornithologists for her work on song sparrows. In 1931 she became the fifth woman elected to membership in the American Ornithologists' Union (AOU). She felt isolated, however, from Columbus's bird scientists because the local Wheaton Ornithological Club excluded women. To strengthen ties with other scientists, she reviewed research papers for the journal *Bird-Banding.* Her language skills allowed her to review European work, bringing the latest in behavior research to American readers. The incisive reviews solidified her reputation as a knowledgeable, tough-minded scientist.

The first major publication of Nice's sparrow work appeared to international acclaim in a German journal, *Journal für Ornithologie.* Several major publications in the United States followed, including a popular book about her work, *The Watcher at the Nest*. Nice hoped that the book would reveal the joys of research, stimulating the public to study nature.

Much of Nice's research site was cleared for public gardens during the Great Depression, and in 1936 her husband took a new job in Chicago, Illinois. Separated from the wild sparrows, Nice shifted her attention to studying captive birds, writing reviews and research reports, and supporting ornithological societies. In 1937 the AOU elected her as a fellow, and in the next year she became the first woman president of the Wilson Ornithological Club.

Respect for Nice's achievements continued to grow in Europe as well as the United States. She worked with Konrad Lorenz and other scientists on trips to Europe in the 1930s and kept in contact during World War II. Appalled by postwar hardships suffered in many countries, she led relief efforts by organizing American ornithologists to supply food, clothing, and research literature to European colleagues and their families.

Despite social concerns and advancing age, Nice's extraordinary research and writing productivity continued. Her daughter Constance assisted on a major study of duck behavior at the Delta Waterfowl Research Station in Manitoba, Canada. Environmental problems also consumed more of her writing energies. She campaigned against pesticide misuse, dam construction in Dinosaur National Monument, and development of the Indiana Dunes outside of Chicago. She battled particularly hard for protection of Oklahoma's Wichita National Wildlife Refuge, a favorite site of her early bird studies.

The many tributes to Nice's long, exemplary career included the AOU's Brewster Medal, an honorary doctorate from Mount Holyoke, and a Toronto women's ornithological society named in her honor. To many, her status was enhanced because she worked without grants or other formal support while managing a full family life. But when praised as an "amateur housewife," Nice retorted, "I am *not* a housewife; I am a *trained zoologist*."[8] Her struggles are described in her autobiography, *Research Is a Passion with Me*. Published after her death at age 91, the book best expresses Margaret Morse Nice's lifelong dedication to science. "The study of nature is a limitless field," she wrote, "the most fascinating adventure in the world."[9]

Notes

1. Margaret Morse Nice, *Research Is a Passion with Me* (Toronto: Consolidated Amethyst Communications, 1979), p. 15.

2. *Ibid.*

3. *Ibid.*, p. 21.

4. *Ibid.*, p. 35.

5. *Ibid.*, p. 41.

6. *Ibid.*, p. 43.

7. Margaret Morse Nice, *The Watcher at the Nest* (New York: Macmillan, 1939), p. 7.

8. Milton B. Trautman, "In Memoriam: Margaret Morse Nice," *Auk* 94 (July 1977): 440.

9. Nice, *The Watcher at the Nest*, p. 264.

Bibliography

Ainley, Marianne G. "Field Work and Family: North American Women Ornithologists, 1900–1950," in *Uneasy Careers and Intimate Lives*, eds. Pnina Abir-Am and Dorinda Outram, pp. 60–76. New Brunswick, N.J.: Rutgers University Press, 1987.

Bonta, Marcia Myers. "Song Sparrow Lady." *Birder's World* 7 (Aug. 1993): 24–28.

———. *Women in the Field: America's Pioneering Women Naturalists.* College Station: Texas A&M University Press, 1991.

Gibbon, Felton, and Deborah Strom. *Neighbors to the Birds: A History of Birdwatching in America.* New York: Norton, 1988.

Kastner, Joseph. *A World of Watchers.* New York: Alfred Knopf, 1986.

Nice, Margaret Morse. *Research Is a Passion with Me.* Toronto, Can.: Consolidated Amethyst Communications, 1979.

———. *The Watcher at the Nest.* New York: Macmillan, 1939.

Trautman, Milton B. "In Memoriam: Margaret Morse Nice." *Auk* 94 (July 1977): 430–441.

JULIE DUNLAP

CHRISTIANE NÜSSLEIN-VOLHARD

(1942–)

Developmental Biologist, Geneticist

Birth	October 20, 1942
1964	[B.A.], biology, physics, and chemistry, Johann-Wolfgang-Goethe-Universität, Frankfurt
1968	Diplom, biochemistry, University of Tübingen
1972–74	Research Associate, Max-Planck Institute for Viral Research, Tübingen
1973	Ph.D., biology/genetics, University of Tübingen
1975–76	EMBO Fellowship with Dr. Walter Gehring, Biozentrum Basel, Switzerland

1977	Postdoctoral DFG Fellowship with Dr. Klaus Sander, University of Freiburg
1978–80	Head Group, European Molecular Biology Lab
1981–85	Group Leader, Friedrich-Miescher-Laboratorium, Max-Planck-Gesellschaft
1986–90	Director, Max-Planck Institute for Developmental Biology
1990	Honorary D.Sc., Yale University
1990–	Director, Department of Genetics, Max-Planck-Gesellschaft
1991	Albert Lasker Public Service Award; honorary doctorates from Princeton University and Utrecht University
1992	Gregor Mendel Medal, Genetical Society, Great Britain; Louis Jeantet Prize for Medicine; Alfred P. Sloan, Jr., Prize; General Motors Cancer Research Prizewinner
1993	Honorary doctorates from Harvard University and the University of Freiburg
1995	Nobel Prize in medicine

Germany's best-known woman scientist was born in Magdeburg in 1942 to Rolf and Brigitte Haas Volhard, the second of their five children. Her father was an architect; her mother was the daughter of a painter from a family of artists and musicians. Christiane was accepted into a pre-eminent girls' school where she enthusiastically embraced many subjects: literature, especially Goethe; history; music (she plays flute and sings); art; even cooking and sewing. She writes: "Of my brothers and sisters, I was the only one with a lasting inclination for the sciences. My parents, while unable to offer practical help, were benevolent and supportive of my interest."[1]

By age 12 Christiane wanted to be a researcher in the natural sciences. In 1962 she gave the graduation speech at her school on communication among animals, "a topic I still find interesting today, as I do evolution— I composed my own theory of evolution at 18, which I have since forgotten."[2] In that same year she began to study biology in Frankfurt, but it didn't last long. "It would be breath-taking—or so I believed—but after I began, I found it dull, flat. Only botany was new and stimulating. I quickly changed majors to physics."[3] In 1963 she happened to learn that a new major in biochemistry was being established at the University of Tübingen. Christiane entered the new program during the winter semester of 1964. "That meant for me the beginning of a hard-to-endure separation from my family and friends, but biochemistry appeared to be the only path for me."[4] Yet even this new topic soon lost its charm because of the emphasis placed on organic chemistry over biochemistry. Christiane attended lectures on molecular biology for comfort and came to know the faculty in biology at both the university and the nearby Max-Planck Institutes. These associations helped her escape from the biochemistry program and do her Diplom work at Max-Planck, the first woman student to do so. She worked there with Heinz Schuller in virology and received a solid education in molecular biology.

Technology at the time was not advanced enough to answer many of the questions biologists had about genetics. For her degree, Christiane chose to work with the Drosophila fly instead of the polyp Hydra, the model system at that time for the study of embryology in complex forms. Using such a simple model to explain such complex operations appealed to her, but she spoke with friends and colleagues before committing to a research "relationship" with the fly that has lasted throughout her career. "Drosophila is so ideal for genetic research ... [because of] the small number of chromosomes, and the existence of giant chromosomes of the salivary glands provided a unique physical measure for the numbers of genes and the analysis of chromosomal aberrations."[5]

Walter Gehring, just back from the United States, was establishing a laboratory in Basel, Switzerland. Christiane gathered her courage and asked him if she could continue her developmental biology studies with Drosophila in his lab as a postdoctoral candidate. He accepted and she applied for an EMBO (European Molecular Biology Organisation) fellowship. Her successful application detailed a plan for studying the embryology and molecular foundation of morphogenesis of Drosophila flies, an area that has become her life's work. While there, she first worked with Eric Wieschaus and began to think of developing techniques that would allow the direct study of the fly's eggs.

After her fellowship expired, Christiane found that there were few opportunities to continue her work. She accepted a DFG (Deutsche Forschungsgemeinschaft) fellowship for one year at the University of Freiburg in the lab of Dr. Klaus Sander. At the end of her time there, she teamed up with Eric Wieschaus again at the European Molecular Biology Laboratory, where she led a work group for three years. Here Nüsslein-Volhard began to implement techniques that uncovered development of Drosophila embryos at the earliest stages. The development of mutant Drosophila and methods to study its embryonic pattern formation revolutionized developmental genetics. "The gains we made were—at first only for us—overwhelmingly interesting, not only as confirmation of the beginnings of the famous Drosophila fly, but also functioning as confirmation to the Director General, who had a tendency to say 'This never goes anywhere!' and 'I only have one life!' "[6]

After the successful new techniques and discoveries were established, Christiane moved back to the Max-Planck Institute, where she has been ever since, first as group leader of a laboratory, then as director of developmental biology, and since 1990 as director of the Department of

Genetics. Dr. Nüsslein-Volhard's research groups have discovered much of what is known about how fertilization takes place and how organisms begin existence, including identification of 120 "pattern genes." Her techniques are used in laboratories worldwide. Her research has contributed new knowledge about segmentation, genes that direct axis determination in the embryo, and the mechanism of concentration-dependent transcriptional activation in the case of Biocoid. Most of this knowledge came from her work with Drosophila, but she has begun to work with zebrafish to determine the genetics that control pattern formation in a vertebrate. "The great property of this organism is its embryonic development rather than its genetics. In a mating, hundreds of eggs are produced. . . . After 24 hours the major events have already taken place. . . . Because the embryos are transparent, elegant lineage tracing and transplantation experiments can be performed."[7]

Nüsslein-Volhard believes Germany needs to give "much more encouragement to women who really dare to be a scientist."[8] Many of Germany's most prominent women scientists, including Nüsslein-Volhard, do not have children, partly because the "Kinder, Kueche, Kirche" (children, kitchen, church) adage still holds sway in German culture and partly because there are very few childcare centers. In the spring of 1992 a daycare facility was opened at the Max-Planck Institute for Developmental Biology, backed by Nüsslein-Volhard and partially funded by a scientific prize she had won. She still worries about German women being pushed from the lab back into the kitchen and advocates a mentoring program for women scientists.

In 1991, Dr. Nüsslein-Volhard received the Albert Lasker Award, second only in prestige to the Nobel Prize in the sciences. She also has been awarded the Louis Jeantet Prize in 1992; the 1993 Albert P. Sloan, Jr., Prize; the Franz-Vogt Prize; the Liebnizpreiz of the Deutschen Forschungsgemeinschaft; and the Rosenstiel Medal from Brandeis University. Then in 1995, she won the Nobel Prize in medicine, shared with Edward Lewis of Cal Tech and Eric Wieschaus of Princeton University.

Notes

1. Christiane Nüsslein-Volhard to Kelly Hensley, October 1994.
2. *Ibid.*
3. *Ibid.*
4. *Ibid.*
5. C. Nüsslein-Volhard, "Of Flies and Fishes," *Science* 266 (Oct. 28, 1994): 572.
6. Christiane Nüsslein-Volhard to Kelly Hensley, October 1994.
7. Nüsslein-Volhard, "Of Flies and Fishes," p. 574.
8. Peter Aldhous, "Women in Science 1994," *Science* 263 (Mar. 11, 1994): 1476.

Bibliography

Aldhous, Peter. "Women in Science 1994." *Science* 263 (Mar. 11, 1994): 1475–80.

"Genetic Pioneers Win Lasker Awards in Research." *New York Times* (Sept. 28, 1991): 9, 26.

"In an Unlikely Romance, Biologists Take the Zebrafish into Their Labs." *New York Times* (Nov. 5, 1991): B5.

International Who's Who 1993–94. London: Europa Publ., 1993.

Nüsslein-Volhard, C. "Determination of Anteroposterior Polarity in Drosophila." *Science* 238 (Dec. 18, 1987): 1675–81.

———. "General Motors Cancer Research Prizewinner Laureates Lecture. Alfred P. Sloan, Jr., Prize. The Formation of the Embryonic Axes in Drosophila." *Cancer* 71 (May 15, 1993): 3189–93.

———. "The 1991 Albert Lasker Public Service Award. From Egg to Organism. Studies on Embryonic Pattern Formation." *Journal of the American Medical Association* 266 (Oct. 2, 1991): 1848–49.

———. "Of Flies and Fishes." *Science* 266 (Oct. 28, 1994): 572–74.

KELLY HENSLEY

RUTH PATRICK

(1907–   )

Limnologist

Birth	November 26, 1907
1929	B.S., Coker College
1931	M.S., University of Virginia; married Charles Hodge IV
1934	Ph.D., University of Virginia
1939–	Academy of Natural Sciences of Philadelphia: Assistant Curator, Microscopy Dept. (1939–47); Chair, Dept. of Limnology (1947–73); Francis Boyer Research Chair (1973–)
1950–	University of Pennsylvania: Lecturer in Botany (1950–70); Professor of Biology (1970–)
1970	Elected to National Academy of Sciences
1975	John and Alice Tyler Ecology Award

Ruth Patrick's work in limnology, the scientific study of freshwater ecosystems, has gained her international renown. Her important discoveries have helped to examine and clean up the environment, and she continues to promote a united stance for a better life for all living things.

Ruth Patrick was born to Frank and Myrtle (Jetmore) Patrick in 1907 in Topeka, Kansas. Patrick expressed her fascination for the ecology of aquatic systems at an early age while accompanying her father and sister on weekend field expeditions. Frank Patrick, an attorney who was also an avid naturalist, gave his daughter a microscope at age 7 and continued to encourage her interest in science throughout her childhood and adolescence.

Ruth attended Coker College in South Carolina, receiving her undergraduate degree in 1929. She continued her education at the University of Virginia, earning an M.S. in 1931 and a Ph.D. in 1934. She spent her summers at the Woods Hole Biological Laboratory, the Cold Spring Harbor Biological Laboratory, and the Biological Laboratory of the University of Virginia at Mountain Lake, where she developed her research focus on the ecology of aquatic organisms. It was at Cold Spring Harbor that Patrick met her husband, Charles Hodge IV. She kept her maiden surname in honor of her father. Patrick and Hodge had one child, Charles Hodge V.

Patrick has spent most of her professional life associated with the University of Pennsylvania and the Academy of Natural Sciences in Philadelphia. She started her career as an assistant curator of the microscopy department of the Academy of Natural Sciences and went on to found the limnology department in 1947. She was appointed Francis Boyer Research Chair in 1973, and in 1976 she became the first woman to chair the Academy's board of directors. Concurrently she has taught botany at the University of Pennsylvania, where she was promoted to full professor of biology in 1970.

The diatom, a microscopic, single-celled alga, has been the focus of much of Patrick's research. This organism is an important link in the food chain of freshwater ecosystems. Patrick invented a device called the diatometer, which plots the size and growth of these microscopic organisms. Patrick's fieldwork has taken her to hundreds of freshwater rivers, streams, and lakes in the Americas. In documenting her work on diatoms, Patrick has published over 100 scholarly papers and collaborated with Charles Reimer on the two-volume *Diatoms of the United States.*

Continued interest in the diatom led Patrick to expand her research to the broader field of limnology. While studying the effects of pollutants in freshwater rivers and streams, she was the first to discover the diatom's usefulness in pollution control. Using her diatometer, Patrick discovered the relationship between the number of diatoms present in a system and the health of that system. By measuring the number and size of diatoms in an aquatic ecosystem, she was able to identify the type and extent of pollutants in a given body of freshwater. Her research has included limnological expeditions to Mexico, Peru, and Brazil.

Patrick achieved another first when her field studies in the Conestoga River Basin established the importance of pattern in the diversity of organisms in healthy aquatic communities. This led future researchers determining the health of an ecosystem to study diversity of species rather than one organism.

Patrick believes that scientists should work in concert with their counterparts in government and industry to find solutions to the world's problems. She is quoted as having said in 1975 that

> my great theme in life is that academia, government, and industry have got to work closely on all the big problems of the world. Unless academics and industry get together there will not be many bright young people trained in the future. We have to develop an atmosphere where the industrialist trusts the scientist and the scientist trusts the industrialist. You've got to trust people.[1]

Putting this philosophy into practice, Patrick has worked with many federal advisory groups to shape national environmental policy. These include the National Academy of Science Committee on Science and Public Policy; the General Advisory Committee of the Environmental Protection Agency; the Advisory Council of the Renewable Resources Foundation; and the Smithsonian Institution Council. She has served on the board of directors of Pennsylvania Power and Light Company and DuPont. In addition, she has advised numerous state-level groups on water policy issues.

Patrick was elected to the National Academy of Sciences in 1970. In 1975 she was awarded the John and Alice Tyler Ecology Award. This prize was established in 1973, honoring a single recipient for work accomplished over the previous decade. At that time, the Tyler award was the most lucrative of scientific prizes, carrying a $150,000 cash award. Patrick has been recognized with many other honors, including the Eminent Ecologist Award of the Ecological Society of America; the Certificate of Merit of the Botanical Society of America; and numerous honorary degrees from colleges and universities throughout the United States.

Note

1. Constance Holden, "Ruth Patrick: Hard Work Brings Its Own (and Tyler) Reward," *Science* 188 (June 5, 1975): 998.

Bibliography

"AIBS Announces Distinguished Service Awards." *BioScience* 30, no. 6 (1980): 432.

American Men and Women of Science. New York: Bowker, 1906– .

Emberlin, Diane. *Contributions of Women: Science.* Minneapolis: Dillon Press, 1977.

Holden, Constance. "Ruth Patrick: Hard Work Brings Its Own (and Tyler) Reward." *Science* 188, no. 4192 (1975): 997–99.

O'Neill, Lois Decker, ed. *The Women's Book of World Records and Achievements.* Garden City, N.Y.: Anchor Press/Doubleday, 1979.

Patrick, Ruth. "A Proposed Biological Measure of Stream Conditions, Based on a Survey of the Conestoga Basin, Lancaster County, Pennsylvania." *Proceedings of the Academy of Natural Sciences of Philadelphia* 10 (1949).

———. *Rivers of the United States.* New York: J. Wiley, 1994– .

Patrick, Ruth, and Charles W. Reimer. *The Diatoms of the United States, Exclusive of Alaska and Hawaii.* Philadelphia: Academy of Natural Sciences, 1966.

Patrick, Ruth, Emily Ford, and John Quarles. *Groundwater Contamination in the United States.* Philadelphia: University of Pennsylvania Press, 1987.

Patrick, Ruth, et al. *Surface Water Quality: Have the Laws Been Successful?* Princeton: Princeton University Press, 1992.

Vare, Ethlie Ann, and Greg Ptacek. *Mothers of Invention: From the Bra to the Bomb: Forgotten Women and Their Unforgettable Ideas.* New York: William Morrow and Co., 1988.

Who's Who in America. Chicago: A.N. Marquis, 1899– .

ANN LINDELL

DEBORAH L. PENRY

(1957–)

Biological Oceanographer

Birth	February 28, 1957
1979	B.A., University of Delaware
1979–82	Research Assistant, Dept. of Invertebrate Ecology, Virginia Institute of Marine Science
1982	M.A., William and Mary College

1982–83	Lab Chemist, Core Labs, Inc., Lake Charles; Research Associate, U.S. Department of Energy Program: Brine Disposal Monitoring, McNeese State University
1988	Ph.D., oceanography, University of Washington
1988–90	Postdoctoral Researcher, School of Oceanography, University of Washington
1990–92	Postdoctoral Researcher, Horn Point Laboratory, University of Maryland
1991–	Assistant Professor, Dept. of Integrative Biology, University of California, Berkeley
1993	Alan T. Waterman Award; Young Investigator Award, National Science Foundation

In 1993, Dr. Deborah Penry, a biological oceanographer, became the eighteenth recipient of the National Science Foundation's Alan T. Waterman Award for outstanding research by a scientist under age 35. Penry received the award for her research on how marine animals feed and digest food. Specifically, she has applied principles of chemical reactor theory used in engineering to develop models of digestion. Penry's theory provides a framework from which the feeding and digestive processes in other animals can also be described.

Penry's interest in animal digestion can be traced to her fishing days as a child growing up near the Chesapeake Bay in Maryland.

> Water-oriented activities were a part of growing up in Maryland. . . . I enjoyed going out on the boat with my father and took a great interest in everything I saw. When it came to fishing, my greatest joy was not in actually catching the fish, but in later cleaning them and examining the contents of their stomachs. I was fascinated by seeing what these fish had been eating.[1]

Not surprisingly, oceanography became a strong interest early in her life.

Like many young people entering college, Penry did not begin her academic career with a completely formed plan of action. Encouraged by her parents to "be anything she wanted to be" and considering her interest in oceanography, she chose to pursue an undergraduate biology major. She soon found that she enjoyed not just biology but many other areas of science as well. Her coursework in mathematics, physics, and other fields awakened an avid interest in studying the interaction of biology with these areas. This interest would serve her well in her later research.

As Penry began graduate school, her focus was on ichthyology, a branch of zoology that deals with fish. She intended to begin by studying the invertebrate animals that serve as food for fish, reasoning that she had to understand what fish eat before moving on to ichthyology. However, she soon found herself immersed in the study of these invertebrates and benthic (bottom of the ocean) ecology. Her research focus never shifted from this area.

When the time came to decide where to do her doctoral work, Penry already knew with whom she wanted to work—Dr. Peter A. Jumars, professor of oceanography at the University of Washington. Through her own research and studies, Penry had become familiar with Jumars's work and liked his research approach. She says:

> You can see patterns of community structures over and over everywhere on the ocean floor, but we don't know why these patterns are repeated. Jumars works from the premise that you can't understand why organisms are found together by looking at the whole community. Instead, you should look at the processes, such as feeding and digestion, going on at the population or individual level.

Penry contacted Jumars and was pleased to find him willing to oversee her doctoral studies.

Working with Jumars, Penry set out to study the digestive processes of deposit feeders, animals that ingest mud and sand from the ocean floor, digest some part of this material, and excrete the rest. These feeders play a small but key role in the carbon-nitrogen cycle of the ocean, and thus their feeding and digestion processes have important physical and chemical consequences for other animals. Penry found that although there were theories for how an animal should forage, or search, for food, there were no comparable theories for how an animal should digest what it finds. Because an animal's digestive process will constrain, or limit, how it forages, a theory for digestion was needed in order to have a more accurate picture of the feeding process. "Because every animal has a different digestive strategy, I wanted a framework where I could generalize the process."[2]

One day Jumars handed Penry a chemical engineering textbook, suggesting that she might find it useful. A section of the book described the chemical reactor theory used in engineering to achieve a maximum production rate of a chemical. A chemical reactor is a vessel or tank in which chemical reactions take place. Chemical reactors can vary in size and shape and are classified on the basis of how the reactants, or chemicals, enter the tank (continuously or at intervals) and how they are brought together in the tank (with or without mixing). Three ideal reactor types are the batch reactor, the plug-flow reactor, and the continuous-flow, stirred-tank reactor. In the batch reactor, chemicals enter the tank and are mixed together. The reaction is allowed to proceed for a set period of time, and then the products are all removed. In the plug-flow reactor, the chemicals flow into and then out of the tank continuously, with no mixing. In the continuous-flow, stirred-tank reactor, the chemicals flow into and out of a tank that is continuously mixing.

Penry and Jumars were struck by the parallels between chemical reactor design and their problem of modeling digestion. After further research, they found that digestion of food in the animal's gut is homologous with conversion of chemicals in a reactor. Penry explains, "Animal guts are like chemical reactors because, essentially, you put material in, chemical reactions occur, and material comes out again."[3] Penry and Jumars used their theory to model digestion in marine deposit feeders, categorizing the animals' guts in terms of the three types of reactors. They also compared digestion in these invertebrates to digestion in different mammals, such as the cow and rabbit.

After finishing her postdoctoral work, Penry applied for a position in the newly formed integrative biology department at the University of California, Berkeley. This department, the first of its kind in the United States, was created by the merger of many small departments such as zoology, paleontology, and botany. By joining these different areas of biology, the department is able to offer a program that integrates all levels of organization from molecules to ecosystems, all taxa of organisms from viruses to higher plants and animals, and all periods of time. For Penry, working in this environment has been a rewarding experience. Her research in digestion has centered on marine animals, but she is also interested in applying models to other animals, plants, and even insects. Her position in the department gives her the opportunity to consult with colleagues working in other areas and in other organisms, allowing her to gain a broader perspective.

Penry's research involves doing a little of everything, from creating models on her computer to field testing to collecting samples from the ocean. But research is only one of her roles in the department. She also teaches both undergraduate and graduate courses, including invertebrate zoology, biological oceanography (a course she developed), and the ecology of marine benthic organisms. In addition, she is a guest lecturer in other departments. For example, she presents her research to chemical engineering classes to illustrate the use of engineering in biology. Much of Penry's time is spent with her students, encouraging them to work on special projects in her laboratory and overseeing their research. The grant money from the Waterman Award was used to support her students in their research and to purchase new equipment for the laboratory.

Although she was only the second woman to win the Waterman Award, which has been presented annually since 1976, Penry has not encountered many difficulties in being a woman in her field. "I haven't found oceanography to be a field where discrimination has been prevalent, and the integrative biology department has been very supportive of my work." Like many women today, she has chosen to combine a career with marriage and a family. This choice naturally leads to compromises, because there are limits to the time she can spend on her work and also devote to her young daughter. But then, her daughter has the unique opportunity to visit Penry's laboratory, making her "probably one of the few 3-year-olds who knows what algae is." Recalling her mother's early fishing trips, there's no telling where that interest may lead.

Notes

1. Material for this entry, unless otherwise noted, comes from a telephone interview with Deborah Penry on May 30, 1995.
2. Ron Kaufman, "Berkeley Oceanographer Is Second Woman to Receive NSF's Alan T. Waterman Award," *Scientist* 7 (June 14, 1993): 22.
3. *Ibid.*

Bibliography

Jumars, Peter A., and Deborah L. Penry. "Digestion Theory Applied to Deposit Feeding." *Lecture Notes on Coastal and Estuarine Studies* 31 (1989): 114–28.
Penry, Deborah L., and Peter A. Jumars. "Modeling Animal Guts as Chemical Reactors." *American Naturalist* 129 (1987): 69–96.

                    PATRICIA MURPHY AND LESLIE O'BRIEN

NAOMI E. PIERCE

(1954–)

Biologist

Birth	October 19, 1954
1976	B.S., biology, Yale University
1976–77	John Courtney Murray Fellowship, Yale University: a traveling fellowship to Southeast Asia
1983	Ph.D., biology, Harvard University
1983–84	Fulbright Postdoctoral Research Fellow, Griffith University, Queensland, Australia

1984–85	NATO Postdoctoral Fellow, Science Dept. of Zoology, Oxford University
1984–86	Research Lecturer, Christ Church, Oxford University; Research Fellow, Dept. of Zoology, Oxford University
1986–90	Assistant Professor of Biology, Princeton University (1986–89); Associate Professor (1989–90)
1987–88	Visiting Research Fellow, University of New England, Armidale, NSW, Australia
1988	MacArthur Fellowship Award
1989–90	Science Scholar, Bunting Institute, Radcliffe College
1991–	Hessel Professor of Biology, Curator of Lepidoptera—Museum of Comparative Zoology, Harvard University

In 1991, Naomi E. Pierce was appointed Hessel professor of biology at Harvard University. Recognized for her effort to uncover the ways in which behavioral ecology influences the evolution of certain kinds of butterflies, she is one of the few tenured women on the faculty. Naomi is in constant demand for committee work, and she recognizes that the careers of women in science can be difficult. Women scientists, she observes, become increasingly isolated as they move up the academic ladder, because there are fewer and fewer women and fewer and fewer opportunities for interaction and discussion the higher one climbs. Naomi's own positive experience early in her career with mentor Charles Remington, with whom she still consults, contributed to her support for mentoring; she encourages women in science to discuss their experiences with others and to foster supportive relationships with their colleagues.

At Harvard, Naomi acquired the funding to set up a molecular biology laboratory for work in phylogenetic studies. Her lab is about 75 percent female and is a supportive atmosphere for the women who work there. She sees her lab as more of an information center than a hierarchical domain. She currently has nine graduate students and four postdoctoral students working in the lab. She finds teaching upper-level undergraduate and graduate students doing independent research very satisfying. Although seminars for graduate students are helpful, she feels that the students would be better served to be out in the field doing research. Naomi also enjoys teaching younger students because they are uninhibited about showing enthusiasm. She prefers to teach students individually and in small groups, avoiding large theater lecture classes, which she did not like as a student.

Naomi began her academic studies as a history, arts, and letters major at Yale University. Born in Denver, Colorado, she had grown up in a home where science was emphasized (her father worked as a geophysicist), but her interests lay in the humanities. She felt influenced by her grandfather, who was a novelist in Japan. Naomi is still an avid reader of novels, particularly contemporary fiction. Despite this interest, Naomi felt compelled to complete a number of introductory science courses in case she decided to attend medical school.

The direction of her academic work became clear during her junior undergraduate year, when she took a course on evolution with Charles Remington. He often lectured about his studies of butterflies at the Rocky Mountain Biological Laboratory, a high-altitude biological station in Colorado, Naomi's home state. Naomi was intrigued and decided to approach Remington about his research on the butterflies. "One day I was walking by his office and he was standing in there by himself and there was no one around. I darted in and introduced myself and asked him [about his work]. His eyes lit up and he made me sit down [so he could] tell me all about [it]."[1] Naomi had no idea about how one could study butterflies in the mountains of Colorado, nor that one could get a job as a field biologist, but her conversation with Remington led to summer courses and work at the Rocky Mountain Biological Laboratory and to her determination to become a field biologist.

In her senior year at Yale University, Naomi changed her major to biology and devoted herself entirely to scientific study. She graduated from Yale with a B.S. degree in biology in 1976.

The year after she completed her undergraduate degree, Naomi traveled throughout Southeast Asia on a John Courtney Murray Fellowship from Yale University to study butterflies. In Australia, she was told about a specific pair of species that cooperate in a mutualistic relationship. The "mistletoe butterflies" are a species in the Lycaenid family. At night, the ants herd the Lycaenid caterpillars high up into eucalyptus trees to feed on hanging mistletoe. At dawn, the caterpillars are herded back down to the base of the tree, where they are protected from predators and parasites by the ants until nightfall. As "payment" for this protection, the caterpillars secrete a protein- and carbohydrate-rich liquid as food for the ants. Although she had not seen the two species, Naomi was so fascinated by this butterfly-ant interaction that she chose it as the focus for her doctoral research at Harvard University. She received her Ph.D. in biology from Harvard in 1983.

A Fulbright Postdoctoral Research Fellowship in Australia and a NATO Postdoctoral Fellowship in England allowed Naomi to continue her study of the mutualistic Lycaenids. During her appointment as a research fellow and research lecturer at Oxford University, she started to pursue ideas that had emerged from her previous work on the Lycaenids. She had studied two species of mutualistic butterflies, the Australian *Jalmenus evagoras* and the North American *Glaucopsyche lygdamus*. Naomi began to consider the potential evolutionary repercussions that could result from the kinds of behavioral ecological interactions that she had observed.

Lycaenid butterflies as a group demonstrate incredible variation in life history—from mutualism to parasitism, from carnivory to herbivory—and Lycaenid caterpillars feed on many different kinds of host plants. While Naomi was still at Oxford, she embarked on a large-scale comparative study of life history evolution of Lycaenid butterflies, in order to attempt to uncover the ways in which behavioral ecology might have influenced the evolution of the entire group. Complete life histories are known for about 1,500 species of the 4,000 to 5,000 species of Lycaenids currently described. Naomi examined the phylogenies that already existed for the group and found that although a very good classification scheme was in place for the Lycaenid butterflies, a rudimentary phylogeny was based only upon that scheme. Thus, it became clear to Naomi that without a more complete phylogeny of the Lycaenids, it would be very difficult to undertake the extensive comparative study that she proposed.

Morphological characters were used to construct the taxonomy of the Lycaenidae, because in many cases the same morphological characters can be used to construct a phylogeny for the same group. However, biomolecular information can build upon the phylogeny based solely on morphological characteristics. Naomi saw the need to reconstruct a phylogenetic history for the group utilizing the newer molecular biology techniques. She retrained herself as a molecular biologist in order to begin this lengthy project. As it develops, the phylogeny will be a significant contribution to the basic scientific research on the Lycaenidae and will be a valuable resource for other researchers.

Naomi held the positions of assistant and associate professor at Princeton University from 1986 to 1990. In 1988 she was one of 31 recipients of the prestigious MacArthur Fellowship Award. The award allowed Naomi to pursue her research unrestricted for five years and to give several graduate students the opportunity to do fieldwork in Australia.

As a graduate student deep in the study of the butterfly-ant interactions that fascinated her, Naomi was caught up in fieldwork and travel and had "so much fun," she says, that she "wouldn't have traded it for anything."[2] Now, as a molecular phylogenetist, she spends most of her time on the administrative duties of running two large laboratories and a collective project. Naomi acknowledges that the academic lifestyle is neither ideal nor easy, noting that there are other options for employment in biology. Above all, Naomi advises women pursuing careers in science to follow their hearts and to persevere.

Notes

1. Naomi Pierce, personal correspondence with contributor.
2. Naomi Pierce, personal correspondence with contributor.

Bibliography

American Men and Women of Science, 1995–96. New Providence, N.J.: Bowker, 1994.

Baylis, M., and N. Pierce. "The Effect of Host Plant Quality on the Survival of Larvae and Oviposition Behaviour of Adults of an Ant-Tended Lycaenid Butterfly, *Jalmenus evagoras.*" *Ecological Entomology* 16 (1991): 1–9.

Pierce, N.E. "The Evolution and Biogeography of Associations between Lycaenid Butterflies and Ants." *Oxford Surveys in Evolutionary Biology* 4 (1987): 89–116.

———. "Lycaenid Butterflies and Ants: Selection for Nitrogen Fixing and Other Protein Rich Food Plants." *American Naturalist* 125 (1985): 888–95.

Pierce, N.E., and S. Eastal. "The Selective Advantage of Attendant Ants for the Larvae of a Lycaenid Butterfly, *Glaucopsyche lygdamus.*" *Journal of Animal Ecology* 55 (1986): 451–62.

Pierce, N.E., and W.R. Young. "Lycaenid Butterflies and Ants: Two-Species Stable Equilibria in Mutualistic, Commensal, and Parasitic Interactions." *American Naturalist* 128 (1986): 216–27.

D. BOSWELL LANE

ELIZA LUCAS PINCKNEY

(1722–1793)

Horticulturist

Birth	December 8, 1722
1738	Moved to South Carolina with her father, mother, and sister
1744	Produced the first truly successful crop of indigo; married Colonel Charles Pinckney
1746	Son Charles Cotesworth Pinckney born
1748	Daughter Harriott Pinckney born
1750	Son Thomas Pinckney born
1753	Moved to England with her husband and children

1758	Returned to South Carolina with her husband and daughter
Death	May 26, 1793

Born in the West Indies and schooled in England, Eliza Lucas Pinckney achieved notoriety in mid-eighteenth-century Charles Town, South Carolina, as an efficient plantation manager, an avid gardener, and a favorite in Charles Town society. Her agricultural interests extended beyond vegetables or flowers. At age 22, Eliza Lucas Pinckney was the first southern planter to successfully cultivate indigo, a plant used as a dye for textiles. Her efforts led to a widespread cultivation of indigo among southern planters and boosted the agricultural economy of the region.

Eliza Lucas was born in the West Indies to George Lucas, a planter and lieutenant colonel in the British Army, and his wife (name unknown) in 1722. She was the oldest of four children. In 1738 she moved to Charles Town (later known as Charleston), South Carolina, with her father, mother, and younger sister. George Lucas owned the Wappoo Creek plantation as well as two other plantations. The family lived on the Wappoo plantation.

Young Eliza had been educated in England and enjoyed her life there very much. However, she found her new home in the Charles Town area to be quite agreeable. In 1740 she wrote in her letterbook, "I prefer England to it [Charles Town], 'tis true, but think Carolina greatly preferable to the West Indies. . . . Charles Town, the principal one in this province, is a polite, agreeable place. The people live very Gentile and very much in the English taste."[1]

After being welcomed into Charles Town society, Eliza Lucas soon found herself in charge of her father's plantations. Her father had returned to the colony of Antigua when war erupted between England and Spain in the War of Jenkins Ear.[2] The property under Eliza's care included "the 600-acre Wappoo Creek plantation with its 20 slaves . . . the 1500-acre Garden Hill plantation, whose products were pitch, tar, and pork, and 3000 scattered acres of rice along the Waccamaw River."[3]

Eliza proved to be disciplined and well able to conduct the business of plantation life. Of her new responsibilities, she wrote to a friend in England:

I have the business of three plantations to transact, which requires much writing and more business and fatigue of other sorts than you can imagine. But least you should imagine it too burthensome to a girl at my early time of life [age 18], give me leave to answer you: I assure you I think myself happy that I can be useful to so good a father, and by rising very early I find I can go through much business.[4]

Eliza also acted as schoolteacher for her younger sister. Although it was illegal in the colonial South at that time, she taught two slave girls on her plantation to read in order that they might teach the other slave children. Even though she was constantly busy with obligations to the plantation and her family, Eliza kept up her social calendar in Charles Town and visited friends and attended other engagements in the city.

While managing the plantations in her father's absence, Eliza tried a number of crops. She began her experiments with cultivating indigo in 1740. She was not the first planter in the Carolinas to attempt to cultivate indigo. However, she was the first to successfully cultivate the plant on a large scale and produce quality dye for the marketplace. The process of producing dye from the indigo plant involved fermenting the leaves of the plant in water, draining the fermented product, and then beating it to produce a blue sediment. The addition of lime and additional draining of the sediment produced particles that were left to dry until they could be cut into cubes.[5]

Eliza Lucas received seeds for the indigo plant from her father in the West Indies. She also had the help of a neighboring planter and a dye-maker whom her father sent to help her. Her first attempts at producing indigo were unsuccessful. She ended up firing the dyemaker sent by her father because the workman sabotaged the process. After hiring another dyemaker, Lucas went on to produce indigo good enough to sell on the market. Magnanimously, she shared indigo seeds with neighboring planters. Indigo cultivation grew in the southern colonial area until, in 1747, the colony exported 138,300 pounds of dye for that year.[6] Unfortunately, her father lost the lands not long after Eliza produced the successful crop.

Soon after her success in indigo production, Eliza Lucas married Charles Pinckney, a wealthy landholder and member of the Governor's Royal Council. Pinckney was almost twice Eliza's age. The two had known each other since Eliza's first arrival in Charles Town. Eliza and Pinckney's late wife were great friends. Eliza and Charles Pinckney had three children together: Charles Cotesworth, Harriott, and Thomas. They led a comfortable and happy life. In 1753 the family sailed to England, where Charles Pinckney secured a position as an agent to South Carolina merchants. Eliza's husband died of malaria in 1758 soon after returning to Charles Town.

Once again, Eliza was left to manage a plantation. She applied the same business acumen to her late husband's land as she had to her father's. She also turned her attention to the upbringing of her sons and daughter. Charles Cotesworth was elected to the Assembly. Harriott married Daniel Horry, whose family was prominent in Charles Town society. Thomas became a lawyer. Both sons fought against the British in the Revolutionary War. Eliza Pinckney also supported the Revolution by loaning money to South Carolina.

Eliza Lucas Pinckney received recognition for her loyalty not only from her fellow South Carolinians but also from George Washington. In fact, Washington visited Pinckney and her daughter after he was elected president. He also served as a pallbearer at her funeral. Eliza Lucas Pinckney died of breast cancer in Philadelphia on May 26, 1793.

Notes

1. Eliza Lucas Pinckney, *The Letterbook of Eliza Lucas Pinckney, 1739–1762*, ed. Elise Pinckney with Marvin R. Zahniser (Chapel Hill: University of North Carolina Press, 1972), pp. 6–7.

2. Constance B. Schulz, "Eliza Lucas Pinckney (1722–1793)," in *Portraits of American Women: From Settlement to the Present* (New York: St. Martin's Press, 1991), p. 700.

3. Caroline Bird, *Enterprising Women* (New York: W.W. Norton, 1976), pp. 33–34.

4. Pinckney, *The Letterbook*, p. 7.

5. Schulz, "Eliza Lucas Pinckney," p. 72.

6. *Ibid.*, p. 73.

Bibliography

Bird, Caroline. *Enterprising Women*. New York: W.W. Norton, 1976.

Nicholas, Edward. *The Hours and the Ages: A Sequence of Americans*. Port Washington, N.Y.: Kennikat Press, 1949.

Pinckney, Elise. "Pinckney, Elizabeth Lucas," in *Notable American Women, 1607–1950*, Vol. 3, ed. Edward T. James. Cambridge, Mass.: Belknap Press of Harvard University Press, 1971.

Pinckney, Eliza Lucas. *The Letterbook of Eliza Lucas Pinckney, 1739–1762*, ed. Elise Pinckney with Marvin R. Zahniser. Chapel Hill: University of North Carolina Press, 1972.

Ravenel, Harriott Horry. *Eliza Pinckney*. New York: Charles Scribner's Sons, 1896.

Schulz, Constance B. "Eliza Lucas Pinckney (1722–1793)," in *Portraits of American Women: From Settlement to the Present*. New York: St. Martin's Press, 1991.

Williams, Frances Leigh. *Plantation Patriot: A Biography of Eliza Lucas Pinckney*. New York: Harcourt, Brace, and World, 1967.

HEATHER MARTIN

ELIZABETH WATIES ALLSTON PRINGLE

(1845–1921)

Horticulturist

Birth	May 29, 1845
1870	Married John Pringle
1885	Began managing a plantation
1913	Published *A Woman Rice Planter*
Death	December 5, 1921

Elizabeth Pringle achieved success as a plantation owner/farmer and later in life as a writer. She chronicled her life as a plantation owner in personal journals from which excerpts were compiled and published in her first book, *A Woman Rice Planter. Chronicles of Chicora Wood*, a detailed account of plantation life under her father's management, was published posthumously in 1922. These accounts of plantation life offer an uncensored glimpse into a woman's acceptance of life based on subordination and hierarchy and the attitudes of southern aristocracy. Among her accomplishments, Pringle practiced a scientific approach to agriculture and employed new techniques in farming, a task she thoroughly enjoyed.

Born Elizabeth Waties Allston to parents with aristocratic bloodlines, Adele (Petigru) Allston and Robert Withers Allston, she was the third of five offspring to survive childhood. Known to family and friends as Bessie, Elizabeth was born at the family summer home at Canaan Seashore near Pawley's Island in South Carolina.

Elizabeth's father was a prominent plantation owner and politician, having served in the state legislature for 28 years and as governor of South Carolina from 1856 to 1858. He was one of the South's largest slave owners, controlling the lives of 630 slaves at one time. He maintained seven plantations in South Carolina, cultivating thousands of acres of rice fields. He wrote articulate publications about rice development and seacoast crops. The Allston family is known to have become established in the Georgetown District as early as 1763 through the procurement of land grants in this South Carolina community.

Elizabeth's mother, Adele, was of Scotch-Irish descent with distant French Huguenot ancestry. The Huguenot name was said to be a social advantage in South Carolina at the time. Adele embodied the expected standard of a plantation mistress: she had beauty, grace, and good social connections. Her brother was the noted Charleston lawyer and outspoken Unionist, James Louis Petigru. The Allston family home was at Chicora Wood on the Peedee River, 14 miles north of Georgetown.

Elizabeth was brought up at a time when women were required to exist in the context of a patriarchal society. They were expected to personify what were considered to be feminine values. They were not to think for themselves, speak out in public, or express their needs and desires. Elizabeth's mother was responsible for instilling in her the proper codes of behavior. At an early age, she required Elizabeth to keep a daily journal. This simple exercise of writing and thinking on paper became a habit she continued throughout her life, and it is particularly responsible for her success. Elizabeth's routine of writing necessitated thinking for one's self, an activity not then encouraged for women.

Young Elizabeth was taught by a governess at Chicora Wood until she was 9 years old, at which time she was sent to Charleston for a formal education at Madame Acelie Togno's boarding school. Her course of study included arithmetic, diction, English, French, history, music, and singing. Elizabeth took particular delight in French and English history and studied at the school until 1863.

In 1863 the sight of Union gunboats on the Peedee River made Elizabeth's father transfer the family to Dairlington County, his farm 80 miles inland, while he stayed behind at Chicora Wood. The women were forced to care for themselves, planting a garden to put food on the table and setting up a loom to weave material for clothing.

Elizabeth's father died the next year, leaving each of his children a plantation and 100 slaves, but none of them ended up with anything from the estate. Debt and the emancipation of the slaves left the estate insolvent. The only land retained by the family was the Chicora Wood plantation and home.

For a time after the war, Elizabeth taught in a school founded by her widowed mother in Charleston. In 1868 she moved back to Chicora Wood with her mother, her sister Jane, and her younger brother Charles. On April 26, 1870, Elizabeth married John Julius Pringle of the White House plantation 8 miles down the Peedee River. When he died in 1877, after years of debt and grief over the loss of an infant son, Elizabeth returned to her family and her childhood home.

In 1880 a bequest gave Elizabeth the means to purchase the White House, where she had spent seven years with her husband. In 1885, at age 40 and despite the skepticism of family and friends, Elizabeth began managing the White House plantation herself while continuing to live at Chicora Wood. She proved to be a capable plantation manager, consistently making a profit until the turn of the century. On the plantation she grew rice; cultivated strawberries, peaches, and scuppernong grapes; and kept livestock and poultry. Elizabeth was open to new techniques in agriculture and farming. She applied the use of an incubator to hatch chicks and employed inoculation of alfalfa seed to improve both quality and production.

After the death of her mother in 1896, Elizabeth acquired Chicora Wood and took over its management. The following 18 years proved to be a continual struggle to fulfill her financial obligations, particularly in regard to taxes and labor costs. Years of bad weather contributed heavily to her inability to turn a profit at either plantation after the turn of the century.

Elizabeth aspired to become an author throughout her life. Her early efforts were met with rejection, but she never gave up. In 1879 she published a translation in a Charleston newspaper, and part of her childhood diary was published in a newspaper series in 1884. From 1904 to 1907 the *New York Sun* published diary excerpts under the pseudonym "Patience Pennington." These excerpts and some others were compiled into the book *A Woman Rice Planter* (1913). This work and *Chicora Wood*, published after her death, provide a compelling portrayal of the death of an old industry and a way of life.

During Elizabeth's later years, the Chicora plantation was restricted to livestock, fodder, and vegetable crops. She was a principal force behind the repair of the Price Frederick Church near Chicora and participated actively in the Mount Vernon Ladies Association of the Union.

Elizabeth Pringle was a friend to all around her at her Chicora home. Neighbors, workers, friends—regardless of race or status—were able to turn to her in times of need. She was a fair employer—at times too fair, often at a cost to herself and her livelihood, yet she always managed. She died in her home at age 76 and was buried at Magnolia Cemetery in Charleston beside her husband and their infant son.

Bibliography

Childs, Margaretta P. "Pringle, Elizabeth Waties Allston," in *Notable American Women, 1607–1950*. Cambridge, Mass.: Belknap Press, 1971.

Pringle, Elizabeth W. Allston. *Chronicles of Chicora Wood*. New York: Charles Scribner's Sons, 1922.

———. *A Woman Rice Planter*. Columbia: University of South Carolina Press, 1992.

Sherr, Lynn, and Jurate Kazickas. *The American Women's Gazeteer*. New York: Bantam Books, 1976.

SYLVIA NICHOLAS

MARGIE PROFET

(1958–)

Biologist

Birth	August 7, 1958
1980	B.A., political philosophy, Harvard University
1985	B.S., physics, University of California, Berkeley
1993	MacArthur Foundation Fellowship

History is made not by important people but by individuals who do important work. This certainly can be said of Margie Profet, a woman who has challenged conventional wisdom to present new theories relating to how humans adapt to their environment. This unlikely researcher came to the scientific community without the standard academic credential of a Ph.D. to challenge accepted theories on allergies, pregnancy sickness, and the cleansing benefits of female menstruation.

Margie Profet was born to Karen and Robert Profet in 1958 in Berkeley, California. Margie was the second child in the Profet household, following a son, James. While Karen and Robert continued to complete their studies at Berkeley, two more girls were born, Julie and Kathleen. In 1965 the Profets moved to Manhattan Beach, California, to raise their four children. Karen, a systems engineer, and Robert, a physicist, worked for the aerospace industry. Even as a young child at age 7, Margie remembers being bored with the routine of life and school. This bright, creative, and curious child was a good student who was constantly asking questions. As a young adult she was not focused on any particular career, but she envisioned herself working in a field that allowed her to ask life's questions and to look for the answers. Margie recalls being close to her brother and sisters and living in a family whose parents "respected individuality and independence."[1]

At age 18, Margie entered Harvard University to study political philosophy. This experience provided the foundation of her research career. The study of philosophy encouraged her to challenge conventional wisdom and search for meaning, when none is thought to exist. Margie admits that "she enjoyed the social life of school and was not the best student." However, during her junior and senior years her situation changed when she took a one-on-one tutorial focusing on Plato and Nietzsche with Dr. Harvey Mansfield. "He saw a spark in me and was willing to take a chance on me and not just see me as another California flake." Margie viewed the tutorial and other work with Dr. Mansfield on her senior thesis on Nietzsche as a valuable process: "Dr. Mansfield in his nondirective style helped to change my life. This thesis provided me with one of the greatest intellectual experiences of my life."

After graduation, Margie headed for Europe and spent two years in Munich, Germany, working as a computer programmer with National Semi-Conductor. Her travels also included climbing Mt. Kilimanjaro during a trip to South Africa. She recalls staying with a German friend so she could practice English in exchange for a place to sleep. She returned to the United States realizing that she needed to further her education to find the answers to her questions. In 1985, Margie earned a B.S. degree in physics from the University of California, Berkeley. However, feeling stifled by the constraints of the classroom, the need to meet deadlines, and the traditional rigors of measuring success, Margie decided not to continue her academic training but simply to think.

With two degrees, one in political philosophy and another in physics, Margie found her real interest in evolutionary physiology. By taking part-time jobs, she provided herself with time simply to ask questions and search for answers. She remembers going into a library and doing what she refers to as "detective hunts" in which she would "start with the journals beginning with the letter A and go all the way through the alphabet reviewing the articles and finding a topic that was of interest and keep on asking why. It was like being at Disneyland and entering the world of discovery."

Her first theory was inspired by her personal experiences of being allergic to various foods and chemicals. Late one night, after scratching because of her allergies to shampoos and soaps, she wondered what the function of these allergies was, knowing that allergies were caused by a highly specialized class of antibody. Margie looked for the evidence to explain why the body reacted to certain substances and what the benefit of these reactions might be. Finally, after much research, she published an article in the March 1991 issue of *Quarterly Review of Biology* entitled "The Function of Allergy: Immunological Defense against Toxins." This publication came after a ranking toxicologist, Bruce Ames, gave Margie a part-time job in his lab.

Margie theorized that humans develop allergic reactions as a means of protecting the body from harmful toxins. She proposed that when the body is exposed to a toxin or when a toxin is ingested, it sends a warning signal through an allergic reaction. Thus, the allergy serves as one of the last defense mechanisms the body can mount against toxins. The body is thereby warned that the particular substance is harmful to it. To support her theory, Margie included research on individuals with a history of allergies and demonstrated that they are less likely to develop cancer than individuals without allergies. Allergies are seen, therefore, as an internal warning device for the body.

This theory was received by the scientific community with great skepticism. How could this woman, without proper scientific credentials, expound a theory on the benefits of allergies to the body's defense system? Although many researchers were willing to examine the new hypothesis, some within the scientific community were very reluctant to accept the concept. The theory has dramatic implications not only for the researcher but also for the medical community. The treatment for allergies would be directed at making the toxin less harmful (such as heating food to which one has developed an allergy to eliminate some of the toxic qualities) as opposed to desensitizing the body to the toxic substance.

The basis for Margie's next theory developed as a result of listening to her sister and sister-in-law discuss the experience of pregnancy. Margie continued to study and examine the relationship between being pregnant and experiencing morning sickness. She believes that the ability of the brain to discern what is toxic becomes recalibrated during pregnancy, so that almost any food or odor can cause an aversion to almost anything. Women are sick in the morning because their digestive system slows during sleep; awakening, they get sick as a means of eliminating toxins from their body. The toxins protect not only the mother but the fetus as well. Margie proposed that the greater the degree of morning sickness, the greater the protection to the unborn infant. Morning sickness occurs primarily during the first trimester, which coincides with the time of the greatest development of the fetus. Women from all over the world presented anecdotal records that supported her work, as did some scientific research as well. Not surprisingly, the scientific community was not uniformly receptive of this new theory; but some did see in it an avenue for future study.

This unconventional scientist continued to question conventional wisdom when she began to re-examine the reason for menstruation. Margie recalls that one night while sleeping, she found herself dreaming of a cartoon about a woman menstruating and what she thought were the ovaries, the red lining of the uterus, and red blood flowing from the cervix. But there were many black triangles impeded in the uterus. The constant noise from her cat, Gelato, woke her up from this dream. However, she could not escape the memory of her dream and what significance it might have had. After reviewing her dream, Margie thought that the tiny black triangles were pathogens containing bacteria within the lining of the uterus. Upon further research, Margie was able to support the idea that sperm carry pathogens into the uterus and that the menstrual flow allows the uterus to rid itself of bacteria and infection. Margie published her new theory in the September 1993 issue of *Quarterly Review of Biology*. It was called "Menstruation as a Defense against Pathogens Transported by Sperm." This theory, too, has significant implications for medicine and how irregular menstrual bleeding is treated by physicians.

In addition to these theories, Margie continues to pursue other areas of interest, including ideas and questions "concerning immunological puzzles" that need further examination. She has also recently authored a book, *Protecting Your Baby To Be: Preventing Birth Defects—First Trimester*. This book contains dietary advice and highlights the need to pay attention to one's own body.

Margie describes her life as one of balance in which her family, friends, and animals play an important role. Her brother, James, is a missionary who is engaged in feeding the poor. Both of her sisters, Kathleen and Julie, have two children. She describes them as being "emotionally close" and a source of support along with her parents. In addition to family and friends, Margie collects wild animals such as squirrels, raccoons, foxes, cats, and wild parrots. She finds squirrels to be very interesting, learning to take food by hand and performing tricks for her.

Recognition finally came to Margie Profet in June 1993 at age 34. She was awarded the MacArthur Foundation "genius award." The $250,000 that accompanied the recognition allows this most unconventional scientist to continue to ask questions and search for answers without being encumbered by bureaucracy and the limitations of the academic world. Margie has challenged and will continue to challenge conventional wisdom and examine how women and men change as required by new and ever-changing environments.

Note

1. All quotations in this entry were taken from a telephone interview with Margie Profet by Joan Lewis on June 26, 1995.

Bibliography

Angier, Natalie. "Radical New View of Role of Menstruation." *New York Times* (Sept. 21, 1993): C1.

Bloch, Hannah. "School Isn't My Kind of Thing." *Time* 142 (Oct. 4, 1993): 72–73.

Kahn, Alice. "A Brand-New Thought about Women's Cycles." *San Francisco Chronicle* (Sept. 27, 1993).

Plumer, William. "A Curse No More." *People Weekly* 40 (Oct. 11, 1993): 75–76.

Profet, Margie. "The Function of Allergy Immunological Defense against Toxins." *Quarterly Review of Biology* 66 (Mar. 1991): 23–62.

————. "Menstruation as a Defense against Pathogens Transported by Sperm." *Quarterly Review of Biology* 68 (Sept. 1993): 335–86.

Rudavsky, Shari, and Tom Zimberoff. "Margie Profet." *Omni* 8 (May 1994): 69–74.

Toufexis, Anastasia. "A Woman's Best Defense." *Time* (Oct. 4, 1993): 72.

JOAN LEWIS

EDITH SMAW HINCKLEY QUIMBY
(1891–1982)
Biophysicist

Birth	July 10, 1891
1912	B.S., Whitman College
1915	Married physicist Shirley Quimby
1919–42	Assistant/Associate Physicist, Memorial Hospital for Cancer and Allied Diseases, New York City
1940	Janeway Medal, American Radium Society
1941	Gold Medal, Radiological Society of North America
1943–61	Associate Professor/Professor of Radiology, Columbia University
1947	Scientific Achievement Medal, International Women's Exposition
1949	American Design Award, Lord and Taylor
1952	Jagadish Bose Memorial Gold Medal, Indian Radiological Society
1957	Medal, American Cancer Society
1958	Gold Medal, Interamerican College of Radiology
1961–78	Professor Emerita, Columbia University
1962	Judd Award, Memorial Hospital for Cancer and Allied Diseases
1963	Gold Medal, American College of Radiology
1973	Scientific Achievement Award, American Medical Association
1976	Distinguished Service Award, Radiological and Medical Physics Society
1977	Coolidge Award, American Association of Physicists in Medicine
Death	October 11, 1982

Edith Smaw Hinckley was born in 1891 in Rockford, Illinois, to Arthur and Harriet Hinckley. Edith's father was an architect and farmer who moved his family from Illinois to Alabama to Idaho and then finally to California. Edith remembers her childhood spent in a "middle-class family . . . in a medium-sized city in the Middle West."[1] Unlike many parents of the time who felt that education, especially of a scientific nature, was of no use to women, Arthur Hinckley encouraged Edith's curiosity and numerous questions, teaching her to find answers for herself. It was this curiosity that nurtured one of the most celebrated and honored pioneers in radiology.

While attending Boise High School in Idaho, Edith was challenged by her science teacher, Mr. Rhodenbaugh, to explore natural phenomena and to use the science laboratory to answer some of the questions that arose in her inquiring mind.[2] At Whitman College in Walla Walla, Washington, where she studied on a full four-year scholarship, stimulating classes in physics taught by B.A. Brown, and in mathematics taught by Walter Bratton, greatly influenced Edith's choice of career.[3] She became the first woman at Whitman College to take the mathematics/physics major. She received her B.S. degree in 1912.[4]

After completing her college education, Edith began teaching physics and chemistry at a high school in Nyssa, Oregon.[5] In 1914 a teaching fellowship in physics at the University of California, Berkley, allowed her to work toward a master's degree in physics. Here she met Shirley L. Quimby, also a graduate student in physics, and they were married the following year.[6] After receiving her M.A. in 1917, Edith and her husband moved to Antioch, California, where Shirley was a science teacher at the high school. Edith—tall, blonde, gray-eyed, good-natured, and home-loving—now seemed content with domestic life. She especially seemed to enjoy cooking, perhaps because her kitchen became another kind of laboratory. When Shirley joined the Navy during World War I, Edith took over his position teaching science at the high school. When the war was over Shirley remained in the military for another year in New London, Connecticut, where he worked on submarine detection. Edith joined her husband in Connecticut and once more seemed content with domesticity.[7] However, she soon began working outside the home again.

After leaving the Navy, Shirley began work on his Ph.D. in physics at Columbia University.[8] Because his part-time instructorship did not pay enough to support them, Edith needed a job. Fortunate in having had a father who encouraged her scientific interests, Edith was again fortunate in her search for employment. She found a job as assistant physicist to the chief physicist, Dr. Giocchino Failla, at New York City Memorial Hospital for Cancer and Allied Diseases. Dr. Failla's work at this hospital involved developing a laboratory for radiological research. Unusual for his time, Failla was not prejudiced against hiring and working with a woman in scientific research. But Edith was qualified for the position, and in 1919 she was one of the few women in America to be engaged in medical research.[9]

Radiological physics as a science did not yet exist because using radiation to treat the ill had just begun. Edith Quimby, only in her twenties, became a pioneer in radiation physics. Her research focused on this completely new science involving the medical use of radiant energy.[10] It was clear that the new technique was useful in exploring inside the body and also in treating some diseases, but the proper and safe dosages had not yet been established.[11] Edith conducted studies that included both live patients and cadavers to measure how much radiation penetrates certain areas of the body.[12] She measured both the generation and penetration of radiation so that dosages of radiation could be given to patients without damaging the healthy body parts. Specifically, she established "how much radiation was emitted from a specific source, how much was delivered in the air, how much to the skin, and how much within the body, under varying conditions of irradiation."[13]

From 1920 to 1940, Edith wrote 50 technical articles on radium and X-ray treatments and the correct dosages for cancer patients.[14] Because of her outstanding work in providing the first practical guidance for physicians, in 1940 she received the highest honor of the American Radium Society, the Janeway Medal. She was the first woman to receive this honor and only the second person without an M.D. to do so. In the same year she received an honorary doctor of science degree from Whitman College. In 1956 she received another honorary doctor of science degree, this time from Rutgers University.[15] In 1941 she was presented with another prestigious award, the Gold Medal of the Radiological Society of North America, for her "continuous service to radiology." Only one other woman, Marie Curie, had been so honored.[16]

One of Quimby's most important contributions was that she made doctors aware of the invaluable help they could get from radiotherapists and radiologists. Radiology became another specialty in medicine, and Quimby trained people in medicine to become specialists in this new area of medicine and science.[17] She was more qualified than most physicians to teach radiology, so she was appointed assistant professor of radiology at Cornell University Medical College in 1941.[18] In 1943 she followed Dr. Failla to the College of Physicians and Surgeons at Columbia University. The fact that she, a nonphysician teacher at a medical school of graduate physicians, was able to gain the respect of physicians not only in the classroom but also in the medical world is a testimony to her scientific knowledge, painstaking research, and personality.[19]

While working part-time on the Manhattan Project (the development of the atomic bomb) during World War II, Quimby added research on radioactive isotopes to her already impressive list of achievements. Her research led to the recommendation of safety measures for those handling isotopes as well as for safe disposal of radioactive wastes. She became an expert in procedures for cleaning up radioactive spills.[20] Because of this work, Quimby served on the Atomic Energy Commission's Committee for the Control and Distribution of Radioactive Isotopes and the National Committee for Radiation Protection. She was made a full professor at Columbia in 1954. In 1960 she retired from Columbia and its Radiological Research Lab, and she was named professor emerita in 1961.

Edith Quimby's research on radiation and its effects on the human body has been invaluable. The results of her work have been published in over 70 articles and four books. In addition to the Janeway Medal and the Gold Medal of the Radiological Society, Quimby received numerous other awards. Among these are the American Cancer Society medal in 1957; gold medal, Interamerican College of Radiology in 1958; Judd Award, Memorial Hospital for Cancer and Allied Diseases in 1962; gold medal, American College of Radiology in 1963; Scientific Achievement Award, American Medical Association in 1973; Coolidge Award, American Association of Physicists in Medicine in 1977.[21]

Certainly Dr. Quimby's research contributed much to the establishment of the field of radiation physics. Equally important, her personality combined with her scientific knowledge, ability, and research skills won professional acceptance from physicians. As a nonphysician, she was able to show physicians that radiotherapy called for highly educated physicists and that collaboration was in the best interests of the physician and the patient. To Edith Quimby goes much of the credit that radiation physicists are accepted as professionally equal to those who practice medicine.[22] She also must be given credit for succeeding in a male-dominated profession at a time when women were not widely accepted as being competent scientists.

Outside the science laboratory Dr. Quimby enjoyed many other activities, especially reading detective stories, playing sports, going to the theater, playing bridge, cooking, and traveling.[23] She belonged to the Democratic Party and the League of Women Voters. Not only did she gain the approval and respect of her colleagues professionally, but those who knew her outside the world of science found that she made friends easily and created for them a warm feeling of good cheer.[24]

Edith Hinckley Quimby died on October 11, 1982, at her home in New York at age 91.[25]

Notes

1. Anna Rothe, ed., *Current Biography: Who's News and Why, 1949* (New York: H.W. Wilson, 1950), p. 492.

2. Edna Yost, *American Women of Science* (Philadelphia and New York: Frederick A. Stokes, 1943), p. 96.

3. Rothe, *Current Biography: Who's News and Why, 1949*, p. 492.

4. Yost, *American Women of Science*, p. 96; Rothe, *Current Biography*, p. 492.

5. Yost, *American Women of Science*, p. 97.

6. Rothe, *Current Biography*, p. 492.

7. Yost, *American Women of Science*, pp. 97–98.

8. Rothe, *Current Biography*, p. 492.

9. Yost, *American Women of Science*, p. 98.

10. Diane Emberlin, *Contributions of Women: Science* (Minneapolis, Minn.: Dillon Press, 1977), p. 153.

11. Iris Noble, *Contemporary Women Scientists of America* (New York: Julian Messner, 1979), p. 14.

12. Janet Podell, ed., *The Annual Obituary, 1982* (New York: St. Martin's Press, 1983), p. 496.

13. Yost, *American Women of Science*, p. 101.

14. Margaret Rossiter, *Women Scientists in America: Struggles and Strategies to 1940* (Baltimore: Johns Hopkins University Press, 1982), p. 261.

15. Allen G. Debus, ed., *World Who's Who of Science* (Chicago: Marquis Who's Who, 1968), p. 1386.

16. Yost, *American Women of Science*, p. 100.

17. Noble, *Contemporary Women Scientists of America*, p. 15.

18. Podell, *The Annual Obituary, 1982*, p. 496.

19. Yost, *American Women of Science*, p. 105.

20. Emberlin, *Contributions of Women: Science*, p. 154.

21. Podell, *The Annual Obituary, 1982*, p. 496.

22. Yost, *American Women of Science*, pp. 104–5.

23. Rothe, *Current Biography*, p. 493; Yost, *American Women of Science*, p. 106.

24. Rothe, *Current Biography*, p. 493.

25. Podell, *The Annual Obituary, 1982*, p. 495.

Bibliography

Debus, Allen G., ed. *World Who's Who of Science*. Chicago: Marquis Who's Who, 1968.

Emberlin, Diane. *Contributions of Women: Science*. Minneapolis: Dillon Press, 1977.

Noble, Iris. *Contemporary Women Scientists of America*. New York: Julian Messner, 1979.

Podell, Janet, ed. *The Annual Obituary, 1982*. New York: St. Martin's Press, 1983.

Rossiter, Margaret. *Women Scientists in America: Struggles and Strategies to 1940*. Baltimore: Johns Hopkins University Press, 1982.

Rothe, Anna, ed. *Current Biography: Who's News and Why, 1949*. New York: H.W. Wilson Co., 1950.

Yost, Edna. *American Women of Science*. Philadelphia and New York: Frederick A. Stokes Co., 1943.

SHIRLEY B. MCDONALD

ELIZABETH SHULL RUSSELL

(1913–)

Geneticist

Birth	May 13, 1913
1933	A.B., University of Michigan
1934	M.A., Columbia University
1935–37	Assistant Zoologist, University of Chicago
1936	Married William Lawson Russell
1937	Ph.D., zoology, University of Chicago
1939–40	Independent Investigator, Jackson Laboratory; Nourse Fellowship, American Association of University Women
1946–57	Resident Associate, Jackson Laboratory
1947	Finney-Howell Fellowship
1957–82	Senior Staff Scientist, Jackson Laboratory
1958–59	Guggenheim Fellowship
1963–	Member, National Academy of Sciences
1982–	Emerita Senior Scientist, Jackson Laboratory

Elizabeth Russell's work with the genetics of Funnyfoot mice pioneered the study of muscular dystrophy and led her into the study of how genes cause cancers. It is fitting that a geneticist of Elizabeth Russell's stature was born to an Ann Arbor, Michigan, family with a scientific predisposition. Her father was a zoologist and geneticist who taught at the University of Michigan. Among her uncles were a geneticist, a plant physiologist, a botanical artist, and a physicist. But it was her mother who provided both nature and nurture, the genes and the example for her daughter. Mrs. Shull earned a master's degree in zoology and taught at the college level before marrying.

Interested in science early, Elizabeth spent one summer vacation cataloguing the plants in a nearby wood. She was only 16 when she entered the University of Michigan, graduating in 1933 with a bachelor of arts degree, majoring in zoology and earning minors in general science and mathematics. Because teaching jobs were so hard to find during the Great Depression, she accepted a scholarship to Columbia University to continue her education. In 1934 she took two classes in genetics that sparked a lifelong curiosity. A paper that disagreed with prevailing thought about genetics at the time heightened her interest. Written by Dr. Sewall Wright of the University of Chicago, it described how genes might do specific tasks in determining physiology.

As soon as she finished her master's degree at Columbia in 1934, Elizabeth went to the University of Chicago to work with Dr. Wright. She taught undergraduate classes as an assistant zoologist while pursuing her doctorate. Her thesis work was on the role of genes in the fur coloring of guinea pigs. She married a fellow student, Dr. William Lawson Russell, in 1936. Upon completing her doctorate in 1937, she followed her husband to the Roscoe B. Jackson Laboratory in Bar Harbor, Maine. The young couple subsequently had three sons and a daughter.

Elizabeth began her own independent research at the Jackson Laboratory in 1939. With the backing of the Elizabeth Pemberton Nouse Fellowship from the American Association of University Women, she investigated tumors in fruit flies. Later she received a Finney-Howell Research Fellowship to study pigmentation in mice. In 1946 she became a regular member of the staff at Jackson.

Jackson Laboratory supplies to researchers worldwide genetically pure, inbred mice with genetic predispositions to certain disorders. In October 1947, a fire destroyed many homes and buildings on Mt. Desert Island, including Jackson Laboratory facilities containing 90,000 mice and their genealogies. Elizabeth was given the task of retracing the breeding ancestry and repopulating the mouse stock. She had been a resident associate for only one year, but her painstaking genetic work was already known. Many customers of the lab offered to return mice, aiding her tremendously in the successful project to rebuild the populations of scientifically valuable mice. Russell wrote of her project:

> In what I hope was a sensible plan, we accepted the offers of mice most closely related to our former stock, added some new special types that were generously offered, built up a common colony from which both in-house research colonies and animal resource colonies were derived, and arranged to carry our recharacterization of the life histories of mice from inbred strains in the common foundation colony.[1]

During this critical period, Russell's marriage was ending and her work became even more valuable to her. Around this time, she noticed one of the mice dragging its rear feet. There had been no injury; and when she mated siblings of this "Funnyfoot" mouse, the dragging feet appeared in succeeding generations. Under Russell's direction, a summer student at the laboratory, Ann Michelson, conducted studies that settled the issue: Funnyfoot mice had a hereditary dystrophy. This discovery allowed researchers to investigate disorders, such as muscular dystrophy, that affect humans. Demand for Funnyfoot mice unfortunately could not be filled because the mice tended to die young, before mating. Elizabeth solved this by transplanting Funnyfoot ovaries into normal, fecund mice.

Russell's generosity with her students did not end with Michelson. Another student, Dave Harrison, returned to the Laboratory to work full-time and began watching a project she was working on. This research involved injecting an anemic strain of mice with the bone marrow of normal mice. The transplanted marrow went to the bones of the anemic mice and began functioning just as it had in the normal mice, curing the hereditary anemia.

Harrison saw an opportunity to study aging as a corollary to this project. The cured anemic mice lived longer than others of their strain who had not received bone marrow transplants, but not as long as completely normal mice. He found that the transplanted white blood cells outlived the mice who donated them. Although the mice died of natural causes, the donated cells did not also age and die inside the anemic mice.

This strain of anemic mouse is sterile, and Russell searched for the cause. She found that the female mice never developed the usual 5,000 oocytes in the embryonic stage. The female anemic mice ovulated a few times, then ran out of "eggs." When this happened, production of the hormone estrogen decreased and the ovary was bathed in too much gonadotrophin, leading to a tendency to develop ovarian tumors.

This continuing research into how genes produce physiological effects and disorders led to international recognition. In 1954 the Jackson Laboratory hosted a conference to assess what had been learned about mammalian genetics and cancer during the 25 years of its existence. The conference yielded an impressive overview of this knowledge. Elizabeth applied for a Guggenheim Fellowship so she could review what was known about mammalian physiological genetics in the same way. In 1957 she became a senior staff scientist at Jackson; a year later, in 1958, she was awarded the Guggenheim Fellowship and headed for the California Institute of Technology to conduct her review.

Much of Russell's work in the 1960s and 1970s dealt with anemias in mice and embryonic development in mice. She has written several reviews of mouse genetics. She is a lively writer and has used this talent generously to pen tributes to scientists who influenced her career and made great strides in her field: William E. Castle, Sewall Wright, Clarence Cook Little. She wrote of Little, a founder of the Jackson Laboratory, during the period after the 1947 destruction of facilities in a fire:

All experiments in progress were lost, and some staff members and assistants had to move to other laboratories for a short period, while those who remained at Bar Harbor first piled on top of each other . . . then sat at desks on the open floor of a half-built animal house which was under construction at the time of the fire, and nursed the minuscule colonies of mice which began to trickle in. . . . Prexy [Little's nickname], with his energy, enthusiasm, power of persuasion, and vision, was the essential driving force [behind the rebuilding].[2]

In 1982 Russell became an emerita scientist at Jackson Laboratory. She now continues her work at a more leisurely pace. As more is known about the role of genes in "switching on" cancers, their location and proximity to each other, and the mapping of human DNA in the Human Genome Project, more of Elizabeth Russell's questions will be answered. She is a member of the National Academy of Sciences, the American Academy of Arts and Sciences, the Genetics Society of America, the American Society of Naturalists, and the Society of Developmental Biology.

Notes

1. Elizabeth S. Russell, "Origins and History of Mouse Inbred Strains: Contributions of Clarence Cook Little," in *Origins of Inbred Mice*, ed. Herbert C. Morse III. (New York: Academic Press, 1978), p. 42.
2. *Ibid.*

Bibliography

American Men and Women of Science 1992–1993, 18th ed. New Providence, N.J.: R.R. Bowker, 1992.

Noble, Iris. *Contemporary Women Scientists of America.* New York: Julian Messner, 1979.

O'Neill, Lois Decker. *The Women's Book of World Records and Achievements.* Garden City, N.Y.: Anchor Press/Doubleday, 1979.

Russell, Elizabeth S. "Analysis of Genetic Differences as a Tool for Understanding Aging Processes." *Birth Defects* 14 (Mar. 1976): 515–23.

———. "Developmental Studies of Mouse Hereditary Anemias." *American Journal of Medical Genetics* 18 (Aug. 1984): 621–41.

———. "A History of Mouse Genetics." *Annual Review of Genetics* 19 (1985): 1–28.

———. "Origins and History of Mouse Inbred Strains: Contributions of Clarence Cook Little," in *Origins of Inbred Mice*, ed. Herbert C. Morse III, pp. 33–44. New York: Academic Press, 1978.

Wright, Sewall. "Physiological and Evolutionary Theories of Dominance." *American Naturalist* 68 (1934): 25–53.

KELLY HENSLEY

Florence Rena Sabin

(1871–1953)

Anatomist

Birth	November 9, 1871
1893	B.S., Smith College; Phi Beta Kappa
1900	M.D., Johns Hopkins University; Sigma Xi
1901	Awarded a fellowship by the Baltimore Association for the Advancement of University Education of Women, under the direction of Franklin Paine Mall; published *An Atlas of the Medulla and Midbrain*
1902	Assistant Instructor in Anatomy, Johns Hopkins Medical School; published a series on the origin of the lymphatic system
1903	Awarded the prize of the Naples Table Association
1903–1905	Associate Instructor in Anatomy, Johns Hopkins Medical School
1905–17	Associate Professor of Anatomy, Johns Hopkins Medical School
1910	Honorary Sc.D., Smith College
1917–25	Professor of Histology, Dept. of Anatomy, Johns Hopkins Medical School
1924	First woman to be President of the American Association of Anatomists
1925	First woman elected to membership in the National Academy of Sciences; accepted position as researcher at Rockefeller Institute for Medical Research, New York City
1926	Honorary Sc.D., University of Michigan
1929	Honorary Sc.D., Mount Holyoke College
1931	Named one of America's twelve most eminent living women; Honorary LL.D., Goucher College
1932	National Achievement Award, Chi Omega Sorority
1933	Honorary Sc.D., New York University; Wilson College
1934	Honorary Sc.D., Syracuse University
1935	Received the M. Carey Thomas Prize at the 50th anniversary of Bryn Mawr College; Honorary Sc.D.,

Florence Rena Sabin. Photo reprinted by permission of the Sophia Smith Collection and College Archives, Smith College.

	Oglethorpe University; Honorary Sc.D., University of Colorado
1937	Honorary Sc.D., University of Pennsylvania; Honorary Sc.D., Oberlin College
1938	Retired from Rockefeller Institute; moved to Denver; Honorary Sc.D., Russell Sage College
1939	Honorary Sc.D., University of Denver
1944	Accepted position on Postwar Planning Commission
1945	Trudeau Medal, National Tuberculosis Association
1947	Jane Addams Medal for distinguished service by an American woman; Medal for Achievement, University of Colorado; American Brotherhood Citation; Friendship

	Award, American Women's Association; Honorary Sc.D., Colorado State College of Education
1947–1951	Chair, Interim Board of Health and Hospitals, Denver; Manager, Denver Department of Health and Welfare
1949	Elizabeth Blackwell Award, Hobart and William Smith Colleges
1950	Honorary Sc.D., Women's Medical College of Pennsylvania
1951	Lasker Award for outstanding achievement in the field of public health administration, American Public Health Association; Chair, reorganized Board of Health and Hospitals
1953	Distinguished Service Award, University of Colorado; Elizabeth Blackwell Citation; New York Infirmary for Women and Children
Death	October 3, 1953
1958	Statue of Florence R. Sabin placed in Statuary Hall in the National Capitol

"My secret, is working for every man and woman. It is belief in and practice of that creed."[1] With these words Florence Rena Sabin illustrated the key to her many firsts and achievements. Sabin had not one but three brilliant careers: teacher and researcher at Johns Hopkins Medical School; research scientist at the Rockefeller Institute of Medical Research in New York City; and public health advocate with the Colorado Department of Health. These careers united her study of anatomy, histology, physiology, embryology, and public health reform.

Florence Sabin was a woman of many "firsts." She was the first woman faculty member at Johns Hopkins Medical School and the first woman to receive full professorship there. She was the first woman president of the American Association of Anatomists; the first woman elected to the National Academy of Sciences, where no other woman was elected for another 20 years; the first woman to become a full staff member of the Rockefeller Institute; and the first recipient of the Jane Addams Medal for distinguished service by an American woman. According to Dr. Simon Flexner, head of medical research at the Rockefeller Institute, Sabin was widely regarded as "the greatest woman scientist and one of the foremost scientists of all time."[2]

Sabin was born in 1871 in Central City, Colorado, to George Kimball and Rena Miner Sabin. The family moved to Denver in 1875, where Mrs. Sabin died three years later in childbirth. In 1885, Florence went to the Vermont Academy at Saxtons River, Vermont; she graduated in 1889.

Early in her studies at Smith College, Sabin thought of teaching and becoming a doctor. She had been told repeatedly by her grandmother, "too bad you're not a boy, you would have made a good doctor."[3] Outraged by this, she vowed to become a doctor anyway. The atmosphere at Smith, which was the first women's school in the United States to grant an advanced degree, was such that women had the opportunity to receive an education equal in quality to that provided in the best men's schools in the country. She was told of the effort being made in Baltimore to establish a medical school at the Johns Hopkins University, and she was encouraged to apply. Sabin graduated with honors from Smith College in 1893, determined to study medicine.

Johns Hopkins Medical School opened that same year, financed by daughters of the trustees of Johns Hopkins University, who attached a stipulation that women be admitted on the same terms as men and that high scholastic standards for admission be maintained. This institution became the first medical school in the country to rival the high standards that were required at the prestigious European universities. These women founded the Women's Fund Committee and were often referred to as the "Women of Baltimore." The reality of the medical school was an extraordinary victory for women. Until this time, women often faced humiliation, severe opposition, and relative nonacceptance in their efforts to obtain medical degrees.

Sabin entered the fourth class of the medical school in 1896. She was one of 15 women in a class of 42. One of her first major research projects was a three-dimensional model of the mid and lower brain. The duplication of this model became the standard for worldwide use in the teaching of anatomy of the nervous system. From this project, Sabin published *An Atlas of the Medulla and Midbrain* (1901), which became a popular textbook in medical schools for the next 30 years. She realized that her real love was the laboratory rather than clinical practice.

Florence Sabin began a lifelong involvement in social issues, principally public health and, as a pioneer herself, women's rights. She became involved with the Just Government League of Maryland to work toward women's rights. She helped with the Maryland *Suffrage News*, took part in letter-writing campaigns, and attended suffragist meetings and parades. In 1914 the League was absorbed into the National Women's Party, which became an influential voice in achieving women's right to vote in 1920.

The issue of sexual discrimination arose throughout Sabin's early career, and her hard work and achievements were not always enough to contest the prejudice against women. In 1901 she was awarded a fellowship at Johns Hopkins, by the Baltimore Association for the Advancement of University Education of Women, where she began her research into the development and structure of the lymphatic system. In 1902 she became an assistant professor in anatomy; in 1917, a full professor of histology. She had originally been passed over to succeed her mentor, Franklin Paine Mall, as professor of anatomy because of her gender. Sabin was deeply disappointed by this and felt she had failed all women in medicine. She refused to give up because of discrimination, claiming "there is research to be done." She was determined to be the best researcher ever.

Sabin was associated with Johns Hopkins, teaching and conducting research, for 25 years. Her study of lymphatics, the origin of blood vessels, and the development of blood cells in small pig embryos led to the discovery that the lymphatics were closed at their collecting ends. This new knowledge of the lymphatic system was considered later to be one of the five most significant contributions in the previous 25 years of scientific research and Sabin's most notable contribution. Sabin said of her research, "When I began my work, it was the accepted theory that lymphatics arose from tissue spaces and then grew toward the veins." But her persistent research proved that the lymphatics develop in a manner opposite to that original theory; rather, they arise as buds from the veins and grow outward as continuous channels.[4]

During this period Sabin traveled to the laboratories in Leipzig, Germany, where she studied a new technique that involved using "living stains," allowing one to work with live tissues. She integrated this method of vital staining in studying normal blood and diseases of the blood, demonstrating the properties of living cells as compared to dead blood cells. The distribution of the dyes enabled researchers to clearly identify cell types and to study their functions under normal and diseased conditions. Among the cells she studied were monocytes, the cells involved in tissue reactions against infectious disease.

Sabin loved teaching and was highly respected by and devoted to her students. She was revolutionary in her teaching ideas, favoring dynamic teaching methods and encouraging complete freedom within the course for the student and teacher. She introduced cooperative learning and group work for science, a philosophy she herself followed by giving fair credit and recognition to those working with her. Sabin reflected on her years of teaching at the end of her career at Johns Hopkins: "I have ceased to be a professional teacher, but remain a professional student."[5] She was greatly interested in younger people who were involved in medical research because she believed that in them and through them lay the hope of the future.

In 1925, Sabin accepted a position at the Rockefeller Institute under director Dr. Simon Flexner. This and similar institutes reflected a change in purpose of medical study from the cure of diseases to preventive medicine. The Institute had become one of the leading research centers in the world, focusing on infectious diseases. Sabin organized the Department of Cellular Studies, concentrating her research on the relationship of her past studies of the monocyte and other white blood cells in the defense of the body against infections to tuberculosis. Her work on the tubercle bacillus launched a new technique for the study of infections, allergies, and immunities. In 1926 she joined the Research Committee of the National Tuberculosis Association, which coordinated all tuberculosis research conducted in the United States. This program was the largest cooperative research project yet undertaken, with universities, research institutes, and pharmaceutical companies combining resources to advance the knowledge of tuberculosis and the causes of the infection.

Sabin retired in 1938, returning to Denver, Colorado, to make a home with her sister Mary. She continued traveling east for research and lectures. In 1944, Sabin was asked by Governor John Vivian of Colorado to chair a subcommittee on health for his Postwar Planning Commission. Shortly thereafter she began her third career, dedicating her remaining working years to the fight for public health reform. She investigated the health situation in Colorado and discovered that the state had one of the highest disease rates and worst sanitation systems in the country. Sabin became chair of the Public Health Subcommittee and was influential in writing a series of reform bills, known as "The Sabin Health Bills." She traveled to every county in the state, campaigning for passage of the bills with her slogan, "Health to Match Our Mountains."

Sabin focused her efforts on the city of Denver, which remained unaffected by the passage of the reform bills because it was governed under a home-rule principle. During her tenure as manager of Denver's Department of Health and Welfare, more was accomplished concerning health issues in four months than had been achieved in the city's entire history. She was instrumental in taking the state health department out of politics, doubling its appropriations, and encouraging the formation of county health departments. She ridded the city of its rat infestation, improved sewage disposal, and saw to it that no unpasteurized milk could enter the city.

Throughout her lifetime, Sabin did more than any other person to open the doors of laboratories, medical schools, and hospitals to women. Through her involvement with the "Women of Baltimore," she agreed that participation in the struggle would not be too high a price to pay for a medical education. She was never allowed to forget that she was a woman in a "man's world" and was always willing to do whatever she could to help women reach their highest potential. Sabin once recalled that "the admission of women into Johns Hopkins Medical School on the same terms as men has opened up to women every opportunity for advanced work in medicine which they have had since."[6] She believed strongly that women should not give up their training when they marry, as was common practice. She claimed, "Indeed, one of the next steps in the feminist movement is for educated married women to claim and carry on a share of professional work."[7]

Florence Sabin dedicated her life to helping others, paving the way for thousands of women who followed her into medicine and research. She was a true pioneer. Sabin ceaselessly advocated equality and equal opportunity for all—scientifically, philosophically, and politically. A bronze statue of Sabin stands in the Statuary Hall of the National Capitol in Washington, with a replica in Denver's Public Health Department. The engraved words "Teacher, Scientist, Humanitarian" signify the richness of Florence Sabin's life.

Notes

1. Janet Kronstadt, *Florence Sabin, Medical Researcher* (New York: Chelsea House, 1990), p. 86.

2. "Florence R. Sabin, Scientist, 81, Dies," *New York Times* (October 4, 1953): 89.

3. Kronstadt, *Florence Sabin*, p. 26.

4. Philip D. McMaster and Michael Heidelberger, "Florence Rena Sabin," *Biographical Memoirs, National Academy of Sciences*, Vol. 34 (New York: Columbia University, 1960), p. 260.

5. Florence R. Sabin, "The Opportunity of Anatomy," *Science* 15 (May 1925): 499.

6. Florence R. Sabin, "Women in Science," *Science* 10 (Jan. 1936): 25.

7. Florence R. Sabin, "The Extension of the Full-Time Plan of Teaching Clinical Medicine," *Science* 11 (Aug. 1922): 150.

Bibliography

Bluemel, Elinor. *Florence Sabin: Colorado Woman of the Century*. Boulder: University of Colorado Press, 1959.

Downing, Sybil. *Florence Rena Sabin, Pioneer Scientist*. Boulder: Pruett Publishing, 1981.

Haber, Louis L. *Women Pioneers of Science*. New York: Harcourt Brace Jovanovich, 1979.

Kaye, Judith. *The Life of Florence Sabin*. New York: Henry Holt and Co., 1993.

Kronstadt, Janet. *Florence Sabin, Medical Researcher*. New York: Chelsea House, 1990.

McMaster, Philip D., and Michael Heidelberger. "Florence Rena Sabin," in *Biographical Memoirs, National Academy of Sciences*, Vol. 34. New York: Columbia University Press, 1960.

Phelan, Mary Day. *Probing the Unknown: The Story of Dr. Florence Sabin*. New York: Crowell, 1969.

Sabin, Florence R. *Franklin Paine Mall: The Story of a Mind*. Baltimore: Johns Hopkins University Press, 1934.
Yost, Edna. *Women of Modern Science*. Westport, Conn.: Grenwood Press, 1984.

GAIL M. GOLDERMAN

KATE OLIVIA SESSIONS
(1857–1940)
Horticulturist

Birth	November 8, 1857
1909	Leader in the founding of the San Diego Floral Association
1935	"K.O. Sessions Day" was celebrated at the California-Pacific International Exposition, San Diego
1939	Meyer Medal, Council of the American Genetic Association
Death	March 24, 1940

Kate Sessions's career and period of influence was long, spanning nearly 60 years. Besides her nursery, her collecting, and her work with the San Diego Floral Association, she was a supervisor of gardening for the San Diego Public Schools and a member of the extension faculty of the University of Southern California. She was a popular public speaker; through her speaking she stimulated interest in civic beautification and home gardens. Her personality was dynamic, sometimes abrupt and overly frank. She was not noted for her patience. Nonetheless, her passion for plants enabled her to overcome any obstacle, and her knowledge attracted horticulturists from afar to learn from her.

Kate Olivia Sessions was born in San Francisco in 1857, the first of two children and the only daughter of Josiah Sessions, a breeder of fine horses, and Harriet (Parker) Sessions. In the following year Josiah Sessions moved his family to a farm in Oakland. Kate grew up riding her pony on the Oakland hills, becoming familiar there and in her mother's garden with flowers. She later said, when asked how she got started in gardening, "I was always started; I grew up in a garden."[1] As a small child, she often "helped" in the gardens of her father and uncle, unfortunately without being able to distinguish between weeds and vegetables.

She was educated in the Oakland public schools. After graduation from high school she made a trip to Hawaii, where she was greatly impressed by the beautiful plant life. In 1877 she entered the University of California, Berkeley, where she was known as the prettiest girl on campus. Apparently her interests extended beyond being a campus belle, for she enrolled in the scientific course, most unusual for a woman at that time, and graduated in 1881 with a specialization in chemistry. Her thesis was entitled "The Natural Sciences as a Field for Women's Labor." She accepted a teaching position at Russ School in San Diego, but soon left teaching to pursue her interest in horticulture—to labor in the natural sciences.

In 1885, Sessions opened a nursery in Coronado with an office and flower shop in San Diego. "I had little knowledge for the work," she later wrote, "although I was a college graduate. Thirty years ago one got but little of botany and natural sciences at college. . . . But I loved plants and trees, and I could always make things grow, so I wasn't afraid."[2] By 1892, she was renting 30 acres of land from San Diego for use as a nursery. Her rent was 300 trees donated to the city and 100 trees a year planted on the property. This property became Balboa Park, one of the premier attractions in San Diego to this day; the trees she paid as rent to the city did much to begin the beautification of San Diego.

Kate Sessions closed her flower shop in 1909 and moved her nursery a number of times, but she continued to operate a nursery business until her death. She was not a particularly good businesswoman, never accumulating much wealth. However, she supported herself with her business and, more important, was able to spend her life doing the work she loved. The petite, feminine young woman who arrived in San Diego put aside her wardrobe of fine dresses to become a gardener. She wore heavy shoes and grew weathered and bent from hours of working out-of-doors. Kate Sessions earned recognition and respect in a man's world. She was noted as a collector, having introduced many plants to the San Diego area and popularized more. These include bougainvilleas, palms, jasmines, tecomas, and eucalypti never before seen in the area until she grew them in her gardens. She has been credited with being the one person responsible for the beauty of the parks and gardens of San Diego and Coronado.

In association with Alfred D. Robinson and other horticulturists, Kate Sessions participated in the founding of the San Diego Floral Association. The organization was intended to promote a knowledge of floriculture, to stimulate an intelligent love of flowers, and to beautify the houses, schools, and public grounds of San Diego. She served as an officer or board member of the Association for over 20 years. A monthly magazine, *California Garden*, is sponsored by the Association. Kate Sessions was a regular contributor to the magazine. She discussed new plants and described her collecting trips to Europe in 1925 and to Hawaii in 1926. Her writing was direct and wonderfully descriptive. The extract that follows is from her writing about ground covers in *Garden and Home Builder*:

> A new prostrate-growing shrub that is but slightly known to the trade and very little in cultivation as yet, although introduced to cultivation over a hundred years ago, is *Thomasia foliosa*, an Australian evergreen requiring a very mild climate. Its sinuous and serrated leaf is a dark glossy green, quite rough, like shriveled leather. It bears small, dainty, white, inconspicuous blossoms with seed pods like nail heads. It is absolutely flat to the ground and a most generous grower. A plant the size of a finger in two years covered a space two by two and a half feet![3]

The esteem in which Sessions was held is exemplified by the honors she received. She was honored by many tree plantings during her lifetime. During the California-Pacific International Exposition of 1935, a "K.O. Sessions Day" was held. In 1939, she was the first woman ever awarded the Meyer Medal by the Council of the American Genetic Association. The medal was established in 1919 by bequest of Frank Meyer, a pioneering plant explorer for the U.S. Department of Agriculture, and honors distinguished service in the field of foreign plant introduction. Sessions earned the award for her outstanding contributions to the horticulture of her native state. The description of the award ceremony stated that

> The people of San Diego owe to Miss Sessions a debt that they can never repay. Her influence has made their town lovely and gracious. . . . She has dedicated her genius and energy to the principle of leaving her community a legacy of beauty and making life itself more serene and beautiful.[4]

In 1956 an elementary school in Pacific Beach was named in her honor, as was a memorial park in the same city in 1957. One year after receiving the Meyer Award, Kate Sessions died at age 82 from bronchial pneumonia following a broken hip. She is buried in Mount Hope Cemetery, San Diego.

Notes

1. Victoria Padilla, *Southern California Gardens* (Berkeley: University of California Press, 1961), p. 168.
2. *Ibid.*, p. 169.
3. Kate Olivia Sessions, "Ground Cover Plants for Southern California," *Garden and Home Builder* 46 (Dec. 1927): 352–53.
4. "Two Major Medals Awarded during 1939," *Journal of Heredity* 30 (Dec. 1939): 531–33.

Bibliography

California Garden (Autumn 1953). Entire issue.

Cockerell, T.D.A. "Kate Olivia Sessions and California Floriculture." *Bios* 14 (Dec. 1943): 167–79.

Notable American Women, 1607–1950. Cambridge and London: Belknap Press, 1971.

Padilla, Victoria. *Southern California Gardens*. Berkeley: University of California Press, 1961.

Payne, Theodore. "History of the Introduction of Three California Natives." *El Aliso* 2 (Mar. 15, 1950): 109–14.

Robinson, Alfred D. "An Appreciation, Not an Obituary." *California Garden* (May 1940).

Sessions, Kate Olivia. "Color Planting for Pacific Coast Gardens." *Garden Magazine* 32 (Dec. 1920): 205–8.

———. "Ground Cover Plants for Southern California." *Garden and Home Builder* 46 (Dec. 1927): 352–53.

"Two Major Medals Awarded during 1939." *Journal of Heredity* 30 (Dec. 1939): 531–33.

CAROL W. CUBBERLEY

LYDIA W. SHATTUCK

(1822–1889)

Botanist

Birth	June 10, 1822
1848–51	Student, Mount Holyoke Seminary
1851–89	Teacher, Mount Holyoke Seminary/College
1873	Attended Anderson School on Penikese Island

Lydia W. Shattuck. Photo courtesy of the Mount Holyoke College Library Archives.

1874	Attended Priestley Centennial
1887	Traveled to Hawaii to see Mauna Loa
Death	November 2, 1889

Lydia Shattuck holds an important place in science and in science education. At Mount Holyoke, where she spent her career, she not only carried on the tradition of the laboratory method of teaching that was begun by Mount Holyoke's founder, Mary Lyon, but also expanded Mount Holyoke's influence in the world of science and helped define its role as a women's college that specialized in science. As a teacher and an inspiration for younger teachers, she had a profound influence on the teaching of biology in the United States and elsewhere. As one of the first women to join and be a founding member of scientific societies, she helped enlarge the sphere of women engaged in scientific research. Her participation in the Penikese Island school helped prove that women were capable of advanced work in science and led the way for the next generation of women scientists to enter graduate schools.

Lydia White Shattuck was born in East Landaff (later called Easton), New Hampshire, in 1822. Her parents, Timothy S. Shattuck and Betsey (Fletcher) Shattuck, were first cousins and had married against the wishes of their families. In order to avoid unpleasantness, the Shattucks had migrated to the wild mountainous regions of New Hampshire, where Lydia was born. She was their fifth child and the first to live to maturity. As a young girl Lydia spent much of her time out-of-doors, climbing the White Hills with her younger brother William, "watching the birds and the ever-changing sky and above all seeking out every variety of the flowers."[1]

Having finished her schooling by age 15, she began teaching in the district school in order to support herself. During the next 11 years Shattuck attended academies in Newbury, Vermont, and Haverhill, New Hampshire, for short periods when she could take time out from her teaching. At age 26, much older than most of her classmates, she entered Mount Holyoke Seminary in Massachusetts, one of the first academies for young women, and briefly studied with Mary Lyon, the founder of the Seminary. To help pay her board and tuition at the Seminary, she worked four hours per day in the domestic hall, where she had charge of bread-making and baking. She found the chemistry of the bread-making process extremely interesting and would later use this knowledge in her chemistry lectures.

Her roommate at the Seminary describes her as having "fair white skin . . . bright blue eyes . . . and peculiarly sunny golden brown hair" and recalls how Shattuck first became interested in botany as a field of study. Although she had always loved flowers, Shattuck had as much trouble with botany as her classmates did. Again her roommate remembered that "on one early Spring day she brought to our room . . . a flower which she found difficulty in analyzing. She worked over it for some time and could not succeed in placing it." Working together, the two roommates finally succeeded, and "from that moment, she dated her love of Botany as a science."[2]

Lydia Shattuck graduated from the Seminary in 1851, remaining there as a faculty member throughout her life and winning recognition chiefly as a botanist of "unquestioned standing in the world of science."[3] She was an early proponent of evolution, seeing no conflict between Darwin's theory and her own liberal views of religion. Her friendship with Harvard University botanist Asa Gray may have encouraged her to hold this view of evolution. The field of chemistry was her second love. She sometimes said that in the winter she liked chemistry best, and in the summer, botany.

Shattuck made every attempt to keep current in her two fields. She was a "corresponding member" of the Torrey Botanical Club of New York City during the 1880s. She was also one of the few women chemists who attended the 1874 Priestley Centennial, which led to the formation of the American Chemical Society. However, when the photograph was taken of the founders of the Society, Shattuck and the other women chemists were asked to stand aside with the male chemists' wives, so none of them appear in the photograph.[4]

Shattuck was also one of the first women to attend the Anderson School of Natural History conducted by Elizabeth and Louis Agassiz on Penikese Island in Buzzard's Bay, Massachusetts, in 1873. Louis Aggasiz, a famous scientist of the times, carefully chose the 50 people, 15 of whom were women, who would attend that first summer session on Penikese Island. The Anderson School played a critical role for women scientists in the United States because it offered a rare opportunity for advanced training in biology at a time when such training was otherwise closed to women. Additionally, Shattuck conscientiously read the British journal *Nature*, one of the few scientific journals published at the time, so that she "knew exactly what was going on in geology, zoology, and physiology."[5]

Like most women scientists of her day, Shattuck never published scientific papers. She did, however, do active research on the classification of plants and corresponded with many scientists at other institutions. An expert on the Connecticut Valley region of Massachusetts, she was at one time president of the Connecticut Valley Botanical Association. She also traveled to Europe, Canada, and Hawaii to pursue her botanical interests. She was thrilled to watch the eruption of the Mauna Loa volcano in Hawaii. Many of her students who became missionaries in foreign lands showed their devotion to Shattuck and her research by sending specimens of plants to her from their new homes. Thus, through Shattuck's work, the Mount Holyoke Seminary herbarium came to house thousands of plants from all over the world.

Her students caught her enthusiasm for biology. **Cornelia Clapp**, a noted biologist, was a student of Shattuck's in the 1860s. Interviewed in 1921, Clapp recalled Shattuck as the "most honored person at Penikese." It was Shattuck who had ensured that Clapp would also attend the Anderson School at Penikese. Clapp also remembered Shattuck's delight in looking through a microscope as she exclaimed, "I never in all my wildest dreams expected to have a microscope to use for my own." Clapp also recalled an incident when she and Shattuck were collecting biological specimens. To look for animals in a pond, "we took all our clothes off and went in." When someone came along, they hid behind a log; but Shattuck didn't mind her body being seen as long as her head didn't show: "Miss Shattuck was not prudish. Miss Shattuck had a streak of the modern."[6]

Although Shattuck's advanced training never went beyond Penikese Island, many of her students did go on to earn doctorates soon after graduate schools opened their doors to women in the 1880s. Shattuck's student, Henrietta Hooker, was awarded one of the first Ph.D.s in botany from Syracuse University in 1888; and her student and friend, Cornelia Clapp, earned two Ph.D.s in zoology, one from Syracuse in 1889 and another from the University of Chicago in 1896. Indeed, so many of Shattuck's students followed the same path that Mount Holyoke soon became one of the leading undergraduate institutions in graduating women who subsequently obtained doctorates in the biological sciences.[7]

When Mount Holyoke Seminary celebrated its fiftieth anniversary in 1887, just before it became a college, Shattuck gave a speech to the alumnae in which she said:

> Our non-resident professors give us much credit for doing good work in science. One of them recently suggested that in the future as a college we give a special scientific direction to our pursuits; for said he, "there is a spirit of study here favorable to scientific research, and science does not tolerate any half-way work." Since, therefore, the instruction of the Seminary has had a scientific trend from the first, without tendency to convert us into agnostics or infidels; since this is a scientific age and we are bound to keep abreast of the times; since every college has its own particular individuality—let us press onward in these lines until we obtain full recognition among the colleges of New England.[8]

Shattuck's last work, before her death on November 2, 1889, involved soliciting funds for a new building to house the chemistry and physics departments of Mount Holyoke College. This building, which opened in 1893, became Shattuck Hall.

Notes

1. Charlotte Mann Paine, in Lydia Shattuck Papers, Mount Holyoke College Library Archives.

2. *Ibid.*

3. Arthur C. Cole, *A Hundred Years of Mount Holyoke College: The Evolution of an Educational Ideal* (New Haven, Conn.: Yale University Press, 1940), p. 156.

4. Margaret W. Rossiter, *Women Scientists in America: Struggles and Strategies to 1940* (Baltimore and London: Johns Hopkins University Press, 1982), p. 78.

5. Cornelia Clapp, oral interview (1921), 56. In Cornelia Clapp Papers, Mount Holyoke College Library Archives.

6. *Ibid.*

7. M. Elizabeth Tidball and Vera Kistiakowsky, "Baccalaureate Origins of American Scientists and Scholars," *Science* 193 (Aug. 20, 1976): 646–52.

8. Emma P. Carr, "The Department of Chemistry: Historical Sketch," *Mount Holyoke Alumnae Quarterly* 2 (Oct. 1918): 159–65.

Bibliography

Carr, Emma P. "The Department of Chemistry: Historical Sketch." *Mount Holyoke Alumnae Quarterly* 2 (Oct. 1918): 159–65.

Cole, Arthur C. *A Hundred Years of Mount Holyoke College: The Evolution of an Educational Ideal.* New Haven, Conn.: Yale University Press, 1940.
Memorial of Lydia W. Shattuck. Boston: Beacon Press, 1890.
Notable American Women, 1607–1950. Cambridge and London: Belknap Press, 1971.
Rossiter, Margaret W. *Women Scientists in America: Struggles and Strategies to 1940.* Baltimore and London: Johns Hopkins University Press, 1982.

CAROLE B. SHMURAK

ELLEN KOVNER SILBERGELD

(1945–)

Biologist

Birth	July 29, 1945
1967	A.B., modern history, Vassar; Leverhulme and Fulbright Fellow, London
1968–70	Secretary and Program Officer, Committee on Geography, National Academy of Sciences, National Research Council, Washington, DC
1972	Ph.D., environmental engineering sciences, Johns Hopkins University
1972–75	Postdoctoral Fellow in Environmental Medicine and Neurosciences, Johns Hopkins University
1975–79	Staff Fellow and Head, Unit on Behavioral Neuropharmacology, Therapeutics Branch, National Institute of Neurological and Communicative Disorders and Stroke (NINCDS), NIH
1979–81	Chief, Section on Neurotoxicology, NINCDS, NIH
1982–91	Chief Toxic Scientist and Director, Toxic Chemicals Program, Environmental Defense Fund
1987–91	Associate Faculty, Dept. of Health Policy and Management, School of Hygiene and Public Health, Johns Hopkins Medical Institutions
1989–	Adjunct Professor, Dept. of Pharmacology and Experimental Therapeutics, University of Maryland, Baltimore

Ellen Kovner Silbergeld. Photo courtesy of the University of
Maryland at Baltimore.

1990–	Affiliate Professor of Environmental Law, University of Maryland Law School
1991–	Professor, Dept. of Pathology, University of Maryland; Adjunct Professor, Dept. of Health Policy and Management and Dept. of Environmental Health Sciences, Johns Hopkins School of Hygiene and Public Health Medical School; Professor, Dept. of Epidemiology and Preventive Medicine, University of Maryland Medical School
1993	MacArthur Foundation Fellowship

Ellen Kovner Silbergeld made key decisions that changed the direction
of her professional career. Each change provided many opportunities to
use her academic background and previous experiences to bring a new
perspective to her work as a researcher and an activist. Silbergeld's

unique combination of talents and interests has enabled her to make important contributions to the field of biology and to public policy.

Silbergeld was born in Washington, D.C., in 1945. Her parents, Joseph and Mary Gion Kovner, influenced her life in many ways. They held strong social, political, and personal beliefs and were willing to take risks. From them, Silbergeld learned about the importance of taking stands. Her decision to attend Vassar was influenced by the fact that her mother had also attended a women's college. Silbergeld's parents encouraged her to pursue her own personal interests.

In 1967, Silbergeld earned a bachelor's degree in modern history from Vassar College. During her undergraduate years she was a history major who hated science. After graduation she received a Fulbright Fellowship and traveled to England to begin studying for a doctorate in economic history at the London School of Economics. Although Silbergeld was interested in quantitative history, she soon discovered that she hated economics. In 1968 she returned to the United States and began working as a secretary and program officer for the Committee on Geography at the National Academy of Sciences of the National Research Council in Washington, D.C. While reading and eventually editing reports, she developed an interest in science.

This experience marked the first major change in the direction of her career. In 1968, Silbergeld decided to return to graduate school to pursue a doctorate in environmental engineering at Johns Hopkins University in Baltimore, Maryland. Because she was the only woman in the program, she had to cope with the prejudices often experienced by women in engineering and science during this time.

Following the advice of her academic advisor, Silbergeld took a course in basic biology as a supplement to her courses in economics, resource management, and the environment. During the first lecture she discovered that she loved biology. Silbergeld decided to change the path of her career for the third time and focused her attention on the study of biochemical toxicology. She earned her doctorate in environmental engineering sciences in 1972. She also began studying lead neurotoxicity, a topic that has become a major focus of her work as a researcher. She became an advocate for changes in public policy that promise to protect individuals, especially children, from the dangers of lead.

In 1975, Silbergeld assumed a position as a scientist at the National Institutes of Health (NIH) in Bethesda, Maryland. At age 35, she became one of NIH's youngest lab chiefs when she was appointed head of the Section on Neurotoxicology of the National Institute of Neurological and Communicative Disorders and Stroke. Her research into the effects of compounds, such as lead and food dye, has earned her the respect of her colleagues. She was among the scientists who worked at the developing edge of the fields of neuroscience and toxicology, two fields that, she points out, "invented themselves" in the 1970s.

The year 1982 brought another change in Silbergeld's career. She became a senior toxicologist at the Environmental Defense Fund in Washington, D.C. She was told that in doing so she was taking a risk that might irreversibly change her reputation as a researcher. Silbergeld states that her work at this well-respected organization was "very exciting" because of the intense interactions among lawyers, economists, engineers, and other individuals from various backgrounds. Silbergeld investigated "very real problems" and experienced a sense of accomplishment because she was working with others in an organization that has tremendous influence in political debates.

As an activist for this environmental group, she provided expert testimony on behalf of the United Auto Workers in its 1982 lawsuit against the battery manufacturer Johnson Controls. Although the lower court ruled in favor of the company and upheld its "fetal protection policy," which barred all fertile women from jobs exposing them to lead, the policy was later overturned by the U.S. Supreme Court.

Her experiences at the Environmental Defense Fund validated Silbergeld's decision to combine scientific research with an advocacy career. For her, science is a passionate pursuit of knowledge. She believes that applying knowledge gained from research in the area of public policy is important for several reasons. First, the application teaches scientists how policymakers view the results of research and forces them to clarify the policymakers' misinterpretations of scientific matters. Second, the application of the results reveals gaps in knowledge that become opportunities for new research.

In 1991, Silbergeld decided to leave the Environmental Defense Fund to return to research, teaching, and advising students. She accepted a position as professor in the Department of Pathology of the University of Maryland Medical School in Baltimore. Her work in the environmental movement was a major incentive for her return to academia. In 1992 she joined the Department of Epidemiology and Preventive Medicine. Silbergeld has also served as a faculty member at Johns Hopkins Medical Institutions, at the University of Maryland in Baltimore, and at the University of Maryland Law School.

In her current role, Silbergeld teaches students that good science can improve public policy and that good public policy can lead to improvements in science. She enjoys working with students who have inquiring and skeptical minds and who present ideas that challenge her own assumptions. Students who develop these ideas, design strategies for investigating these ideas, and then show that their ideas are right are, in her words, "the way to the future." Working with these students as they develop our knowledge beyond what we now know is "terrifically exciting" for her.

During one of her lectures about lead toxicity, Silbergeld shared her knowledge of the history of this basic element, discussed the toxic effects of lead, and introduced unanswered questions about its effects on humans. For example, she asked her students, "If lead is just an element, then why can't we figure out how it is toxic?" Silbergeld also discussed her recent studies of the possible toxic effects of the mobilization of bone lead during pregnancy and lactation.

Throughout her career, Silbergeld has made numerous contributions to the field of biology. She does not boast about her triumphs despite the fact that she has received numerous honors and awards, including a prestigious MacArthur Foundation Fellowship. She has served on various committees and advisory boards and has helped organize many symposia and workshops. Silbergeld has been a consultant for the government of Bermuda and an exchange fellow in Yugoslavia. She has reviewed grants for the National Foundation of the March of Dimes, the National Science Foundation, and many other groups. She has served on the editorial board of numerous journals, including the *American Journal of Industrial Medicine* and *Chemical and Engineering News*.

Throughout her life, Silbergeld made decisions that changed the direction of her career. When asked to reflect on these decisions, she commented that doors were opened and that she wanted to do "something that seemed very compelling." At these turning points she did not perceive the decisions as risks; rather, she thinks that other people were willing to take risks on her behalf. For example, being accepted to the doctoral program at Johns Hopkins when she did not have undergraduate work in biology and chemistry was a risk for the university. Silbergeld believes she is fortunate that these doors were opened to her and that it was not impossible for her to go down the paths that she chose. She made each choice at the time because she was convinced that each was the right thing to do.

Silbergeld does not know if the world has changed sufficiently so that young women can pursue their interests without facing the same problems that she encountered. She has a teenage daughter and believes that girls now seem "tremendously more confident about themselves and much less prone to evaluate themselves in terms of how boys evaluate them." She believes that individuals are very fortunate if they love their work, if they have the ability to do the work that they love, and if they can make a living doing this work. Silbergeld lives according to these guidelines. For her, science is "selfish enjoyment" because she is pursuing her own ideas and because her work is rewarding.

Silbergeld's career demonstrates how a unique combination of talents and interests can provide a rich background for contributing to one's profession in ways that offer personal satisfaction. This former history major has proved that becoming a scientist, an environmental activist, and a university professor is worth the risks and challenges.[1]

Note

1. This entry is based on personal interviews between Dr. Hamberger and Dr. Silbergeld held at the University of Maryland campus, November 4, 1994.

Bibliography

American Men and Women of Science, 18th ed. New Providence, N.J.: R.R. Bowker Co., 1992.

Florini, Karen L., and Ellen K. Silbergeld. "Getting the Lead Out." *Issues in Science and Technology* 9 (Summer 1993): 33–39.

Hening, Robin Marantz. "Ellen Silbergeld: The Making of a Biochemist." *SciQuest* 54 (Mar. 1981): 22–24.

Silbergeld, Ellen K. "Lead in Bone Implications for Toxicology during Pregnancy and Lactation." *Environmental Health Perspectives* 91 (1991): 63–70.

———. Review of *The Burning Season: The Murder of Chico Mendes and the Fight for the Amazon Rain Forest,* by Andrew Revkin. *New England Journal of Medicine* 324 (Mar. 1991): 930.

———. "Talking Point: How Research and Activism Can Mix." *New Scientist* 132 (Nov. 1991): 12.

Who's Who in America, 46th ed., 1990–1991. Wilmette, Ill.: Marquis Who's Who, 1990.

Who's Who in the East, 22nd ed., 1989–1990. Wilmette, Ill.: Marquis Who's Who, 1988.

NAN MARIE HAMBERGER

MAUD CAROLINE SLYE

(1869–1954)

Pathologist

Birth	February 8, 1869
1899	A.B., Brown University
1914	Received Gold Medal for science exhibit, American Medical Association
1905–1908	Graduate Student, University of Chicago
1908–11	Postgraduate Fellow, University of Chicago

Maud Caroline Slye. Photo courtesy of the Library of Congress.

1911–43	Staff Member, Sprague Memorial Institute, Chicago
1919–22	Instructor in Pathology, University of Chicago
1922	Gold Medal, American Radiological Society
1922–26	Assistant Professor of Pathology, University of Chicago
1926–45	Associate Professor of Pathology, University of Chicago
1937	D.Sc., Brown University
Death	September 17, 1954

Although Maud Slye's theories were later shown to be an over-simplification of the genetics of cancer, her studies were instrumental in establishing the role that heredity plays in human susceptibility to cancer. However, her scientific credentials were unquestionable. She held honorary memberships in the Seattle Academy of Surgery and the Southern California Medical Society. She held memberships in the American

Association for the Advancement of Science, the American Medical Association, and the Association of Cancer Research, among others. By the 1950s, owing to her work, physicians were asking patients whether there was any history of cancer in their family.

Maud Slye was born in Minneapolis in 1869, the middle child of three children. Her father, James Alvis Slye, was a lawyer and author. Her mother, Florence Alden (Wheeler) Slye, was the granddaughter of the founder and first president of Allegheny College, the Reverend Timothy Alden. She was interested in poetry. Her father died shortly after she graduated from high school, so Maud worked as a stenographer in St. Paul for nearly a decade, from 1886 until 1895, to assist with family finances.

When she did begin college at the University of Chicago in 1895, she worked her way through school as a clerk in the president's office to pay her tuition and board while taking a full course load. After three years of nearly full-time work as well as full-time student status, Maud suffered a "nervous breakdown." She withdrew from college and recovered with relatives in Massachusetts. There she took courses at the Woods Hole Marine Biological Laboratory, which intensified her interest in biology. She finally completed her bachelor's degree at Brown University in 1899. Her first professional position was as a high school psychology teacher, where she was first exposed to genetics.

Maud began her graduate studies at the University of Chicago and became a graduate assistant of Professor Charles Whitman, whom she had met in Woods Hole. Her research initially involved studying the nervous disorders of Japanese waltzing mice. She planned to breed these mice with normal mice to track inheritance patterns of the nervous disarrangement: moving in a circular waltzing motion, tossing their heads, and demonstrating general hyperactivity. Maud cared for the mice herself, buying their feed from her graduate assistant wages with very little money left for her own food.

While waiting for the mice to breed, she became interested in the work of Leo Loeb, a pathologist studying the endemic incidence of cancer that persisted in certain communities. At the time, there were three animal examples of possible inheritance of cancer (eye cancer in cattle from a ranch in Wyoming, thyroid cancer in rats, and lung tumors in certain strains of mice). Maud realized that a careful breeding program was needed to determine whether cancer was inherited. She decided to change her research direction and use her carefully recorded breeding colonies to track the inheritance of spontaneous tumors. Specifically, she wanted to determine the inheritance patterns in litters born to female mice with active, spontaneous breast cancer. The project she envisioned would involve breeding and housing all mice until their natural deaths and autopsying every mouse, examining tissues for evidence of malignancy or other causes of death.

Maud Slye sought financial support for a cancer inheritance project, but her requests were often dismissed. Two factors influenced her fundraising efforts: first, it was a time when there were very few women in the university environment and the attitude of most men was that a woman's place was in the home; and second, her theory of cancer inheritance was not accepted or understood by scientists at the time. She finally gained approval for her project from the pathology department chair, Dr. Ludwig Hektoen, but the promised funding never materialized through the university channels. Maud continued to feed her mouse colonies from her own meager salary.

In 1911 the Sprague Memorial Institute was established at the University of Chicago with H. Gideon Wells, M.D., an eminent pathologist, appointed as director. At the Institute, Maud was given a salary, funding, and a laboratory large enough to breed and raise colonies of mice, but not enough money to hire an assistant. She enlisted the volunteer assistance of Harriet Holmes to prepare the tissue slides and of the director, Dr. Wells, to confirm readings of the tissue slides and lend credibility to her conclusions. Maud personally recorded all events in the lives of her mice. She knew each mouse by name/number and heredity. When her aged mother was dying in California and required her presence for an extended time, Maud shipped the mice and the records by train boxcar to California so she could continue to look after them. She kept the mouse cages painstakingly clean to eliminate any deaths by infection or other diseases. For many years she took no vacation because she refused to leave her mice in the care of anyone but herself.

Maud Slye's paper, "The Incidence and Inheritance of Spontaneous Cancer in Mice," was presented to the American Society of Cancer Research in May 1913. In this paper, Maud described the 298 mice out of 5,000 raised that had spontaneously developed cancer. From this work she reached these conclusions: (1) all mice bred of noncancerous parents died without cancer by autopsy; (2) if cancerous mice were bred with noncancerous mice, cancer-free young were produced; (3) when the second-generation cancer-free young were bred with other second-generation cancer-free young, they produced third-generation cancerous mice; (4) if both parents were cancerous, the majority of the offspring would develop cancer; (5) cancer was not contagious, that is, it could not be transmitted by housing noncancerous mice in proximity to cancerous mice; (6) the cancer toxin theory was dispelled because mice with enormous tumor burdens remained alive; and (7) the presence of cancer did not interfere with reproduction. In June 1914, the American Medical Association presented Maud Slye an award for her efforts. Her exhibit on the hereditary transmission of cancer in mice won a gold medal as the best scientific exhibit. In recognizing her work, the committee acknowledged the "personal sacrifice" she had made.[1]

However, many scientists engaged in cancer research at the time refused to accept Maud's work and her conclusions: namely, that cancer can be selectively bred in or bred out, that susceptibility to cancer is a recessive trait, and that resistance to cancer is a dominant trait. Maud took each criticism of her work as a personal assault and responded by letter. In 1936 she proposed the theory, based on her years of research with mice, that there seemed to be three types of genes associated with cancer: one that determines its location, one that determines its type, and one that determines the degree of malignancy (how rapidly it will proliferate and kill). She also espoused the theory that two factors are necessary for cancer to develop: a hereditary predisposition, and a trauma or chronic irritation that can result in cancer in susceptible tissues. Her work was vehemently attacked by Dr. Clarence Cook Little, first as controversial but also as having been done by a woman. Dr. Little's theory of cancer susceptibility was that it was due to a dominant, heterozygous trait; but his research had been done primarily by grafting or inducing tumors, not by studying spontaneous tumors as Slye had done. The two researchers never did resolve their disputes.

Throughout her career, Slye repeatedly made pleas for a human cancer registry to track the hereditary aspects of human cancer similar to the way she had tracked cancer in 140,000 mice over her lifetime as a research scientist. But her pleas fell on deaf ears. Dr. Little and others publicly denounced her efforts to equate human cancer to cancer in mice. Although scientific success eluded her, Maud took solace in poetry, which her mother had enjoyed. She wrote two books of poems: *Songs and Solaces* (1934) and *I in the Wind* (1936). In 1944 all funds for her mouse studies ceased. She made the decision to let the remaining mice die a natural death, so she cashed in her own insurance policy to purchase food for the mice. In the same year she was automatically retired from the University of Chicago at age 65. Ten years later Maud Slye died of a heart attack.

Note

1. J.J. McCoy, *The Cancer Lady* (Nashville: T. Nelson, 1977).

Bibliography

Current Biography, 1940, pp. 743–45. New York: H.W. Wilson, 1940.
Current Biography, 1954. Obituary. New York: H.W. Wilson, 1954.
McCoy, J.J. *The Cancer Lady: Maud Slye and Her Heredity Studies*. Nashville: T. Nelson, 1977.
The National Cyclopaedia of American Biography, Vol. F (1932–1942). New York: J.T. White and Co., 1942.

Parascandola, John. *Notable American Women: The Modern Period.* Cambridge, Mass.: Harvard University Press, 1980.

KAREN JAMES

JOAN ARGETSINGER STEITZ
(1941–)
Geneticist

Birth	January 26, 1941
1963	B.S., chemistry, Antioch College
1967	Ph.D., biochemistry and molecular biology, Harvard University
1970–1978	Assistant Professor, Dept. of Molecular Biophysics and Biochemistry, Yale University
1975	Young Scientist Award, Passano Foundation
1976	Eli Lilly Award in Biological Chemistry
1978–	Professor, Dept. of Molecular Biophysics and Biochemistry, Yale University
1981–	Fellow, American Association for the Advancement of Science
1982	U.S. Steel Foundation Award in Molecular Biology
1982–	Member, American Academy of Arts and Sciences
1983	Lee Hawley, Sr., Award for Arthritis Research (shared with J.A. Hardin and M.R. Lerner)
1983–	Member, National Academy of Sciences
1986	National Medal of Science; Member, Connecticut Academy of Science and Engineering
1987	Radcliffe Graduate Society Medal for Distinguished Achievement
1987–	Associate Member, European Molecular Biology Organization
1988	Dickson Prize for Science, Carnegie-Mellon University
1989	Fritz Lipmann Lecturer, American Society for Biochemistry and Molecular Biology; Warren Triennial Prize (co-recipient with Thomas R. Cech)

Joan Argetsinger Steitz. Photo courtesy of Joan Argetsinger
Steitz.

1992	Christopher Columbus Discovery Award in Biomedical Research; Member, American Philosophical Society; Fellow, American Academy of Microbiology
1993	Rebecca Rice Award for Distinguished Achievement, Antioch College Alumni Association
1994	Weizmann Women and Science Award

Joan Argetsinger Steitz, born in 1941 in Minneapolis, Minnesota, is one of the most prominent scientists in the field of molecular genetics. When she was in high school, her field did not even exist. She received her bachelor of science degree in chemistry from Antioch College in Yellow Springs, Ohio, in 1963. While at Antioch, Steitz took courses in classical genetics, finding the idea that molecules could explain what was going on in genetics to be incredibly intriguing. She has continued to pursue her interest in genetics throughout her career.

Dr. Steitz obtained her Ph.D. in biochemistry and molecular biology from Harvard University in 1967. She studied the test-tube assembly of the ribonucleic acid (RNA) antibacterial virus R17 for her thesis at Harvard under James D. Watson, who, with Francis Crick, had discovered the double helix structure of DNA. After leaving Harvard, she went to England to pursue postdoctoral studies at Cambridge University. There she worked with Francis Crick. While at Cambridge, she researched how bacterial ribosomes recognize where to start protein synthesis on messenger RNA (mRNA).

In 1970, Professor Steitz accepted a position as assistant professor of molecular biophysics and biochemistry at Yale University in New Haven, Connecticut. She became a full professor at Yale in 1978. Since arriving at Yale, Steitz has made many significant contributions in the field of molecular genetics. She considers the most important of these to be the discovery of small nuclear ribonucleoproteins, or snRNPs (pronounced "snurps"). There are several types of snRNPs. The best known of these particles are involved in the processing of messenger RNA (mRNA) in the cell nucleus of mammals. Double-stranded deoxyribonucleic acid (DNA) is first transcribed into single-stranded RNA in the cell nucleus. Then snRNPs mediate the removal of sections of the RNA called "introns" from the RNA strand. Next the snRNPs coordinate the rejoining of the sections that code for proteins (exons), connecting them in the same order that they occurred on the DNA molecule. This process is termed "splicing." The mRNA is then transported to the cell cytoplasm, where it is translated into amino acids and the amino acids are folded into proteins. The discovery of snRNPs was possible because some patients with rheumatic diseases make antibodies (autoantibodies) against their own snRNPs. Knowing which autoantibodies a patient has can help physicians diagnose these diseases. Steitz often makes the point in her speaking engagements that even though in most cases basic research leads to improvements in clinical medicine, clinical medicine has also benefited basic research.

The discovery of snRNPs, however, is only one of Dr. Steitz's many professional accomplishments. She has published more than 160 research articles related to her subject area. At Yale, she was appointed an investigator at the Howard Hughes Medical Institute and the Henry Ford II professor of molecular biophysics and biochemistry. Also at Yale, Steitz established a laboratory devoted to the study of RNA structure and function. Her work has brought her various honors, including the Eli Lilly Award in Biological Chemistry in 1975, the National Medal of Science in 1986, and the Weizmann Women and Science Award in 1994. She has also been awarded five honorary degrees, including one from her alma mater, Harvard. Although Steitz is an active researcher, she also devotes much energy to her teaching duties. Advising and mentoring students play an important role in her professional life.

Dr. Steitz, like any prominent scientist, has an arduous life style. She serves on committees at Yale as well as on national and international committees in her field. She edits the papers of other scientists for journals in molecular genetics and referees articles or reads them for inclusion in specific journals. She is frequently asked to lecture all across the country as well as throughout the world. She attends scientific meetings and is a board member of a variety of organizations. Many of her work days are long, stretching into the evening, and work weeks often continue past Friday. In the summer when she is not teaching, she travels for lectures and meetings and concentrates on her research projects.

Early in her life, Dr. Steitz found that it was "fun to find out things that no one had ever known before."[1] She believes that her father's support of her interest in science was critical to her becoming a scientist. She stresses the importance of being interested in what you are doing regardless of what occupation is chosen. She advises future scientists to "always take on the most challenging problems or situations that you can find and pursue them to the best of your ability."[2] She considers the ability to communicate with other people and get along with other researchers important skills for a scientist.

Of all her honors, Steitz considers the Weizmann award among the most gratifying because it promotes "women scientists."[3] A strong believer in the importance of role modeling, Steitz felt the lack of female scientist role models in her own professional development. She notes that although conditions have improved for young women seeking careers in the sciences, such pursuits are still often male-dominated and the presence of successful women in a given field can be an inspiration to women students. Despite the shortage of role models, Steitz has thrived in a challenging discipline. Joshua Lederberg, professor at Rockefeller University in New York, has stated, "She is one of the country's outstanding scientists, male or female."[4]

Dr. Steitz lives in Branford, Connecticut. She is married to another scientist and has a son who is interested in math, science, and baseball. Her current research focuses on RNA molecules and the AIDS virus; this endeavor is likely to advance the frontiers of science as well as benefit AIDS patients.

Notes

1. Joan A. Steitz, telephone interview by Carla Shilts, May 26, 1995.
2. *Ibid.*

3. Joan A. Steitz, "First Winner of New Award for Women Scientists Sets High Standard" [interview by Barbara Spector], *Scientist* 8 (June 13, 1994): 3.
4. *Ibid.*

Bibliography

American Men and Women of Science. New York: Bowker, 1989/90–.

Black, D. L., B. Chabot, and J. A. Steitz. "U2 as Well as U1 Small Nuclear Ribonucleoproteins are Involved in Pre-Messenger RNA Splicing." *Cell* 42 (1985): 735–50.

"A Fitting First." *Scientist* 8, no. 12 (1994): 3, 11, 14.

Lerner, M. R., J. A. Boyle, S. M. Mount, S. L. Wolin, and J. A. Steitz. "Are snRNPs Involved in Splicing?" *Nature* 283 (1983): 220–24.

Steitz, J. A. "Polypeptide Chain Initiation: Nucleotide Sequences of the Three Ribosomal Binding Sites in Bacteriophage R17 RNA." *Nature* 224 (1969): 957–64.

Steitz, J. A., and K. Jakes. "How Ribosomes Select Initiator Regions in Messenger RNA: Base Pair Formation between the 3″ Terminus of the 16S rRNA and the mRNA during Initiation of Protein Synthesis in *E. coli*." *Proceedings, National Academy of Science* 72 (1975): 4734–38.

Wassarman, D.A., and J.A. Steitz. "Interactions of Small Nuclear RNAs with Precursor Messenger RNA during In Vitro Splicing." *Science* 257 (1992): 1918–25.

Who's Who in Frontier Science and Technology. Chicago: Marquis Who's Who, 1984–.

Who's Who in Technology. Chicago: Marquis Who's Who, 1984–.

Who's Who of American Women. Chicago: Marquis Who's Who, 1970/71–.

CARLA J. SHILTS

NETTIE MARIA STEVENS
(1861–1912)
Geneticist

Birth	July 7, 1861
1899	B.A., physiology, Stanford University
1900	M.A., Stanford University
1902–1904	Research Fellow in Biology, Bryn Mawr College
1903	Ph.D., Bryn Mawr College

1905   Awarded the Ellen Richards Prize, given to promote
                scientific research by women
Death   May 4, 1912

In 1905, Nettie Maria Stevens published a paper based on her research
in which she concluded that the type of chromosome in sperm determined the sex of embryos.[1] This breakthrough study laid the ground-
work for what is now one of the key principles of genetic theory.
However, only in the past decade or so has Stevens been given full credit
for her discovery. A fellow biologist, Edmund Wilson, has been given
credit historically for this theory, although he and Stevens reached some
of the same conclusions through independently conducted research. Today, Stevens's work is considered as presenting the stronger evidence
and conclusions for the theory of sex determination by chromosomes.

Although Stevens continued working on this theory until her death in
1912, she never received national recognition or support for her findings
during her lifetime. What, then, drove this woman who came of age in
the nineteenth century to pursue research and study in a field in which
few women were recognized? Nettie Maria Stevens's life was characterized by quiet determination, collaboration, and a commitment to the
fields of cytology and genetics.

The daughter of Ephraim and Julia (Adams) Stevens, Nettie Maria was
born in 1861 in Cavendish, Vermont. Her mother and father had three
other children, but only one, Nettie's sister Emma, survived past childhood. The girls' mother died in 1863. Nettie and Emma were raised by
their father and stepmother, Ellen Thompson, in Westford, Massachusetts.[2] The Westford public schools provided Nettie with a solid education, and she proved to be an excellent student. She also excelled at the
private Westford academy, which her sister, Emma, also attended. After
graduating from Westford Academy in 1880, Nettie "taught Latin, English, math, physiology, and zoology at the high school in Lebanon, New
Hampshire."[3] She continued her education at Westfield Normal School in
Massachusetts and then returned to teaching, first at the Minot's Corner
school in Westford, then at Westford Academy, which she had attended
earlier. Stevens taught at Westford from 1885 to 1892. She later worked as
a librarian. The portrait of Stevens's personality comes mainly from her
acquaintances from schools that she attended and at which she taught. She
received great praise from her co-workers and was a model teacher.

In 1896, Stevens began attending Stanford University under special
enrollment until she was officially admitted as a freshman in 1897. She
graduated with a bachelor's degree in physiology in 1899 and a master's
degree in 1900. She pursued a Ph.D. at Bryn Mawr College and earned
this degree in 1903.

By the time she received her doctorate from Bryn Mawr College, Stevens was involved in studying the behavior of chromosomes in animals. She received a grant from the Carnegie Institution of Washington in 1904 that allowed her to continue her research on chromosomes and sex determination.

Stevens studied insects in her research. Her breakthrough discovery came with experiments conducted on the common mealworm, whose scientific name is *Tenebrio molitor*. Stevens discovered two kinds of sperm produced by the *Tenebrio*. She identified sperm that carried an X chromosome and sperm that carried a Y chromosome. The unfertilized eggs of the mealworm all contained two X chromosomes. Stevens theorized that an egg fertilized by sperm carrying an X chromosome would produce a female embryo and that an egg fertilized by Y chromosome–carrying sperm would produce a male embryo. After making this connection between chromosomes and sex determination in the mealworm, Stevens studied other insects and searched for evidence to support her theory. She found this evidence in the study of aphids. Hans Ris writes that "in the complex life cycle of aphids she [Stevens] found a perfect correlation between chromosome composition and sex."[4] Stevens's theory is commonly accepted today. However, when she first published her findings, she met with skepticism. Even Edmund Wilson, a biologist with whom Stevens had worked, did not totally support her theory. However, Wilson made similar discoveries after Stevens had published her work and is often given credit for her discovery.

Although Nettie Maria Stevens never received the recognition she deserved for her theory on chromosomes and sex determination, she was well respected as a biologist and teacher. In answer to a question from a student, Stevens once said, "How could you think your questions would bother me? They never will, so long as I keep my enthusiasm for biology; and that, I hope, will be as long as I live."[5]

Stevens died of breast cancer in 1912 in Baltimore. During her lifetime, she achieved the rank of associate professor at Bryn Mawr College and held the title of research fellow in biology at that institution from 1902 to 1904. She also conducted research at the Zoological Institute at Wurzburg, Germany, and at other laboratories outside the United States. These were great accomplishments for a woman of the early 1900s. Nettie Maria Stevens's accomplishments and her dedication to her research can serve as inspiration for future women scientists.

Notes

1. Nettie Maria Stevens, *Studies in Spermatogenesis with Especial Reference to the "Accessory Chromosome,"* Carnegie Institution Publication No. 36, part 1 (Washington, D.C.: Carnegie Institution, 1905).

2. Marilyn Bailey Ogilvie and Clifford J. Choquette, "Nettie Maria Stevens

(1861–1912): Her Life and Contributions to Cytogenetics," *Proceedings of the American Philosophical Society* 125 (August 1981): 294.

3. *Ibid.*, p. 295.

4. Hans Ris, "Stevens, Nettie Maria," in *Notable American Women, 1607–1950*, vol. 3, ed. Edward T. James (Cambridge, Mass.: Belknap Press of Harvard University Press, 1971), p. 372.

5. Ogilvie and Choquette, "Nettie Maria Stevens," p. 303.

Bibliography

Brush, Stephen G. "Nettie M. Stevens and the Discovery of the Sex Determination by Chromosomes." *Isis* 69 (June 1978): 163–72.

Morgan, T[homas] H[unt]. "The Scientific Work of Miss N. M. Stevens." *Science* (Oct. 11, 1912): 468–70.

Morgan, Thomas Hunt, and Nettie Maria Stevens. "Experiments on Polarity in Tubularia." *Journal of Experimental Zoology* 1 (1904): 559–85.

Ogilvie, Marilyn Bailey, and Clifford J. Choquette. "Nettie Maria Stevens (1861–1912): Her Life and Contributions to Cytogenetics." *Proceedings of the American Philosophical Society* 125 (Aug. 1981): 292–311.

Ris, Hans. "Stevens, Nettie Maria," in *Notable American Women, 1607–1950*, Vol. 3, ed. Edward T. James. Cambridge, Mass.: Belknap Press of Harvard University Press, 1971.

Stevens, Nettie Maria. *Studies in Spermatogenesis with Especial Reference to the "Accessory Chromosome."* Carnegie Institution Publication no. 36, part 1. Washington, D.C.: Carnegie Institution, 1905.

HEATHER MARTIN

HELEN BROOKE TAUSSIG
(1898–1986)
Physician

Birth	May 24, 1898
1921	B.A., University of California, Berkeley
1927	M.D., Johns Hopkins University School of Medicine; Archibald Fellow in Medicine
1928	Internship in Pediatrics, Johns Hopkins Hospital
1930–46	Instructor in Pediatrics, Johns Hopkins School of Medicine
1930–63	Physician-in-Chief, Harriet Lane Home Cardiac Clinic, Johns Hopkins Hospital

1946–59	Associate Professor of Pediatrics, Johns Hopkins School of Medicine
1947	Chevalier Legion d'Honneur, France
1953	Honorary Medal, American College of Chest Physicians
1954	Feltrinelli Award, Rome, Italy
1957	American Heart Association Award of Merit
1959	Gairdner Foundation Award of Merit, Canada (shared with A. Blalock, M.D.)
1959–63	Professor of Pediatrics, Johns Hopkins School of Medicine
1960	American College of Cardiology Honorary Fellowship
1963	Woman of Achievement Award, American Association of University Women
1963–86	Professor Emerita of Pediatrics, Johns Hopkins School of Medicine
1964	Medal of Freedom of the United States; Dedication of the Helen B. Taussig Cardiac Clinic, University of Göttingen, Göttingen, West Germany
1965	President, American Heart Association
1968	VII Interamerican Award of Merit, Lima, Peru
1970	Pediatric Cardiac Clinic at Johns Hopkins was renamed the Helen B. Taussig Children's Pediatric Cardiac Center
1971	Howland Award, American Pediatric Society
1972	American College of Physicians' Mastership
1976	Milton S. Eisenhower Gold Medal, Johns Hopkins University
Death	May 21, 1986

Helen B. Taussig was the original pediatric cardiologist. She was known for her outstanding powers of deductive reasoning. She is most famous for proposing the surgical procedure now known as the Blalock-Taussig operation, which allowed thousands of babies with congenital heart disease to survive to live useful adult lives.

Helen was born in Cambridge, Massachusetts, in 1898. She was the fourth child of Edith (Guild) Taussig and Frank William Taussig. Her mother was one of the first graduates of Radcliffe College, and her father was an eminent Harvard economist. Although she grew up in a relatively prosperous family, Helen's childhood was troubled. When she was 9 years old, her mother developed tuberculosis and died within two years. Helen also developed a mild case of tuberculosis and for several years was able to attend school only in the mornings. Her academic progress in her early years was further slowed by dyslexia. In spite of

Helen Brooke Taussig. Photo courtesy of The Alan Mason Chesney Medical
Archives of The Johns Hopkins Medical Institutions.

these struggles, she graduated from high school only one year behind
her peers. She attended Radcliffe for two years, where she was a tennis
champion; in 1921 she completed her bachelor of arts degree at the University of California, Berkeley.

In the early 1920s, professional choices for women were few. Helen's
interests included public health and medicine. Studying medicine was
not an easy choice. At Harvard Medical School, women were permitted
to study but were not admitted as candidates for degrees. When she
studied histology there, Helen was not allowed to speak to the male
students or even look at slides in the same room with them, so that she
would not "contaminate" them.

After taking bacteriology and histology at Harvard, she transferred to
Boston University to study anatomy. Dr. Alexander Begg, professor of
anatomy, asked her to study the muscle bundles of the ox heart. This
study resulted in her first publication in 1925 and also marked the be-

ginning of her interest in cardiology. With Begg's encouragement, Helen applied to Johns Hopkins University School of Medicine, where women were admitted as degree candidates. She received her M.D. degree in 1927.

While she was a medical student at Johns Hopkins, Dr. Taussig worked at the Heart Station for Dr. Edward P. Carter. She was not given a medical internship, having been nudged out by another woman by two-tenths of a point. Only one woman was taken on the house staff at a time then, so Dr. Carter offered her a fellowship in cardiology. Following her fellowship, Dr. E.A. Park, chief of pediatrics, put Helen in charge of the new cardiac clinic to deal with the many cases of rheumatic fever.

Dr. Taussig served as chief of the Cardiology Clinic of the Harriet Lane Home, the children's section of the Johns Hopkins Hospital, from 1930 to 1963. Although it wasn't easy to get referrals of patients with rheumatic fever, many cyanotic ("blue") babies, for whom it was thought that nothing could be done, were referred for study.

Taussig was the first to recognize that some of the complex malformations of the heart produced very distinctive features. By studying many "blue babies" with cardiology's new tool, fluoroscopy, she discovered that at birth many cyanotic infants had an abnormal opening in the wall between the ventricles of their heart and a partial blockage of the artery to the pulmonary circulation. These defects resulted in a lack of oxygenation of the blood and hence a cyanotic, or "blue," appearance. From 1941 to 1944, Taussig and Dr. Alfred Blalock, chief surgeon at Johns Hopkins, developed the operation that enabled a more normal circulation to be established. It was first performed in 1946 after two years in the laboratory and 200 experiments on dogs.

The idea for the Blalock-Taussig operation, a cornerstone of modern cardiac surgery, was Helen's; the actual surgery was performed by Blalock. The procedure had wide implications for children with cyanotic congenital heart disease. It was not only a specific, life-saving operation, but also it demonstrated that these kinds of children were physically able to tolerate major surgical interventions without excessive mortality. It prompted surgeons to undertake other critical interventions that they had been hesitant to attempt.

Had Taussig's only accomplishment been the concept of the "blue baby" operation, it would have been enough to justify her numerous accolades. Her recognition as an important contributor to modern cardiology might have faded with the advent of other procedures. However, her influence has been broader and more permanent. Had it not been for her book *Congenital Malformations of the Heart* (1947), the methods of identifying congenital heart defects would have been unknown to many physicians around the world. The book served as the stimulus that influenced many to enter the field of pediatric cardiology and provided the basis on which the discipline was built. In addition, Taussig trained many pediatric cardiologists and was one of the founders of the Board of Pediatric Cardiology.

Her accomplishments in the field of cardiology are particularly impressive because she had severe hearing problems. Although her hearing had probably been deteriorating for several years, it dropped abruptly between 1930 and 1931, just at the time she was put in charge of the Pediatric Cardiology Clinic at Johns Hopkins. She had to learn to lip read, and she worked with a special amplified stethoscope. She also developed the ability to "listen with her hands," feeling, more than listening to, the hearts of babies with cardiac disease. By 1951 her hearing loss was extensive. After an operation in 1971 her hearing was much improved, but she later developed nerve deafness and needed hearing aids.

In addition to the important discoveries related to her study of the heart, Helen's scientific curiosity led her to examine the outbreak of phocomelia, or "seal limb" deformity, in Germany in the early 1960s. One of the doctors whom she had trained, who was visiting from Germany, spoke to her about the record number of babies being born in his country with this terrible deformity. Within two weeks Helen was on her way to Germany as a medical detective. Her investigations helped to keep the drug thalidomide out of the United States and halted its use in Europe.

Helen Taussig possessed a keen mind that remained sharp until her death at age 87. Forty-one of her 129 scientific articles were written after her retirement. Her last paper was completed just a few days before her death. She died in an automobile accident on her way home from the Delaware Museum of Natural History, where she was doing research on avian hearts. Dr. Taussig had proposed that isolated cardiac malformations were not errors in development, but rather genetic variants. To document her hypothesis, she performed a gross examination of the hearts of more than 5,000 birds over a three-year period. This was an important distinction to Taussig, because distraught parents of children with heart defects often ask physicians about events during pregnancy and blame themselves for whatever they fear led to the cardiac anomaly in the child. She was also an outspoken advocate for socialized medicine, malpractice reform, and liberalization of abortion laws, and she argued against prolonging life in hopeless situations.

In recognition of her many achievements, Helen Taussig was the recipient of more than 20 honorary degrees and more than 30 major international awards, including the presidential Medal of Freedom in 1964, the highest civilian award a U.S. president can give. Although she was widely acclaimed, Taussig was very aware of the discrimination of which she was the victim more than once. For example, in recognition of the

Blalock-Taussig procedure, Dr. Blalock was elected to membership in one of America's most prestigious groups, the National Academy of Sciences, in 1945; Taussig was not chosen for membership until 1973. Blalock was appointed to a full professorship when he joined the Hopkins faculty in 1941; Taussig was made a full professor of pediatrics in 1959, four years before her retirement. She was the first woman to achieve this rank at Hopkins. But it appears that she was able to overcome her bitterness. When interviewed, she observed: "I think that it is [important] to work for the love of working and not for fame."[1] She added that "whatever field you choose, just work quietly and steadily to make this world a better place to live and your life will be worthwhile."[2]

The November 1978 issue of *Medical Times* was dedicated to Dr. Taussig. It was the inauguration of what was to be an annual custom: the Physician of Excellence Award. One entire issue is devoted to an individual whose long-term contributions to the field of medicine have been outstanding. Helen Taussig certainly fit this description. As described by Dr. Edward B. Clark, "Dr. Taussig was loved by her patients, revered by her fellows, and world acclaimed for her pioneering care of children with heart disease."[3] Helen B. Taussig did leave the world a better place—especially for "blue babies."

Notes

1. Joyce Baldwin, *To Heal the Heart of a Child: Helen Taussig, M.D.* (New York: Walker Publishing, 1992), p. 113.

2. *Ibid.*, p. 85.

3. Edward B. Clark, "The Origin of Common Cardiac Defects," *Journal of the American College of Cardiology* 12 (1988): 1087.

Bibliography

Baldwin, Joyce. *To Heal the Heart of a Child: Helen Taussig, M.D.* New York: Walker Publishing, 1992.

Clark, Edward B. "The Origin of Common Cardiac Defects." *Journal of the American College of Cardiology* 12 (1988): 1087.

Engle, Mary Allen. "Blue Babies." *Medical Times* 106 (1978): 45–60.

Harken, Dwight Emary. "The Emergence of Cardiac Surgery." *Journal of Thoracic and Cardiovascular Surgery* 98 (1989): 805–13.

Harvey, W. Proctor. "A Conversation with Helen Taussig." *Medical Times* 106 (1978): 28–44.

McNamara, Dan G. "Helen B. Taussig: The Original Pediatric Cardiologist." *Medical Times* 106 (1978): 23–27.

McNamara, Dan G., et al. "Historical Milestones: Helen Brooke Taussig: 1898–1986." *Journal of the American College of Cardiology* 10 (1987): 662–71.

Neill, Catherine A. "Profiles in Pediatrics: Helen Brooke Taussig." *Journal of Pediatrics* 125 (1994): 499–502.

O'Neill, Lois Decker, ed. *The Women's Book of World Records and Achievements.* Garden City, N.Y.: Anchor Press/Doubleday, 1979.

Taussig, Helen B. *Congenital Malformations of the Heart.* New York: Commonwealth Fund, 1947. (2nd revised ed.: Cambridge, Mass.: Harvard University Press for the Commonwealth Fund, 1960.)

————. "Difficulties, Disappointments, and Delights in Medicine." *Pharos* 42 (1979): 6–8.

————. "Evolutionary Origin of Cardiac Malformations." *Journal of the American College of Cardiology* 12 (1988): 1079–88.

————. "How to Adjust to Deafness." *Medical Times* 109 (1981): 38s–43s.

————. "Little Choice and a Stimulating Environment." *Journal of the American Medical Women's Association* 36 (1981): 43–44.

————. "The Thalidomide Syndrome." *Scientific American* 207 (1962): 29–35.

Wanzer, Sidney H., et al. "The Physician's Responsibility toward Hopelessly Ill Patients." *New England Journal of Medicine* 310 (1984): 955–59.

CLARA A. CALLAHAN

TROTULA

(10??– ca. 1097)

Physician

The historical existence of a person named Trotula is a matter of debate. Most authorities believe, however, that Trotula was indeed a real person whose writings on obstetrics and gynecology influenced medicine for 500 years. She is often identified as the pre-eminent member of the Mulieres Salernitanae (Ladies of Salerno).

During the eleventh century the education of women was not at all popular in most parts of the world. Fortunately, this was not the case in Salerno, Italy. The University of Salerno was the first university to open its doors to women. As a result, a number of women attended the university and one group in particular became prominent eleventh-century medical scholars and practitioners. These women were known as the Ladies of Salerno. They were most skilled in the treatment of obstetric and gynecological disorders. Unfortunately, many of their names and the works attributed to them have been lost. Only the works of Trotula have been preserved to the present day.

Despite the lack of definitive factual information concerning Trotula and the Ladies of Salerno, a detailed biography of Trotula has emerged and is generally accepted. She was a member of the noble family di Ruggiero and the wife of physician Johannes Platearius. Trotula and Johannes had two sons, Mattias and Johannes the Younger.[1] Although there is no record of Trotula having held a chair at the University of Salerno, her husband did hold a chair there. It is assumed that the entire family was involved in teaching at the university. Trotula was a woman of great influence; her titles included Magistra (Directress) and Mulier Sapiens (Woman of Wisdom). Trotula is believed to have died in the year 1097.

The premise for Trotula's existence is based on the existence of many manuscripts bearing her name. *Practica Brevis* (Brief Handbook) and *De Compositione Medicamentorum* (On the Preparation of Medicines) are works written by her entire family.[2] The former is a medical encyclopedia and the latter is a guide to the preparation of medicines. However, *Passionibus Mulierum Curandorum* (Cures of the Diseases of Women) and *Ornatu Mulierum* (Women's Cosmetics) are treatises attributed to her alone, and they are her most prestigious writings. These two works became known as *Trotula Maior* and *Trotula Minor*, respectively. *Trotula Maior* addresses reproductive issues including problems with menstruation, conception, childbirth, and delivery. *Trotula Minor* discusses cosmetics and skin diseases. *Trotula Minor* was eventually incorporated into *Trotula Maior*. Both works had a profound and lasting effect on medical practice for many years. These manuscripts were studied at medical schools through the sixteenth century.[3] The manuscripts were copied, amended, and compiled with other works over the 500-year period. As a result, there are numerous variations of both writings.[4]

Trotula advocated the use of a variety of medical treatments. Many of her cures involved holistic remedies and included the use of herbs. Some of her other cures involved more invasive techniques such as the bleeding of a patient and the sewing of skin together after childbirth.

Her treatises reveal that her knowledge of reproductive medicine was quite advanced for the time. For example, she indicated that the fetus is like a fruit on a tree and that women's menstrual flow is like a flower that needs to be pollinated. She indicated that impotence in males may be due to a spiritual disorder. (We would use the term "psychological disorder" today.) Some of her treatments were practical. Trotula cited lack of exercise as a contributing factor to many conditions.

Some of her cures were less scientific, and others were downright bizarre. For example, if a patient was not menstruating, she advocated draining blood from her feet. If a woman could not conceive a child with her husband, Trotula recommended the following test to see if one or the other was sterile:

take two jars and into each put bran. Into one put the urine of the man and into the other the urine of women and let the jars be left for nine or ten days. If barrenness be from a defect of a woman you will find many worms and the bran foul in her jar. On the other hand you will have similar evidence from the other urine if the barrenness be through the defect of the man. But if you have observed such signs in neither urine neither will be the cause of the barrenness and it is possible to help them to conceive by the use of medicines.[5]

To promote conception, Trotula recommended the following treatment: "Take the livers and testicles of a pig . . . and let them be dried; make a powder of this and give it in a drink to the man and woman who cannot conceive and they will procreate."[6]

Despite the fact that many of her cures were not what we would now consider scientifically sound, Trotula was extremely wise and learned for her time. It is also important to note that her cures were rarely based on superstition or called for the astrological methods that were common during this time.

Trotula's writings also reveal some of her personal qualities. One modern historian has observed that Trotula was knowledgeable, practical, and understanding.[7] Her treatises are filled with common sense. She recommended cures that were simple to follow and affordable to the poor. She understood the role that stress plays in disease and advocated good hygiene, a balanced diet, and exercise. She was a compassionate healer. She understood what was accepted at the time as a woman's modesty: that a woman may feel uncomfortable or embarrassed during a gynecological examination, especially if it is performed by a male doctor.

The uncertainty surrounding Trotula's existence and the authorship of the works that bear her name may be due in part to prejudicial treatment. Prejudice against women in science and medicine, or against Trotula and her works in particular, is nothing new.[8] Writing in the sixteenth century, Karl Sudhoff downplayed the significance of the Ladies of Salerno and their medical findings. He lowered their status to midwives or nurses. Likewise, Charles Singer sought to discredit Trotula's work by calling it pornographic in nature.

Despite these attacks, a lack of historical information, and various refutations, most scholars support the historical reality of Trotula. Italian medical historians have always been steadfast in their support of her authenticity and in their praise of her contribution to the medical field.[9]

Notes

1. Margaret Alic, *Hypatia's Heritage* (Boston: Beacon Press, 1986), p. 51.
2. Caroline L. Herzenberg, *Women Scientists from Antiquity to the Present: An Index* (West Cornwall, Conn.: Locust Hill Press, 1986), p. xxi.

3. Alic, *Hypatia's Heritage*, p. 53.

4. See Kate Campbell Hurd-Mead, *A History of Women in Medicine* (New York: AMS Press, 1977), pp. 127–41.

5. Howard Grant, *A Source Book in Medieval Science* (Cambridge, Mass.: Harvard University Press, 1974), p. 765.

6. *Ibid.*

7. Alic, *Hypatia's Heritage*, p. 51.

8. *Ibid.*, p. 55.

9. *Ibid.*

Bibliography

Alic, Margaret. *Hypatia's Heritage.* Boston: Beacon Press, 1986.

Brooke, Elisabeth. *Women Healers through History.* London: Women's Press Ltd., 1993.

Grant, Edward. *A Source Book in Medieval Science.* Cambridge, Mass.: Harvard University Press, 1974.

Harington, John, Sir. *The School of Salernum.* New York: Augustus M. Kelley, 1970.

Herzenberg, Caroline L. *Women Scientists from Antiquity to the Present: An Index.* West Cornwall, Conn.: Locust Hill Press, 1986.

Hughes, Murial Joy. *Women Healers in Medieval Life and Literature.* Freeport, N.Y.: Books for Libraries, 1968.

Hurd-Mead, Kate Campbell. *A History of Women in Medicine.* New York: AMS Press, 1977.

Mozans, H. J. *Women in Science.* Cambridge, Mass.: MIT Press, 1913.

MICHAEL WEBER

MARY EDWARDS WALKER
1832–1919
Physician

Birth	November 26, 1832
1853	Admitted to Syracuse Medical College
1855	Graduated with a medical degree; married Albert Miller
1860	Marriage ended (divorce made final in 1869)
1861	Moved to Washington to volunteer in the Civil War as a surgeon
1864	Captured and imprisoned in Richmond, VA; received back pay for her services as a civilian surgeon

1866	Congressional Medal of Honor
1868	Began to campaign for women's rights
1881	Announced herself a candidate for the U.S. Senate
1890	Declared herself a candidate for Congress
1917	Medal of Honor rescinded
Death	1919
1977	Medal of Honor restored

Mary Walker was born to Vesta and Alvah Walker in 1832 on a farm in Oswego, New York. She was the youngest of four daughters and had one younger brother. Alvah Walker was a carpenter by trade and an avid reader. When he was unable to recover from an illness that lingered for many weeks, he began to read medical books to search for a cure. This led him to develop three strongly held beliefs about health. He was convinced that liquor, tobacco, and tight clothing were very harmful to health. He preached these beliefs to his children and did not require any of his daughters to wear corsets at home.

Mary's father also felt that girls should be educated. He encouraged each of his daughters to pursue a professional career, which they did. Throughout her life, Mary was greatly influenced by her father. This, coupled with her own ambitious, energetic, and determined nature, caused her life to be quite out of the ordinary. Her first move out of the mainstream was to apply for admission to Syracuse Medical College, the first coeducational medical school in the country. She was accepted just after her twenty-first birthday and was the only woman in her class.

The first woman to graduate with a medical degree in the United States was **Elizabeth Blackwell**, in 1849. When Mary graduated six years later in 1855, women physicians were still quite rare. Consequently, Mary faced great prejudice. It was a common belief that women were too delicate, modest, and innocent to learn about the human body and were subject to "uncontrollable hysteria." In addition, many people believed that women were born with limited intelligence. It was felt that women were unable to do mathematics and were deficient in judgment and courage.

The few patients Mary saw expected to pay less for her services than for those of a man. Mary's answer to them was clear: "My education had been quite as elaborate and occupied the same time, and cost as much. I can do the work I profess as well as any man. The obstacles I had to surmount were greater. Therefore I am justified in charging not only as much, but more."[1]

Her father's influence surfaced again when Mary stopped wearing the cumbersome floor-length dresses (with a wooden hoop underneath) of the day. Instead, she wore bloomers covered by a coat-dress. She even wore pantaloons at her wedding. Mary had married Albert Miller, a fellow classmate, a few months after graduation. Her wedding was quite unorthodox for the times. Not only did she not wear a wedding dress, but she had the part about obeying her spouse removed from their wedding vows and insisted on retaining her maiden name. Their marriage ended after five years when her husband confessed that he had been unfaithful.

In 1861 the Civil War suddenly created a demand for physicians and surgeons. Mary immediately headed to Washington, where she petitioned to become a surgeon in the army. Eager to use her skills, she volunteered to help wherever she was needed without awaiting the outcome of her petition. Her help was gratefully accepted by the sick and wounded.

She applied for a commission as a surgeon in the army several more times, but her application was rejected each time. She was asking for something that had never been done before; namely, there had never been a woman officer in the army. The surgeon-general was not prepared to break this tradition. Mary continued to work as an assistant physician and surgeon, but without title or pay. However, whenever she cared for soldiers on the battlefield she wore the blue union officer's uniform consisting of pants with gold stripes, a felt hat encircled by a golden cord, and the green sash of a surgeon.

In the autumn of 1863 she arrived at Chattanooga, Tennessee. While she was there, the assistant surgeon for the fifty-second Ohio infantry died unexpectedly and Mary was ordered to report as his replacement. She traveled many miles on horseback to make her calls. On April 10, 1864, Mary took the wrong road and was captured as a prisoner of war. She was transported 700 miles by train to Richmond, Virginia, where she was incarcerated in Castle Thunder, a political prison. She spent a total of four months in the dirty prison infested with lice, bedbugs, and mice. She was released from prison on August 12, 1864, when she was traded for a Confederate officer.

In January 1864 she appealed directly to President Lincoln for appointment as a surgeon. In less than a week Mary received a reply, which was another rejection, handwritten by President Lincoln. Surprisingly, in September 1864 she was awarded a contract as acting assistant surgeon, United States Army, with a salary of $100 a month plus back pay. Although she was the only woman assistant surgeon in the history of the army, she was never commissioned as an officer.

Even at the end of the war, Mary lobbied for a promotion to major for her services. Edwin M. Stanton, secretary of war, would not grant her request, reasoning that it was impossible to give her a higher rank because she has never been an officer in the first place. Once again, Mary persisted and sent letter after letter to President Johnson requesting a promotion. President Johnson finally discussed with Stanton some other way to recognize her service. On January 24, 1866, she received from President Andrew Johnson the Congressional Medal of Honor for meritorious service, dated November 11, 1865. The citation read as follows:

> Whereas it appears from official reports that Dr. Mary E. Walker, a graduate of medicine, "has rendered valuable service to the Government, and her efforts have been earnest and untiring in a variety of ways," and that she was assigned to duty and served as an assistant surgeon in charge of female prisoners at Louisville, Ky., upon the recommendation of Major-Generals Sherman and Thomas, and faithfully served as contract surgeon in the service of the United States, and has devoted herself with much patriotic zeal to the sick and wounded soldiers, both in the field and hospitals, to the detriment of her own health, and has endured hardships as a prisoner of war four months in a southern prison while acting as contract surgeon; and Whereas by reason of her not being a commissioned officer in the military service a brevet or honorary rank can not, under existent laws, be conferred upon her; and Whereas in the opinion of the President an honorable recognition of her services and sufferings should be made; It is ordered. That a testimonial thereof shall be hereby made and given to the said Dr. Mary E. Walker, and that the usual medal of honor for meritorious services be given her. Given under my hand in the city of Washington, D.C. this 11th day of November, A.D. 1865.
>
> Andrew Johnson, President
By the President:
Edwin M. Stanton, Secretary of War[2]

After the war Mary all but abandoned her medical practice, which was never very successful. She spent her time on the lecture circuit and lived on her government pension. She campaigned heavily for women's rights. She was at the crest of the women's rights and suffrage movement and often shared the stage with Susan B. Anthony. Her lectures had themes reminiscent of her father: dress reform for women, and the evils of smoking and drinking. She was probably the most ardent about dress reform. Moreover, she practiced what she preached and was arrested more than once for "impersonating a man." At one trial, she asserted her right "to dress as I please in free America on whose tented field I have served for four years in the cause of human freedom."[3]

Mary suffered the greatest disappointment of her life in 1917. Congress revised the Medal of Honor standards to include only "actual combat with an enemy," and the army board rescinded Mary's medal. At 85 years of age, she went before the board to plead her case and remind them of her accomplishments, but the board would not reconsider. She vowed to go on wearing her medal (even though they informed her this was a crime) and refused to give it up.

Determined to the end, Mary continued to appeal to congressmen and war department officials to have her medal restored, but she was never able to reverse their decision. It was on one of these visits that she had a bad fall on the Capitol steps. She never fully recovered. She died on February 21, 1919, at age 86.

In 1977, at the urging of a descendant, the army restored her Medal of Honor. Mary Edwards Walker is listed in the "Register of Recipients" as "Walker, Contract Surgeon (Civilian) Mary (Bull Run, Va & Chattanooga, Tenn, etc.)."[4] To this day, she is the only woman ever to receive this award.

Notes

1. Charles McCool Snyder, *Dr. Mary Walker: The Little Lady in Pants* (New York: Arno Press, 1974), pp. 66–67.

2. *Ibid.*, pp. 53–54.

3. *Above and Beyond: A History of the Medal of Honor from the Civil War to Vietnam*, by the editors of Boston Publishing Company; produced in cooperation with the Congressional Medal of Honor Society of the United States of America (Boston: Boston Publishing Company, 1985), p. 39.

4. *Ibid.*, p. 322.

Bibliography

Above and Beyond: A History of the Medal of Honor from the Civil War to Vietnam, by the editors of Boston Publishing Company; produced in cooperation with the Congressional Medal of Honor Society of the United States of America. Boston: Boston Publishing Co., 1985.

Faust, Patricia L. *Historical Times Illustrated Encyclopedia of the Civil War.* New York: Harper and Row, 1986.

Hall, Marjory. *Quite Contrary: Dr. Mary Edwards Walker.* New York: Funk & Wagnalls, 1970.

Levin, Beatrice S. *Women and Medicine.* Metuchen, N.J.: Scarecrow Press, 1980.

Malone, Dumas. *Dictionary of American Biography.* New York: Charles Scribner's Sons, 1936.

Morton, Leslie T. *A Bibliography of Medical and Biomedical Biography.* London: Scolar Press, 1989.

Snyder, Charles McCool. *Dr. Mary Walker. The Little Lady in Pants.* New York: Arno Press, 1974.

Walker, Mary E. *Hit.* New York: American News Co., 1871.

———. *Unmasked, or the Science of Immorality.* Philadelphia: William H. Boyd, 1878.

LAURIE A. POTTER

ANNA WESSELS WILLIAMS

(1863–1954)

Bacteriologist

Birth	March 17, 1863
1883	Graduated from New Jersey State Normal School
1891	M.D., Woman's Medical College of New York
1891–93	Instructor in Pathology, New York Infirmary
1891–95	Assistant to Department Chair, Pathology and Hygiene, New York Infirmary
1894	Volunteer in diagnostic laboratory, New York City Department of Health; Isolated Park-Williams #8 strain of *Corynebacterium diphtheriae*
1895–1905	Assistant Bacteriologist for the laboratory, New York City Department of Health
1896	Researcher, Pasteur Institute, Paris
1898	Produced rabies vaccine in New York
1902–1905	Consulting Pathologist, New York Infirmary
1905	Published new method of detecting "Negri bodies"; named Assistant Director of New York City Research Laboratories; co-authored (with William H. Park) the second edition of *Pathogenic Micro-Organisms Including Bacteria and Protozoa*
1907	Chair, Committee on the Standard Methods for the Diagnosis of Rabies, American Public Health Association
1915	President, Woman's Medical Association
1929	Co-authored (with William H. Park) *Who's Who among the Microbes*
1931	Vice Chair, Laboratory Section, American Public Health Association
1932	Chair, Laboratory Section, American Public Health Association
1934	Retired from New York City Department of Health Research Laboratory
Death	November 20, 1954

Anna Wessels Williams. Photo reprinted by permission of
The Schlesinger Library, Radcliffe College.

Although during her lifetime Anna Wessels Williams was often over-
shadowed by her male counterparts, she is now widely recognized for
her contributions to the understanding of infectious diseases, diphtheria
immunization, and rabies diagnosis and control. Her work saved countless lives. The course of her life as a bacteriologist and pathologist was
set at age 12, when she first looked under a teacher's "wonderful microscope" at the State Street Public School.[1] Until then, Williams—who was
born in Hackensack, New Jersey, in 1863 as the second child and second
daughter of the six children of William and Jane Williams—was educated at home by her English father, who was a private school teacher,
and her mother, who was a zealous supporter of the mission program
of the First Reformed (Dutch) Church.

Williams graduated from the New Jersey State Normal School in Tren-
ton in 1883, and she taught from 1883 to 1885. After her sister almost

died during delivery of a stillborn child, Williams convinced her parents that she should enter the Woman's Medical College of the New York Infirmary for Women and Children. While she was a student there, she served as an assistant and gained knowledge and experience in experimental therapeutics, pathology, and hygiene. She obtained her M.D. degree in 1891 and took a post as an instructor in pathology and hygiene at the infirmary. In addition, she assisted Dr. Annie S. Daniel in the children's clinic and in the out-practice. During 1892 and 1893 she continued her education in Europe, studying at the universities of Vienna, Heidelberg, and Leipzig, and serving as intern in the Royal Fräuen Klinik of Leopold in Dresden.

Williams's pioneering efforts in the prevention and cure of diphtheria began in 1894 when she volunteered at the newly established diagnostic laboratory of the New York City Department of Health. The laboratory, which was housed in three small rooms of the old police headquarters on Mott Street, had opened in the previous year and was the first municipal laboratory to apply bacteriology to the study of public health problems. At the time, a rampant diphtheria epidemic was the primary cause of death in children. Working together, Williams and the laboratory director, William Hallock Park, searched for a diphtheria antitoxin to replace the existing ineffective, low-yield antitoxin. During her first year at the laboratory Williams isolated in pure culture the now famous "Park-Williams #8," or Park strain of *Corynebacterium diphtheriae*, from a case of mild tonsillar diphtheria. This strain, the first to produce a strong toxin, became the stock strain for producing commercial diphtheria toxin. It rapidly became available to physicians without charge for destitute patients. Over the years it has maintained its reputation as a strong toxin producer and is still manufactured today. Although there was a question as to who the discoverer of the bacillus was, Williams stated in a speech: "I insisted that I alone was the discoverer, since I happened to find it while Dr. Park was on a vacation."[2] Williams's initial bacteriological research, which almost totally eradicated diphtheria in much of the world, was detailed in two articles published in 1894.

In 1895, Williams became a full-time assistant bacteriologist for the laboratory. She progressed in 1905 to become the assistant director, a position that she held until her forced retirement in 1934. Throughout her tenure at the laboratory, she developed smooth organization, teamwork, and a staff that included many women. Her attitude and efforts to consider "always the worker, man or woman" were instrumental in the growth and productivity of the Research Laboratories.[3]

After conquering diphtheria, Williams investigated the bacteriology of streptococcal and pneumococcal infections between 1894 and 1896. In 1896 her search for a scarlet fever toxin took Williams and her now famous "Diphtheria Bacillus #8" to the Pasteur Institute in Paris. There she studied the streptococcus with Alexander Marmorek and revealed through a series of studies on this group of pathogenic bacteria that several toxins are involved in streptococcal infections. Her work in this area occupied much of her 40-year career and culminated in 1932 in her definitive monograph *Streptococci in Relation to Man in Health and Disease.*

Another important goal of Williams's trip to Paris was to perfect rabies diagnosis by working with Emile Duclaux and Marmorek. After studying their methods for diagnosis of rabies and preparation of a vaccine, she returned to New York with the first culture of the rabies vaccine virus. By 1898 she was producing enough vaccine from the culture for 15 persons to be inoculated with injections prepared in her laboratory. Large-scale production of this vaccine in the United States followed. In 1902 she was joined by Alice G. Mann, who labored with her on diagnostic methods for rabies, streptococcus, variola, and vaccinia. Meanwhile, many bite victims continued to die because of the length of time necessary to diagnose an animal as rabid. While painstakingly conducting research on rabies, Williams was the first to see the brain cell peculiar to a rabid animal. At the same time, however, an Italian physician, Adelchi Negri, also discovered the cell and published his results before Williams could publish her findings. Consequently, these brain cells became known as "Negri bodies."

In 1905, however, Williams made a major breakthrough in rabies diagnosis and control by publishing a new method of detecting Negri bodies in prepared and stained brain tissue. Her method took only minutes. Previously, diagnosis of a rabid animal took ten days. Williams's staining method was not improved on until 1939. In 1907 the American Public Health Association acknowledged the importance of her work by naming her chair of the newly created Committee on the Standard Methods for the Diagnosis of Rabies.

Williams collaborated with others on many public health problems, including influenza, venereal diseases, pneumonia, meningitis, poliomyelitis, and smallpox. Among her more important investigations were those concerned with trachoma, a chronic inflammatory eye infection, which was thought to be scarring and blinding the eyes of underprivileged schoolchildren in New York City. According to Williams, the knowledge of trachoma was at the time in a "most confused state."[4] She studied over 3,000 cases and determined that trachoma was often mistakenly diagnosed. Other more benign infections than trachoma were causing the damage to schoolchildren's eyes. As a result, the "trachoma hospitals," which were situated throughout the city, were closed.[5]

Because of her uncanny ability to find the influenza bacillus whenever she looked for it, Williams became known as "the influenza picker."[6] Her research in this area afforded her the opportunity during World War I to serve on the Influenza Commission, which was a joint research project of the Research Laboratory of the Department of Health of New York City, the U.S. Public Health Service, the Divisions of Preventive Medicine of Harvard and Chicago Universities, and the Metropolitan Life Insurance Company. During the raging influenza epidemic, she was sent by the Federal Health Service to Minnesota to help determine the incidence of the influenza bacillus there and its relationship to epidemic influenza. She later aided in determining that this bacillus had nothing to do, etiologically, with the epidemic. During 1917 and 1918, at New York University, she also directed a teaching-training project for American and European medical laboratory workers for war service and worked in a program to detect meningococcal carriers in the military.

Reporting her research, Williams published extensively in the bulletins and publications of the New York City Department of Health and in the *Journal of Experimental Medicine.* In 1905 she co-authored with Park the second edition of *Pathogenic Microorganisms including Bacteria and Protozoa: A Practical Manual for Students, Physicians and Health Officers.* The respected text went through 11 editions, with the final one published in 1939. Williams also collaborated with Park on *Who's Who among the Microbes* (1929), a book for the nonscientist.

On March 31, 1934, after 39 years of continuous service in the Department of Health's laboratory, Anna Williams was retired by Mayor Fiorello La Guardia, who was anxious to maintain the principle of age regulation even in the face of numerous protests and petitions from scientists, physicians, and organizations. Williams lived for ten years in Woodcliff Lake, New Jersey, before moving to Westwood to live with her sister, Amelia Wilson. There she died of heart failure in November 1954.

Williams was an active participant in the professional world. She held memberships in the American Medical Association; the Society of American Bacteriologists; the New York Pathological Society; the Society of Experimental Biology; the American Association of Immunologists; the American Social Hygiene Association; and the New York Academy of Medicine. In addition, she served as president of the Woman's Medical Association in 1915 and as vice chair in 1931 and chair in 1932 of the Laboratory Section of the American Public Health Association. She was the first woman to hold an elected office in that section. The New York Women's Medical Society in 1936 honored her service to the city and her work in advancing the cause of the woman doctor. She accepted the honor as "a tribute largely to the work of my associates."[7] Naming her colleagues—many of whom were women—as partners in her own success, she stated, "without the aid of my associates, little of this work could have been accomplished."[8]

Notes

1. Elizabeth D. Robinton, "Williams, Anna Wessels," in *Notable American Women: The Modern Period, A Biographical Dictionary*, eds. Barbara Sicherman and Carol Hurd Green (Cambridge, Mass.: Belknap Press of Harvard University Press, 1980), p. 737.

2. Anna W. Williams, "Group Work in Public Health Laboratories," *Medical Woman's Journal* 43, no. 6 (June 1936): 152.

3. Wade W. Oliver, *The Man Who Lived for Tomorrow: A Biography of William Hallock Park* (New York: E.P. Dutton, 1941), p. 456.

4. Anna Wessels Williams, "A Study of Trachoma and Allied Conditions in the Public School Children of New York City," *Journal of Infectious Diseases* 14 (1914): 261.

5. Williams, "Group Work in Public Health Laboratories," p. 153.

6. *Ibid.*

7. *Ibid.*

8. *Ibid.*, p. 151.

Bibliography

"Anna W. Williams, Scientist, Is Dead." *New York Times* 104 (Nov. 21, 1954): 86.

Barley, Martha J. *American Women in Science: A Biographical Dictionary.* Denver: ABC-CLIO, 1994.

"Dr. Anna W. Williams." *Medical Woman's Journal* 43, no. 6 (June 1936): 160.

Lovejoy, Esther Pohl. *Women Doctors of the World.* New York: Macmillan, 1957.

Oliver, Wade W. *The Man Who Lived for Tomorrow: A Biography of William Hallock Park.* New York: E.P. Dutton and Co., 1941.

Park, William Hallock, and Anna W. Williams. *Who's Who among the Microbes.* New York: Century Co., 1929.

Robinton, Elizabeth D. "A Tribute to Women Leaders in the Laboratory Section of the American Public Health Association." *American Journal of Public Health* 64, no. 10 (Oct. 1974): 1006–7.

————. "Williams, Anna Wessels," in *Notable American Women: The Modern Period; A Biographical Dictionary*, eds. Barbara Sicherman and Carol Hurd Green, pp. 737–39. Cambridge, Mass.: Belknap Press of Harvard University Press, 1980.

Siegel, Patricia Joan, and Kay Thomas Finley. *Women in the Scientific Search: An American Bio-Bibliography, 1724–1979.* Metuchen, N.J.: Scarecrow Press, 1985.

Vare, Ethlie Ann, and Greg Ptacek. *Mothers of Invention: From the Bra to the Bomb; Forgotten Women and Their Unforgettable Ideas.* New York: William Morrow, 1988.

Williams, Anna W. *Streptococci in Relation to Man in Health and Disease.* Baltimore: Williams and Wilkins Co., 1932.

CASSANDRA S. GISSENDANNER

CICELY DELPHIN WILLIAMS
(1893–1992)
Physician

Birth	December 2, 1893
1929	B.A., Oxford University; Diploma, tropical medicine and hygiene, London University; became the first woman appointed by the British Colonial Medical Service in the Gold Coast (now Ghana)
1933	Described kwashiorkor
1936	M.D., King's College, Oxford University
1946	First woman to hold a superscale post in the Colonial Service as Head, Maternity and Child Welfare, British Colonial Office
1948–51	Chief, Maternal and Child Health Section, World Health Organization
1955–57	Senior Lecturer in Nutrition, London University
1959	Professor of Maternal and Child Health, American University, Beirut
1964–67	Advisor in Training Programs, Family Planning Association, United Kingdom
1965	James Spence Memorial Gold Medal, British Paediatric Association
1967	Joseph Goldberger Award, American Medical Association; AMA Council on Foods and Nutrition Award
1968	Professor, International Family Health, Tulane University School of Public Health; Companion of the Order of St. Michael and St. George
1969	Honorary D.Sc., University of the West Indies
1971	Martha May Eliot Award, American Public Health Association
1972	Dawson Williams Prize in Paediatrics, British Medical Association
1973	Honorary Member, American Academy of Pediatrics
Death	July 13, 1992

Dr. Cicely Williams spent her life saving the lives and promoting the health of women and children throughout the world. A woman physician of firsts, a prisoner of war, a saver of lives from the diseases kwa-shiorkor and Jamaican vomiting disease, she has led an extraordinary life of personal heroism and dedication to others.

Cicely was born in Jamaica in 1893. Her father, James Towland Williams, was director of education in Jamaica and was Oxford-educated. Cicely always had the desire to graduate from Oxford too, but there was no money for her schooling. She thought about becoming a nurse but was told she didn't have the stamina for it. For a time after high school, Cicely stayed at home with her family, a pleasant but unchallenging life.

Because of a shortage of doctors during World War I, 41 women were admitted to Oxford, Cicely among them. She found medicine to be everything she had hoped and was privileged to study with Sir William Osler during his last year of teaching. She passed her finals in 1923 but still needed to complete a doctoral thesis.

By the time Cicely finished at Oxford, the war and the shortage of doctors were over. She mailed out 70 applications before finding a gynecological surgery residency at South London Hospital for Women and Children. Cicely found that she enjoyed working with children.

Because of the lack of positions for women and because of her upbringing in Jamaica, Cicely was interested in working overseas for the British Colonial Office. She was told there were no openings, but she badgered the Colonial Office over the two years of her residency and finally received a posting to the Gold Coast (now Ghana) of Africa. Cicely was welcomed there with open arms by her superiors, who told her that they had been asking for women doctors for years. She immediately set out to conquer the local language, learning in a year what usually took four. Cicely was impressed by the love that Ghanaians had for their little ones. If a child had to stay in the hospital, the mother or another family member was always there by the bedside. She was amused and exasperated by the number of people who returned daily to see her. It was believed that the more one saw the doctor, the healthier one could become.

Cicely started to notice a disorder in her youngest patients that she could not identify. Babies and older children seemed to be immune, but she often saw toddlers with swollen legs and bellies who later developed scaly rashes and a reddish tinge to the hair. She found no parasitic organism and concluded that it was a nutritional disease. Other doctors in the Colonial Service insisted that she was misdiagnosing pellagra. She was also told that local burial customs were strict and she would not be allowed to do autopsies. Cicely found instead that families were rushing their dying children home not for burial but because they feared they could not pay the high cost of transporting a body. Cicely offered to pay for moving the bodies and performed postmortem examinations. She found that there was a native name for the condition that children develop when a new baby is born: the weaning disease, or kwashiorkor.

The children she placed on a protein diet became as healthy as children who were still breast-fed or were old enough to eat adult food. Cicely was convinced that she had identified a new form of protein malnutrition. She included her findings in the Gold Coast Colony Annual Medical Report for 1931–32, then published a description of kwashiorkor in *Archives of Disease in Children* in 1933. A later trip to Louisiana stilled any question in her mind about pellagra. Her 1935 article in *Lancet* detailed the differences in the two diseases.

During this period, Cicely nearly died after contracting blood poisoning during an autopsy. After a slow recovery, she was transferred to another city, Kumasi. Disappointed by this unwanted move, Cicely decided to use the year she must spend in Kumasi to write her long-delayed doctoral thesis. "Child Health in the Gold Coast" was accepted without reservation, and in 1936 she graduated from Oxford. Her only disappointment was that her father did not live to see her attain this goal.

Hoping to go back to Africa, Cicely was nonplussed when she was transferred in 1936 to Trengganu, Malaya (now Malaysia). She found the population to be much different from the patients she had known in Ghana. Local religions accepted all that happened as fate. Infant mortality was higher than in Africa, partly because Western companies were shaming new mothers into using canned milk with little nutritional value. Cicely fought futilely with these companies, who would send out "nurses" to new mothers with samples of their products. She took down notes for a paper on rickets, but publication was delayed until after World War II.

The news that all Westerners feared, but knew would come, was broadcast at the end of 1941: the Japanese had invaded Malaya. Cicely made a nightmarish trek across mountains to reach the fortress of Singapore. This move was only a short reprieve; the Allied forces left in 1942 and the Japanese army occupied the bombed-out city. Cicely had begun working with the children in Singapore hospitals during the siege. They moved from place to place, each worse than the last, as bombs damaged makeshift hospitals. When the Japanese troops entered Singapore, Cicely deliberated, then offered children in the hospital to anyone who would take them. Word got out that the hospital was giving away children, and the youngsters were all placed. Thankful for their relative safety, Cicely prepared herself for the occupation.

After the long siege, Cicely was suffering from dysentery and was spared a forced march. After her recovery, she found herself in Changi prison camp in Singapore. The prisoners tried to lead as normal a life as possible and to resist in small ways, but daily life was daunting. There was little food and even less news of the war. After Cicely served for a while as head of the women's side of the camp, the suspicious Kempaitai, the Japanese Secret Police, arrested her. On October 23, 1943, she was questioned and placed in a tiny prison cell with seven men. On December 2 she "celebrated" her fiftieth birthday. A friend, the only other woman in the Secret Police jail, bribed a guard to give her a sliver of soap as a birthday present.

On March 25, 1944, she was sent back to Changi "temporarily," never sure whether she might be reclaimed by the Kempaitai. The dysentery she contracted in prison improved, but recovery from the psychological and physical cruelty was slower. In November 1944 the prisoners saw Allied planes overhead, their first sign that the war was ending. Weak from dysentery and malaria, but alive when the Japanese surrendered, she learned that no one had expected survivors at Changi.

Cicely had to be carried aboard a ship going back to England. When she was well again, she went back to Malaya to be in charge of Maternity and Child Welfare Services, the first superscale post in the Colonial Service ever given to a woman. Cicely was now 52 years old. In 1948 the newly created World Health Organization recruited her to be the first head of the Section on Child and Maternal Health. She left in 1951 to return to Jamaica when the government asked her to head a research team to figure out the cause of Jamaican vomiting sickness. Within a year, the team isolated the harmful substance in spoiled or green ackee fruit and detailed a life-saving treatment of glucose therapy.

Williams returned to academia in 1955 to become the senior lecturer in nutrition at London University, then professor of maternal and child health at the American University in Beirut in 1960. She stayed for four years, then went to London to become an advisor to the Family Planning Association. Cicely never found any irony in this new position. Her philosophy recognizes that children must be wanted before they will be given a healthy life. "If we look after the quality of a population, the quantity will look after itself."[1]

A tart and witty speaker, Cicely said of her years in medicine, "I had some doubts about being a doctor during my early purely scientific years at Oxford, but since I started training at King's College Hospital in 1921, I don't think I have really ever had a moment's boredom."[2] She urged young health professionals to get away from centralized hospitals. "[T]o my mind, it's the general practitioner who is badly needed in medicine and in nursing—people who will look after the people, not just the disease."[3]

When she tallied up the countries she had worked in, she found the number was 70, on five continents—a fine record for someone who was once thought to be too frail for nursing. After a stint as a professor at Tulane University, she settled in Oxford for retirement. Typical of Cicely, it was an active retirement: she was an honorary Fellow at both Somerville and Green Colleges of Oxford University. She died at age 98 on July 13, 1992. One obituary summarized the true mission in Williams's life succinctly: "The world's children are indebted to her."[4]

Notes

1. "A Special Number Marking the Eightieth Year of Cicely D. Williams," *Nutrition Reviews* 31 (Nov. 1973): 377.
2. Beatrice Levin, *Women and Medicine* (Lincoln, Neb.: Media Publishing, 1988), p. 208.
3. *Ibid.*, p. 211.
4. David Morley. "C.D. Williams [Obituary]," British Medical Journal 305 (August 1, 1992): 307.

Bibliography

"Cicely Williams at 90." *Lancet* 3 (Dec. 1983): 1289.

Craddock, Sally. *Retired Except on Demand: The Life of Cicely Williams.* Oxford: Green College, 1983.

Cruickshank, Eric. "Cicely D. Williams: Grand Lady of Medicine." *Nutrition Reviews* 31 (Nov. 1973): 378–81.

Dally, Ann. *Cicely: The Life of a Doctor.* London: Gollancz, 1968.

Darby, William. "Cicely D. Williams: Her Life and Influence." *Nutrition Reviews* 31 (Nov. 1973): 330–33.

Levin, Beatrice. *Women and Medicine.* Lincoln, Neb.: Media Publishing, 1988.

Lovejoy, Esther Pohl. *Women Doctors of the World,* pp. 258, 379–80. New York: Macmillan Co., 1957.

"A Special Number Marking the Eightieth Year of Cicely D. Williams." *Nutrition Reviews* 31 (Nov. 1973). [Includes reprints of many of Dr. Williams's articles.]

Williams, Cicely Delphin. "Kwashiorkor: A Nutritional Disease of Children Associated with a Maize Diet." *Lancet* 16 (Nov. 1935): 1151–52.

———. "A Nutritional Disease of Childhood Associated with a Maize Diet." *Archives of Disease in Childhood* 8 (1933): 423–33.

                                                                        KELLY HENSLEY

HEATHER WILLIAMS

(1955–   )

Ornithologist, Neuroethologist

Birth	July 27, 1955
1977	A.B., biology, Bowdoin College
1977–78	Thomas J. Watson Fellow, Hebrew University, Israel

1985	Ph.D., behavioral neuroscience, The Rockefeller University; Finalist for Lindsley Award, Society of Neuroscience
1986	Air Force Office for Scientific Research (two-year award); married Patrick D. Dunlavey
1986–88	Assistant Professor, The Rockefeller University Field Research Center, NY
1988	National Institutes of Health (NIH), five-year grant
1993	MacArthur Foundation Fellow
1994–	Associate Professor, Williams College, MA; Adjunct Professor, The Rockefeller University

Dr. Heather Williams's life and work are prime examples of the challenge and excitement of discovery. An international youth, athletic pursuits, and scholarly accomplishments all find expression in her professional career. A theme of "exploration" ties together the unusual facets of this notable woman.

Dr. Williams is a researcher and professor of biology. More specifically, she is an ornithologist and neuroethologist: she studies birds (ornithology) and the relationship between their nervous system and their behavior (neuroethology). Her research has focused on the neurological patterns that are related to the song patterns of the zebra finch, *Taeniopygia guttata*. Her work has been recognized as distinctive and instrumental in bringing together several different areas of scientific study. In 1993 her exceptional talent and research was recognized with a prestigious MacArthur Foundation Fellowship award and grant.

Heather Williams was born in 1955 in Spokane, Washington. Her father was in the foreign service with the U.S. government and, therefore, the family moved among different countries when Heather was a youngster. At various times they lived in Turkey, Bolivia, and Laos. This provided the young Williams with a variety of experiences: "Because we moved to such different places and climates, I was always seeing different animals, which probably led to my interest in biology."[1] She recalls that she was always interested in animals, and wherever she lived she liked to collect small specimens.

Like Dr. Williams, many science professionals indicate that their curiosity was turned toward scientific investigation and discovery when they were young. Over the years, Williams's childhood interests matured and were integrated into her career goals. She recalls, "The first thing I can remember wanting to be was an astronaut, followed by [a] geologist and then a biologist, and I have always known I wanted to explore— find out new things."[2]

With these vocational aspirations, Heather Williams finished high school and continued her education at Bowdoin College in Maine. Her college studies concentrated on biology. While at Bowdoin, Heather distinguished herself as a highly motivated and promising scientist. She graduated *summa cum laude* from her department and was the recipient of several other eminent collegiate honors.

Immediately following completion of her A.B. in biology, Heather Williams was accepted as a Thomas J. Watson Fellow at the Hebrew University of Eilat, Israel. She was involved in study and research in a marine biology laboratory during this year abroad. After returning to the United States, Heather entered the doctoral program in behavioral neuroscience at The Rockefeller University, New York. There she studied under Dr. Fernando Nottebohm, an ethologist whose work concentrated on the canary. As is often the case in academia and scientific research, Williams greatly values the past and present role of her mentor in her own research. Her graduate work extended Nottebohm's work into a new area, the zebra finch. Williams's dissertation, "A Motor Theory for Bird Song Perception," was foundational for the many and varied inquiries of her future research.[3]

Williams was granted the Ph.D. in 1985 and continued first as a postdoctoral fellow and then as an assistant professor at The Rockefeller University Field Research Center for Ethology and Ecology. Having done her doctoral and now her postdoctoral work at Rockefeller, she was building a strong professional relationship with several scholars in her area: Dr. Nottebohm, Dr. Jeffrey Cynx, and Dr. David Vicario. These associations have been important: the researchers serve as guides and "sounding boards" for one another's ideas and projects.

In the midst of these many scholarly pursuits, Williams engaged in national competitions of female orienteers. (Orienteering is cross-country running with a compass and a map.) She says that orienteering was a welcome relief from the hours in the research lab. "The years of grad school in Manhattan made me need a diversion that [would get] me out into the countryside, and I have always been active athletically."[4] However, orienteering has been a serious pursuit as well. She ranked third in the United States in this sport for the years 1980–1989. In 1993 she served as a member of the course-setting team for the 1993 World Orienteering Championship.

Besides the physical and mental discipline involved, Williams sees another important link between orienteering and neuroanatomy: "They require you to interpret two-dimensional representations of three-dimensional realities."[5] In orienteering, one needs to interpret a topographical map of the course and surrounding terrain. In neuroanatomy, the researcher tries to determine how the data relates to the structure and function of the brain and nervous system. In this way, orienteering could serve as both a break from the research routine and a stimulant of creative abilities.

In 1986, Dr. Williams married Patrick D. Dunlavey. She had just finished her doctoral studies and was already embarked on her career in research and education. She says that she did not consciously delay marriage and family because of her career, but that the timing worked out well for her.

In 1988 she was invited to join the faculty of the biology department at Williams College in Williamstown, Massachusetts. She made the move to Williams College in part because of the small-town setting, a smaller campus for her research, and the ability to maintain contact with research at The Rockefeller University Field Research Center. Now an associate professor at Williams College and an adjunct professor at The Rockefeller University, she has continued her research on the details of song learning in the zebra finch. Dr. Williams's work has contributed greatly to the body of research and literature on neuroscience, animal behavior, and comparative psychology. In addition to her own publishing credits, she is a reviewer for many related scientific journals.

Dr. Williams's work first focused on a "motor theory" to describe how a bird "hears" the song of other birds. As her work has progressed, she has seen that there are both sensory and motor aspects of song perception. Williams and her colleagues have been able to carefully describe the exact nerves and parts of the brain that appear to be involved in this very complex process. Her findings have helped link the behavior of the bird to the neurological pathways through which the sounds must be transmitted and "translated." She has detailed some of the numerous subtleties that the zebra finch can detect in song patterns. In addition, she has been exploring how this activity appears to be lateralized (or divided among parts) in the brain of the bird, how the new song is made up of parts (syllable chunks) of what of the bird has heard, and the differences in song perception and preference among the sexes (sexual dimorphism).

Some of these findings have been used to propose theories as to how animals perceive and filter sounds in their communication. Much of her work continues to provide intriguing questions for her.[6] Just how much can these birds be selectively influenced by the surrounding bird songs? What are the conditions? Are there constraints? Usually the bird's song is fixed within a certain amount of time, but Williams has found that this can be reversed under certain conditions. Much of this has to do with a brain and nervous system that is very different, but in many ways similar to, that of humans. It is expected that her work will help us better understand neuroscience, psychology, and communication as they relate to humans as well.

From Williams's childhood throughout her years in academia and research, the theme of "exploration" gives form to her diverse biography. A young girl travels to many countries, discovering new ways and forms of life. An athlete trains her body in order to meet the challenge of cross-country orienteering. A determined student disciplines her mind and life in order to achieve her academic goals. And a promising scientist seeks to uncover the implications and limits of new theories in novel areas. The result: a dedicated scientific "explorer" who is willing to doggedly question, investigate, and theorize about the hard questions of neuroanatomy.

Notes

1. Heather Williams to Kathleen King, August 26, 1994, personal e-mail correspondence.
2. *Ibid.*
3. Heather Williams, "A Motor Theory for Bird Song Perception," Ph.D. dissertation, The Rockefeller University, 1984.
4. Heather Williams to Kathleen King, August 30, 1994, personal e-mail correspondence.
5. *Ibid.*
6. *Ibid.*

Bibliography

Cynx, Jeffrey, and Heather Williams. "Hemispheric Differences in Avian Song Discrimination." *Proceedings of the National Academy of Sciences of the USA* 89 (Feb. 15, 1992): 1372–75.
Williams, Heather. "Models for Song Learning in the Zebra Finch: Fathers or Others?" *Animal Behaviour* 39 (Apr. 1990): 745–57.
———. "Sexual Dimorphism of Auditory Activity in the Zebra Finch Song System." *Behavioral and Neural Biology* 44 (1985): 470–84.
Williams, Heather, and Fernando Nottebohm. "Auditory Responses in Avian Vocal Motor Neurons: A Motor Theory for Song Perception in Birds." *Science* 229 (1985): 279–82.
Williams, Heather, and Kristen Staples. "Syllable Chunking in Zebra Finch (*Taeniopygia guttata*) Song." *Journal of Comparative Psychology* 106 (Sept. 92): 278–86.
Williams, Heather, Jeffrey Cynx, and Fernando Nottebohm. "Timber Control in Zebra Finch (*Taeniopygia guttata*) Song Syllables." *Journal of Comparative Psychology* 103 (Dec. 1989): 366–80.
Williams, Heather, Kerry Kilander, and Mary Lou Sotanski. "Untutored Song, Reproductive Success and Song Learning." *Animal Behaviour* 45 (Apr. 1993): 695–705.

KATHLEEN PALOMBO KING

JANE COOKE WRIGHT

(1919–)

Physician, Cancer Researcher

Birth	November 20, 1919
1942	B.A., Smith College
1945	M.D., New York Medical College
1947	Married David Jones, Jr.
1949–52	New York City school physician and visiting physician at Harlem Hospital
1952	Director, Cancer Research Foundation; *Mademoiselle* magazine Merit Award for outstanding contribution to medical science
1955	Director of Cancer Chemotherapy Research, New York University Medical Center
1961	Adjunct Professor of Research Surgery, New York University Medical Center
1965	Spirit of Achievement Award, Women's Division, Albert Einstein College of Medicine
1967–87	Professor of Surgery, New York Medical College; Hadassah Myrtle Wreath Award
1968	Smith Medal, Smith College
1987–	Professor Emerita, New York Medical College

This eminent surgeon and researcher in cancer chemotherapy was born in 1919 in New York City. She was the eldest of two daughters born to Louis Tompkins Wright, a surgeon and pioneer in cancer chemotherapy, and Corinne (Cooke) Wright.

Professor Wright grew up on 139th Street in New York City and attended private schools. Her hobbies were painting in watercolors and reading mystery novels. She was a bright student and excellent swimmer who went to Smith College on a four-year scholarship. At Smith she was also an excellent student and member of the varsity swim team. She graduated in 1942.

Her first career choice was painting, but her father discouraged her because of the uncertainty of that profession. Her second career choice was medicine, so she applied to New York Medical College. She was

Jane Cooke Wright. Photo reprinted by permission of the Sophia Smith Collection and College Archives, Smith College.

accepted and received a four-year scholarship. In medical school she was again an exceptional student, vice president of her class, president of the honor society, and literary editor of the yearbook. Wright graduated from medical school with honors in 1945, the third in a class of 95 students.

She completed her internship and assistant residency at Bellevue Hospital in New York City. An article in a 1953 issue of *Crisis* quoted one supervisor as saying she was by far the most promising intern working with him.[1] Dr. Wright completed her residency in 1948 at Harlem Hospital, where her father had established the Cancer Research Foundation.

On July 27, 1947, Jane Cooke Wright married David Jones, Jr., a graduate of Harvard Law School. Away from the hospital she preferred to be known as Mrs. David Jones or Jane Jones. They had two daughters, Jane and Allison.

Her first professional position in 1949 was as a New York City school physician and visiting physician of Harlem Hospital. When her father died in 1952, Dr. Wright took over as director of the Cancer Research Foundation he had started. The Foundation studied the effects of drugs on tumors and other abnormal growth. She was quoted in 1968 about her work at the Foundation: "There's no greater thrill than in having an experiment turn out in such a way that you make a positive contribution."[2]

In September 1955, Wright joined the faculty of New York University Medical Center as director of cancer chemotherapy research and instructor of research surgery in the Department of Surgery. By 1961 she was an adjunct professor of research surgery. She dedicated herself to analyzing the efficacy of a wide range of drugs in treating cancer to gain a greater understanding of the relationship between patient, tissue culture, and animal response.

Dr. Wright returned to her medical school alma mater, New York Medical College, in July 1967 as a professor of surgery and joined the staff of the affiliated hospitals, Flower 5th Avenue, Metropolitan, and Birds Coler Memorial. She continued her research, was an administrator of the medical college, and was responsible for the development of a program to study cancer, heart disease, and stroke.

In 1952, Dr. Wright received the first of many awards. *Mademoiselle* magazine awarded the Merit Award for her outstanding contribution to medical science through her evaluations of the efficacy of drugs in cancer treatment. On receiving her award, Dr. Wright said, "My plans for the future are to continue seeking a cure for cancer, be a good mother to my children and a good wife to my husband."[3]

Dr. Wright was awarded the Spirit of Achievement Award of the Women's Division of Albert Einstein College of Medicine in 1965 for her deep commitment as a scientist and teacher in advancing medical knowledge and research. In 1967 she was honored with the Hadassah Myrtle Wreath Award for her outstanding contribution to her field. She received the Smith Medal from Smith College in 1968. With other exceptional black scientists, she was featured on a Ciba Geigy poster. She was saluted on the cover of *Cancer Research* with seven other women members of the American Association of Cancer Research for her contributions to research in clinical cancer chemotherapy.

In 1983, Dr. Wright delivered the Surgery Section Distinguished Lecture at the Annual Convention and Scientific Assembly of the National Medical Association. Her paper was entitled "Cancer Chemotherapy: Past, Present, Future." In 1984 she contributed a series of articles on chemotherapy in the *Journal of the National Medical Association*.

In 1987, Dr. Wright became an emerita professor at New York Medical College. This gives her more time for painting, reading, and sailing.

Notes

1. "Young Woman of the Year," *Crisis* 60, no. 1 (Jan. 1953): 4–5.
2. "Jane Cooke Wright," *Current Biography Yearbook* 29 (1968): 443–45.
3. "Young Woman of the Year," pp. 4–5.

Bibliography

"Jane Cooke Wright." *Current Biography Yearbook* 29 (1968): 443–45.
Notable Black American Women, ed. Jessie Carney Smith. Detroit: Gale Research, 1992.
O'Neill, Lois D. *The Women's Book of World Records and Achievements.* Garden City, N.Y.: Anchor Press/Doubleday, 1979.
"Young Woman of the Year." *Crisis* 60, no. 1 (Jan. 1953): 4–5.

HELEN-ANN BROWN

ROSALYN SUSSMAN YALOW

(1921–)

Endocrinologist, Medical Physicist

Birth	July 19, 1921
1943	Married Aaron Yalow
1945	Ph.D., University of Illinois
1946	First woman engineer at Federal Telecommunications Laboratory; joined faculty at Hunter College
1947	Joined staff at Bronx VA Hospital
1950	Began collaboration with Solomon Berson
1959	First use of radioimmunoassay (RIA)
1975	Elected to National Academy of Sciences; A. Cressy Morrison Prize in Natural Sciences, New York Academy of Sciences
1976	Albert J. Lasker Basic Medical Research Award
1977	Nobel Prize in Medicine or Physiology
1978	President, Endocrine Society
1988	National Medal of Science
1991	Retired from Bronx VA Hospital

On October 13, 1977, at 7:00 A.M., Rosalyn Yalow was at work in the Veterans Administration Hospital in the Bronx when she learned she had been awarded the Nobel Prize in Physiology or Medicine. Yalow was the second woman to win the Nobel Prize in Medicine and the first American-educated woman to win a Nobel in science. The prize was awarded for the development of radioimmunoassay (RIA) of peptide hormones. Yalow received half the prize; the other half was shared by Roger C. Guillemin of the Salk Institute and Andrew V. Schally of the Veterans Administration Hospital in New Orleans for their related work on hormones in the brain.

RIA is a biological assay technique used to screen for very tiny amounts of chemicals in biological tissues and fluids. In comparison, it is sensitive enough to detect the presence of a sugar cube in a lake. The use of RIA revolutionized the study of endocrinology and the treatment of hormonal disorders such as diabetes. Blood in blood banks is screened for hepatitis virus, newborns are tested for underactive thyroid secretion to prevent mental retardation, athletes can be checked for drug abuse, and a new science—neuroendocrinology—has come into being through the use of RIA.

Yalow's mother, Clara Sussman, contended that Yalow showed her tenacity and combative spirit early in life. When Rosalyn was 3 years old, her mother visited an egg store with her and her older brother, Alexander, on the way home from a theater visit. Following the egg purchase, Rosalyn insisted that they return home by a different route than her mother intended. Rosalyn sat down on the sidewalk; unable to carry Rosalyn and the eggs, Mrs. Sussman gave in after a crowd began to gather.

When Alexander was in first grade, he was humiliated by having his hand smacked by his teacher. He wept and threw up over the incident. When Rosalyn entered the first grade and had her hand smacked by the same teacher, she struck back. She told the principal she had avenged her older brother after years of waiting. She kept an old photograph of herself as a 5-year-old in huge boxing gloves standing over her supine brother, and she claims that it was that attitude that enabled her to go into physics.

Rosalyn was born in 1921 and grew up in the South Bronx, where she has lived almost all her life. Few people in the Jewish neighborhood— including her parents—had schooling past the elementary level, although education was highly valued. When she graduated from high school in 1937 at age 15, she was admitted to Hunter College. She had already set her sights on medical research as a career. She chose physics as a major because of the excitement surrounding the field at the time, particularly nuclear physics. In 1938, Eve Curie published a biography of her mother, Marie Curie, whom Yalow took as her professional role model.

Rosalyn would have preferred to go to medical school after graduation, but at the time American medical schools were not admitting Jewish men, much less Jewish women. She took a part-time job as a secretary at the Columbia medical school in January 1941, with the understanding that she could take some science courses as long as she also learned stenography. A few months later she received an assistantship in physics from the University of Illinois and tore up her steno books. World War II was imminent, and male graduate students were being lost to the draft. She was the first woman in the University of Illinois engineering school since 1917, when World War I was under way.

Rosalyn met her future husband, Aaron Yalow, on her first day of graduate school at Illinois. Both did Ph.D. research in nuclear physics under Maurice Goldhaber. They were married on June 6, 1943, and remained partners until Aaron's death on August 8, 1992. In January 1945, Yalow received her Ph.D. (the second woman to receive a Ph.D. in physics from Illinois) and returned to New York City as the first woman engineer in the Federal Telecommunications Laboratory of International Telephone and Telegraph. She returned to teaching at Hunter College after a year, but there were no research facilities there. In 1947 she joined the Bronx VA Hospital in order to set up a radioisotope service while still teaching full-time at Hunter.

In January 1950 she resigned from Hunter to devote all her attention to medical physics. Yalow began looking for a collaborator to handle the medical aspects of her research while she focused on the engineering and physics aspects. She met Solomon A. Berson, a young resident at the VA, in the spring of 1950 and formed a partnership that lasted 22 years until his death in April 1972. Yalow and Berson worked 80-hour weeks, speaking to each other in a sort of scientific shorthand. There was unvarnished logic between them and each of them understood the other's meaning thoroughly; although when they subjected other scientists to this type of questioning in meetings, it was sometimes viewed as hostility instead of honesty.

RIA was developed as a side issue of insulin research. Adult diabetics were injected with radioactively tagged insulin to measure how rapidly it would disappear from their system. The adult diabetics retained the insulin longer than the normal controls; after further investigation, Yalow and Berson concluded that people who had taken insulin developed antibodies to it. This flew in the face of the conventional wisdom that the insulin molecule was too small to stimulate the production of antibodies. The *Journal of Clinical Investigation* refused to publish their article until the words "insulin antibody" were removed from the title.

In developing RIA, Yalow and Berson wanted a technique that did not require injecting radioactive material into a living human being. They fixed on a method of labeling the antigenic substance to be measured and mixing it with a specific antibody to the substance. After mixing, the degree of binding of the antibody to the labeled substance could be calculated and used to estimate the amount of the substance present in a biological fluid by comparing it to a known standard. Despite the enormous commercial potential of RIA, Yalow and Berson decided not to patent it.

During the next few years, Yalow and Berson began using RIA to rethink all of endocrinology and made it one of the hottest areas of medical research. During these years, Yalow also bore and reared her two children. Benjamin was born in 1952 and Elanna in 1954. Both have earned doctorates. Benjamin directs the computer center at the City University of New York, and Elanna manages a company that establishes daycare centers.

In 1968, Berson decided to take a position as professor in charge of internal medicine at the Mount Sinai School of Medicine of the City University of New York. Yalow was extremely disappointed at the news, but Berson assumed he could put in a full day's work at Mount Sinai and then work through the night twice a week at the VA lab. But the pace began to catch up with him, and after a while Berson began to work only one night a week at the lab. In March 1972, Berson had a minor stroke. He died a month later of a heart attack at age 54.

Yalow was told she could not win a Nobel without Berson, although RIA was unquestionably an accomplishment worthy of the prize. Nobels are not given posthumously and had not been given to surviving partners in research projects. Yalow increased her work week to 100 hours, and the lab published 60 articles between 1972 and 1976. She was elected a member of the National Academy of Sciences in 1975. In 1976 she was the first woman to win the Albert Lasker Basic Medical Research Award, which is often a forerunner of the Nobel. Following her receipt of the Nobel Prize in Medicine in 1977, she received the nation's highest science award, the National Medal of Science, in 1988.

Yalow retired from the VA Hospital in 1991 and became a science activist. She lectures on causes such as instituting better-quality child care for families, requiring more science in American education, and promoting the role of women in science.

Bibliography

Gilbert, Lynn, and Gaylen Moore. *Particular Passions: Talks with Women Who Have Shaped Our Times.* New York: Crown, 1981.

Gleasner, Diana C. *Breakthrough: Women in Science.* New York: Walker, 1983.

McGrayne, Sharon Bertsch. *Nobel Prize Women in Science: Their Lives, Struggles, and Momentous Discoveries.* Secaucus, N.J.: Carol Pub. Group, 1993.

Moritz, Charles, ed. *Current Biography Yearbook.* New York: H.W. Wilson, 1978.

Opfell, Olga S. *The Lady Laureates: Women Who Have Won the Nobel Prize.* Metuchen, N.J.: Scarecrow Press, 1986.

Rayner, William P. *Wise Women.* New York: St. Martin's Press, 1983.

Stone, Elizabeth. "A Mme. Curie from the Bronx." *New York Times Magazine* (Apr. 9, 1978): 29.

Wasson, Tyler, ed. *Nobel Prize Winners: An H.W. Wilson Biographical Dictionary.* New York: H.W. Wilson, 1987.

Yalow, Rosalyn S. "A Physicist in Biomedical Investigation." *Physics Today* 32 (Oct. 1979): 25–29.

———. "Remembrance Project: Origins of RIA." *Endocrinology* 129, no. 4 (Oct. 1991): 1694–95.

MARGARET SYLVIA

Appendix I: Scientists by Field

Agrostologists
Mary Agnes Meara Chase

Anatomists
Elizabeth Caroline Crosby
Susanna Phelps Gage
Alessandra Giliani
Margaret Adaline Reed Lewis
Florence Rena Sabin

Anesthesiologists
Virginia Apgar

Bacteriologists
Alice Catherine Evans
Rebecca Craighill Lancefield
Barbara Moulton
Anna Wessels Williams

Biologists
Rachel Carson
Mary Agnes Meara Chase
Jewel Plummer Cobb

Rosalind Elsie Franklin
Elizabeth Dexter Hay
Ruth Hubbard
Evelyn Fox Keller
Mimi A.R. Koehl
Elizabeth Fondal Neufeld
Deborah L. Penry
Naomi E. Pierce
Margie Profet
Ellen Kovner Silbergeld

Biomechanics
Mimi A.R. Koehl

Biomedical Researchers
Gladys Rowena Henry Dick

Biophysicists
Edith Smaw Hinckley Quimby

Botanists
Rachel Littler Bodley
Elizabeth Gertrude Knight Britton

Alice Eastwood
Sophia Hennion Eckerson
Katherine Esau
Margaret Clay Ferguson
Ynes Mexia
Lydia W. Shattuck

Cancer Researchers
Mary Jane Guthrie
Ariel Cahill Hollinshead
Jane Cooke Wright

Cell Biologists
Jewel Plummer Cobb
Marilyn Gist Farquhar
Elizabeth Dexter Hay

Comparative Anatomists
Susanna Phelps Gage

Cytologists
Katherine Foot

Developmental Biologists
Christiane Nüsslein-Volhard

Ecologists
Ann Haven Morgan

Embryologists
Susanna Phelps Gage
Rita Levi-Montalcini

Endocrinologists
Rosalyn Sussman Yalow

Entomologists
Maria Sibylla Merian

Environmentalists
Rachel Carson
Ruth Patrick

Geneticists
Madge Thurlow Macklin
Barbara McClintock

Christiane Nüsslein-Volhard
Elizabeth Shull Russell
Joan Argetsinger Steitz
Nettie Maria Stevens

Horticulturists
Louisa Boyd Yeomans King
Martha Daniell Logan
Eliza Lucas Pinckney
Elizabeth Waties Allston Pringle
Kate Olivia Sessions

Icthyologists
Rosa Smith Eigenmann

Limnologists
Ruth Patrick

Medical Physicists
Rosalyn Sussman Yalow

Microbiologists
Rita Rossi Colwell
Charlotte Friend

Microchemists
Sophia Hennion Eckerson

Midwives
Marie-Louise Lachapelle

Marine Ecologists
Jane Lubchenco

Molecular Biologists
Elizabeth Fondal Neufeld

Mycologists
Elizabeth Lee Hazen

Natural Philosophers
Heloise
Herrad of Landsberg

Naturalists
Maria Sibylla Merian

Nature Writers
Florence Merriam Bailey

Neurologists

Roberta Lynn Bondar

Neuroethologists

Heather Williams

Oceanographers

Deborah L. Penry

Ornithologists

Florence Merriam Bailey

Margaret Morse Nice

Heather Williams

Pathologists

Dorothy Hansine Andersen

Edith Jane Claypole

Elise Depew Strang L'Esperance

Maud Caroline Slye

Pediatricians

Mary Ellen Avery

Pharmacologists

Ariel Cahill Hollinshead

Frances Oldham Kelsey

Physicians

Hattie Elizabeth Alexander

Dorothy Hansine Andersen

Elizabeth Garrett Anderson

Aspasia

Mary Ellen Avery

Elizabeth Blackwell

Marie Anne Victoire Gallain Boivin

Lin Ch'iao-chih

Gladys Rowena Henry Dick

Cornelia Bonté Sheldon Amos Elgood

Alice Hamilton

Hildegard of Bingen

Aletta Henriette Jacobs

Sophia Jex-Blake

Anandibai Joshee

Chung-Hee Kil

Helen Brooke Taussig

Trotula

Mary Edwards Walker

Cicely Delphin Williams

Jane Cooke Wright

Physiologists

Edith Jane Claypole

Ida Henrietta Hyde

Margaret Adaline Reed Lewis

Plant Physiologists

Sophia Hennion Eckerson

Zoologists

E. (Estella) Eleanor Carothers

Cornelia M. Clapp

Mary Jane Guthrie

Ethel Nicholson Browne Harvey

Hope Hibbard

Libbie Henrietta Hyman

Helen Dean King

Agnes Mary Claypole Moody

Ann Haven Morgan

Appendix II: Scientists by Awards Received

A. Cressy Morrison Prize in Natural Sciences, New York Academy of Sciences

Rosalyn Sussman Yalow

Alan T. Waterman Award

Deborah L. Penry

Albert A. Michelson Award

Barbara McClintock

Albert Lasker Basic Medical Research Award

Rita Levi-Montalcini

Barbara McClintock

Rosalyn Sussman Yalow

Albert Lasker Clinical Medicine Research Award

Elizabeth Fondal Neufeld

Albert Lasker Public Service Award

Christiane Nüsslein-Volhard

Alcon Award for Vision Research

Elizabeth Dexter Hay

Alfred Sloan Award in Cancer Research

Charlotte Friend

Alfred P. Sloan, Jr., Prize

Christiane Nüsslein-Volhard

Alice Evans Award, American Society of Microbiology

Rita Rossi Colwell

American Association of University Women Achievement Award

Rachel Carson

Elizabeth Caroline Crosby

Barbara McClintock

Helen Brooke Taussig

American Brotherhood Citation

Florence Rena Sabin

American Cancer Society
Award

    Charlotte Friend

American Cancer Society
Medal

    Edith Smaw Hinckley Quimby

American Design Award, Lord
& Taylor

    Edith Smaw Hinckley Quimby

American Heart Association
Achievement Award

    Rebecca Craighill Lancefield

American Heart Association
Award of Merit

    Helen Brooke Taussig

American Medical Association
Council on Foods and
Nutrition Award

    Cicely Delphin Williams

Borden Award for Research in
Nutrition

    Dorothy Hansine Andersen

Botanical Society of America
Certificate of Merit

    Mary Agnes Meara Chase

    Barbara McClintock

    Ruth Patrick

Brewster Medal, American
Ornithologists' Union

    Florence Merriam Bailey

    Margaret Morse Nice

Cameron Prize, University of
Edinburgh

    Gladys Rowena Henry Dick

Charles Leopold Mayer Prize,
Académie des Sciences,
Institut de France

    Barbara McClintock

Chemical Pioneer Award,
American Institute of
Chemists

    Elizabeth Lee Hazen

Chevalier Legion d'Honneur,
France

    Helen Brooke Taussig

Christopher Columbus
Discovery Award in
Biomedical Research

    Joan Argetsinger Steitz

Clement Cleveland Medal,
New York City Cancer
Committee

    Elise Depew Strang L'Esperance

Companion of the Order of
St. Michael and St. George

    Cicely Delphin Williams

Congressional Medal of
Honor

    Mary Edwards Walker

Coolidge Award, American
Association of Physicists in
Medicine

    Edith Smaw Hinckley Quimby

Daniel Giraud Medal,
National Academy of Sciences

    Libbie Henrietta Hyman

Dawson Williams Prize in
Paediatrics, British Medical
Association

    Cicely Delphin Williams

Department of Health and
Human Services Service
Award (Scientific)

    Frances Oldham Kelsey

Dickson Prize for Science,
Carnegie-Mellon University

    Joan Argetsinger Steitz

Distinguished Publication Award, Association for Women in Psychology

Evelyn Fox Keller

Distinguished Scientist Award, Society for Experimental Biology and Medicine

Ariel Cahill Hollinshead

Distinguished Service Award, American Institute of Biological Sciences

Ruth Hubbard

Distinguished Service Award, National Home Planting Bureau

Louisa Boyd Yeomans King

Distinguished Service Award, Radiological and Medical Physics Society

Edith Smaw Hinckley Quimby

E.B. Wilson Medal, American Society for Cell Biology

Marilyn Gist Farquhar
Elizabeth Dexter Hay

E. Mead Johnson Award for Research in Pediatrics

Hattie Elizabeth Alexander
Dorothy Hansine Andersen

Eli Lilly Award in Biological Chemistry

Joan Argetsinger Steitz

Elizabeth Blackwell Award

Dorothy Hansine Andersen
Madge Thurlow Macklin
Florence Rena Sabin

Ellen Richards Prize

E. (Estella) Eleanor Carothers
Helen Dean King
Nettie Maria Stevens

Eminent Ecologist Award, Ecological Society of America

Ruth Patrick

Excellence in Science Award, Federation of American Societies for Experimental Biology

Elizabeth Dexter Hay

Feltrinelli Award, Rome, Italy

Helen Brooke Taussig

Fisher Award, American Association of Microbiologists

Rita Rossi Colwell

Friendship Award for Eminent Achievement, American Women's Association

Elise Depew Strang L'Esperance
Florence Rena Sabin

Gairdner Foundation Award of Merit, Canada

Helen Brooke Taussig

General Motors Cancer Research Prize

Christiane Nüsslein-Volhard

George Robert White Medal, Massachusetts Horticultural Society

Louisa Boyd Yeomans King

George Westinghouse Science Writing Award

Rachel Carson

Gold Medal, American College of Radiology

Edith Smaw Hinckley Quimby

Gold Medal, American Museum of Natural History

Libbie Henrietta Hyman

Gold Medal, American
Radiological Society

    Maud Caroline Slye

Gold Medal, Interamerican
College of Radiology

    Edith Smaw Hinckley Quimby

Gold Medal, International
Biotechnology Institute

    Rita Rossi Colwell

Gold Medal, Linnean Society
of London

    Libbie Henrietta Hyman

Gold Medal, Radiological
Society of North America

    Edith Smaw Hinckley Quimby

Gregor Mendel Medal,
Genetical Society of Great
Britain

    Christiane Nüsslein-Volhard

Henry Gray Award in
Neuroanatomy, American
Association of Anatomists

    Elizabeth Caroline Crosby

    Elizabeth Dexter Hay

Homer Smith Award,
American Society of
Nephrology

    Marilyn Gist Farquhar

Honorary Medal, American
College of Chest Physicians

    Helen Brooke Taussig

Howland Award, American
Pediatric Society

    Helen Brooke Taussig

Interamerican Award of
Merit, Lima, Peru

    Helen Brooke Taussig

International Botanical
Congress Medal

    Katherine Esau

Jagadish Bose Memorial
Medal, Indian Radiological
Society

    Edith Smaw Hinckley Quimby

James Spence Memorial Gold
Medal, British Paediatric
Association

    Cicely Delphin Williams

Jane Addams Medal

    Florence Rena Sabin

Janeway Medal, American
Radium Society

    Edith Smaw Hinckley Quimby

John and Alice Tyler Ecology
Award

    Ruth Patrick

John Burroughs Medal

    Rachel Carson

Joseph Goldberger Award,
American Medical Association

    Cicely Delphin Williams

Kimber Genetics Award,
National Academy of Sciences

    Barbara McClintock

Lasker Award, American
Public Health Service

    Alice Hamilton

    Elise Depew Strang L'Esperance

    Florence Rena Sabin

Lee Hawley, Sr., Award for
Arthritis Research

    Joan Argetsinger Steitz

Lewis S. Rosensteil Award for Distinguished Work in Basic Medical Research

Barbara McClintock

Louis and Bert Freedman Foundation Award for Research in Biochemistry

Barbara McClintock

Louis Jeantet Prize for Medicine

Christiane Nüsslein-Volhard

Louisa Gross Horwitz Prize for Biology or Biochemistry

Barbara McClintock

M. Carey Thomas Prize, Bryn Mawr College

Florence Rena Sabin

MacArthur Foundation Fellowship

Evelyn Fox Keller

Mimi A.R. Koehl

Jane Lubchenco

Naomi Pierce

Margie Profet

Ellen Kovner Silbergeld

Heather Williams

MacArthur Prize Fellow Laureate

Barbara McClintock

Mademoiselle **Magazine Merit Award**

Jane Cooke Wright

Martha May Eliot Award, American Public Health Association

Cicely Delphin Williams

Medaélle de la Recomissance Française

Cornelia Bonté Sheldon Amos Elgood

Médaille de l'Excellence, L'Association des Médecins de Langue Française du Canada

Roberta Lynn Bondar

Medal of Honor, Garden Clubs of America

Louisa Boyd Yeomans King

Medal of Freedom (U.S.)

Rachel Carson

Helen Brooke Taussig

Mercer Award, Ecological Society of America

Jane Lubchenco

Meyer Medal, Council of the American Genetic Association

Kate Olivia Sessions

Mickle Prize, University of Toronto

Gladys Rowena Henry Dick

Milton S. Eisenhower Gold Medal, Johns Hopkins University

Helen Brooke Taussig

Mina Shaughnessy Scholars Award, Fund for the Improvement of Postsecondary Education

Evelyn Fox Keller

Naples Table Association Prize

Florence Rena Sabin

NASA Space Medal

Roberta Lynn Bondar

National Audubon Society
Medal

    Rachel Carson

National Book Award

    Rachel Carson

National Institutes of Health
Merit Award

    Marilyn Gist Farquhar

National Medal of Science

    Mary Ellen Avery

    Elizabeth Caroline Crosby

    Rita Levi-Montalcini

    Barbara McClintock

    Elizabeth Fondal Neufeld

    Joan Argetsinger Steitz

    Rosalyn Sussman Yalow

National Medical Women's
Association Award

    Elizabeth Caroline Crosby

National Women's Hall of
Fame Award

    Barbara McClintock

New York Academy of
Medicine Medal

    Rebecca Craighill Lancefield

Nobel Prize (Physiology or
Medicine)

    Rita Levi-Montalcini

    Barbara McClintock

    Christiane Nüsslein-Volhard

    Rosalyn Sussman Yalow

Order of Canada, Officer

    Roberta Lynn Bondar

Order of Ontario

    Roberta Lynn Bondar

Order of the British Empire

    Cornelia Bonté Sheldon Amos
    Elgood

Oscar B. Hunter Memorial
Award, American Therapeutic
Society

    Hattie Elizabeth Alexander

Paul Karrer Medal, University
of Zurich

    Ruth Hubbard

Peace and Freedom Award,
Women's International
League for Peace and
Freedom

    Ruth Hubbard

Phi Sigma Service Award,
American Chemical Society

    Rita Rossi Colwell

Presidential Young
Investigator Award

    Mimi A.R. Koehl

President's Award for
Distinguished Federal Civilian
Service

    Frances Oldham Kelsey

Public Service Award,
Association of Federal
Investigators

    Frances Oldham Kelsey

Purkinje Gold Medal for
Achievement in Science,
Czechoslovakian Academy of
Sciences

    Rita Rossi Colwell

Ralph M. Waters Award,
American Society of
Anesthesiology

    Virginia Apgar

Research Achievement
Award, journal *Medicine*

> Rebecca Craighill Lancefield

Rhoda Benham Award,
Medical Mycological Society
of the Americas

> Elizabeth Lee Hazen

Salute to Contemporary
Women Scientists Award,
New York Academy of
Sciences

> Elizabeth Dexter Hay

Sault Ste. Marie Medal of
Merit

> Roberta Lynn Bondar

Scientific Achievement
Award, American Medical
Association

> Edith Smaw Hinckley Quimby

Scientific Achievement
Medal, International
Women's Exposition

> Edith Smaw Hinckley Quimby

Silver Medal, Scholar Speciale
Medicina, Italy

> Ariel Cahill Hollinshead

Silver Medal, Society of
Medicine of Paris

> Marie Anne Victoire Gallain
> Boivin

Silver Medal, Union des
Femmes de France

> Cornelia Bonté Sheldon Amos
> Elgood

Sigma Chi Annual
Achievement Award

> Rita Rossi Colwell

Sigma Chi Research Award

> Rita Rossi Colwell

Smith Medal, Smith College

> Jane Cooke Wright

Squibb Award in
Chemotherapy

> Elizabeth Lee Hazen

Star of Europe Medal

> Ariel Cahill Hollinshead

Stevens Triennial Prize,
Trustees of Columbia
University

> Hattie Elizabeth Alexander

T. Duckett Jones Memorial
Award

> Rebecca Craighill Lancefield

Thomas Hunt Morgan Medal,
Genetics Society of America

> Barbara McClintock

Trudeau Medal, National
Tuberculosis Association

> Florence Rena Sabin

U.S. Steel Foundation Award
in Molecular Biology

> Joan Argetsinger Steitz

Virginia Apgar Award,
American Academy of
Pediatrics

> Mary Ellen Avery

Warren Triennial Prize

> Joan Argetsinger Steitz

Weizmann Women and
Science Award

> Joan Argetsinger Steitz

William Wood Gerhard Gold
Medal, Pathological Society of
Philadelphia

> Margaret Adaline Reed Lewis

Wolf Prize in Medicine

> Barbara McClintock

> Elizabeth Fondal Neufeld

**Woman's Award for Work in
Consumer Protection**

> Barbara Moulton

**Young Investigator Award,
National Science Foundation**

> Deborah L. Penry

**Young Scientist Award,
Passano Foundation**

> Joan Argetsinger Steitz

Index

Page numbers in **bold** refer to main entries.

About the Editors
and Contributors

Editors

BENJAMIN F. SHEARER is Vice-President for Student Affairs, Planning, and Program Development at Alvernia College in Reading, Pennsylvania. He and his wife, Barbara, are the authors of *State Names, Seals, Flags, and Symbols* (Greenwood, 1987, rev. ed. 1994), as well as several other books published by Greenwood Press. They are currently preparing *Notable Women in the Physical Sciences* for Greenwood Press (forthcoming 1996).

BARBARA S. SHEARER is Director of Public Services and External Relations at the Scott Memorial Library, Thomas Jefferson University, in Philadelphia. She is co-editor with her husband, Benjamin, of *State Names, Seals, Flags, and Symbols* (Greenwood, 1987, rev. ed. 1994), as well as several other books published by Greenwood Press. She is co-author (with Geneva Bush) of *Finding the Source of Medical Information: A Thesaurus-Index to the Reference Collection* (Greenwood, 1985). She has also published several articles on medical database searching and on bibliometrics.

Contributors

BARBARA I. BOND is a freelance writer living in South Elgin, Illinois.

HELEN-ANN BROWN, M.S., M.L.S., is Head of Library Relations at the

Samuel J. Wood Library, C.V. Starr Biomedical Information Center, Cornell University Medical College.

STEFANIE BUCK, M.A., M.L.S., is Reference Librarian at the Thomas Cooper Library, University of South Carolina, Columbia.

JUDY F. BURNHAM, M.L.S., is University Medical Center Site Coordinator for the Biomedical Library at the University of South Alabama in Mobile. Her research interests lie in information management and bibliometrics, on which she has authored and co-authored several articles.

DIANE M. CALABRESE, Ph.D., is an entomologist and writer based in Columbia, Missouri.

CLARA A. CALLAHAN, M.D., is Clinical Associate Professor of Pediatrics and Associate Dean for Student Affairs at Jefferson Medical College in Philadelphia, Pennsylvania.

FAYE A. CHADWELL, M.A., M.L.S., is the collection development librarian at the University of Oregon. She has published several bibliographies on women in science for books by Dr. Sue V. Rosser.

MARIA CHIARA holds a doctoral degree in anthropology. She lives in Evanston, Illinois, where she is building a research and consulting practice devoted to understanding the career motivations and work experiences of professional women. This knowledge will help employers restructure jobs to make them more compatible with women's personal and professional goals.

NATHANIEL COMFORT has been Science Writer since 1991 at Cold Spring Harbor Laboratory. He is also pursuing a Ph.D. in the history of science. Among his current projects is a history of Cold Spring Harbor Laboratory, co-authored with noted scientist and historian Bentley Glass.

CAROL W. CUBBERLEY, M.S.L.S., D.A.I.S., is Director of Technical Services, University Libraries, University of Southern Mississippi, Hattiesburg. She has reviewed books on gardening for *Library Journal* since 1989. She is also co-author with Anne Lundin of *Teaching Children's Literature: A Resource Guide.*

F. ELAINE DE LANCEY, Ph.D., is Director of African American Studies at Drexel University in Philadelphia, Pennsylvania.

JULIE DUNLAP is a freelance writer and environmental educator living in Columbia, Maryland. She has a Ph.D. in forestry and environmental studies. She has written several books for children about scientists and environmentalists.

MARILYN GIST FARQUHAR, Ph.D., is Professor of Pathology and Co-ordinator of the Division of Cellular and Molecular Medicine at the University of California, San Diego, in La Jolla.

TERESA R. FAUST, M.L.S., is Science Reference Librarian at Wake Forest University in Winston-Salem, North Carolina.

CASSANDRA S. GISSENDANNER, M.S.L.S., is Catalog Management Librarian at the Thomas Cooper Library, University of South Carolina, Columbia.

GAIL M. GOLDERMAN, M.L.S., is Reference Librarian and Acting Systems Librarian at the Schaffer Library, Union College, Schenectady, New York.

NAN MARIE HAMBERGER, Ed.D., is Associate Professor of Education at Alvernia College in Reading, Pennsylvania. Previously she taught for sixteen years at the secondary level. Her research interests include studies of the teaching of literature and values education. She has given presentations about values education at the local, state, and national levels.

JUDITH E. HEADY, Ph.D., is Associate Professor of Biology at the University of Michigan, Dearborn. She has taught gender and science classes and also leads workshops for teachers on active learning in labs and classrooms. She is a regular contributor to professional publications.

KELLY HENSLEY, M.S.L.S., is Media and Computing Services Librarian at the James H. Quillen College of Medicine at East Tennessee State University in Johnson City. She is active in the Medical Library Association.

HELEN HOFFMAN, M.L.S., is Biological Sciences Resource Librarian at the Library of Science and Medicine at Rutgers, the State University of New Jersey. For many years she was Librarian at the Waksman Institute of Microbiology. In addition to her library degree, Ms. Hoffman has degrees in biological and zoological science. Before becoming a librarian, she was a medical writer for a major pharmaceutical company.

JILL HOLMAN, M.I.L.S., is a reference librarian at the Science Library, University of Oregon. Her graduate work specialized in research, technology, and management. Her recent work has included working on a

World Wide Web page for the American Library Association's Feminist Task Force and contributing to the bibliography for the seventh annual University of South Carolina Systemwide Women's Studies Conference.

MARGARET A. IRWIN, M.L.S., is Special Collections Librarian in the Historical Research Center (HRC) at the Houston Academy of Medicine—Texas Medical Center Library. She was the winner of the AMIGOS Bibliographical Inc. 1994 Fellowship and is currently working on a major project for the HRC's Atomic Bomb Casualty Commission Collections.

DEAN JAMES is Director of Cataloging at the Houston Academy of Medicine—Texas Medical Center Library in Houston, Texas. He holds a Ph.D. in medieval history and an M.S. in library science. He is the co-author, with Jean Swanson, of *By a Woman's Hand: A Guide to Mystery Fiction by Women.*

KAREN JAMES, Ph.D., MT (ASCP), is an immunologist who spent more than twenty-five years working in hospital clinical pathology laboratories. She and her husband now live in the mountains of North Carolina, where she does consulting in hospital laboratory management and clinical immunology, medical writing, and manuscript preparation for authors in the fields of immunology and clinical laboratory science.

GRACE FOOTE JOHNS, M.S. in communication theory, is Assistant to the Chair of the Department of Physics at Illinois State University (ISU). She is on the Board of Directors and Organizers of ISU's "Expanding Your Horizons in Science and Math Conference," which encourages young women to prepare for science careers, and holds career conferences for young women with women professionals in science and technology.

DIANE A. KELLY is a Ph.D. candidate in zoology. Her research focuses on hydrostat theory and the functional morphology of reproductive systems.

CARA KENDRIC, M.L.S., is Cataloging Librarian at the Scott Memorial Library, Thomas Jefferson University, Philadelphia, Pennsylvania.

SANGDUK KIM, M.D., Ph.D., is Professor of Biochemistry at the Fels Institute for Cancer Research and Molecular Biology, Temple University School of Medicine, Philadelphia, Pennsylvania.

KATHLEEN PALOMBO KING is Instructor in Computers, Engineering, and Science at the Pennsylvania Institute of Technology, a computer and autoCAD consultant, and also an Ed.D. candidate. She is actively involved in encouraging her students to pursue scientific and technical careers. She is the author of several publications and acts as technical editor of the *Journal of Afro-Latin American Studies & Literatures*.

SHARON SUE KLEINMAN is currently pursuing a master's degree in communication. Her research interests include the sociology of science with an emphasis on gender issues in science pedagogy.

KIMBERLY J. LAIRD is Technical Services Librarian at the Quillen College of Medicine Library, East Tennessee State University, Johnson City.

D. BOSWELL LANE is completing her M.S. degree in zoology. Her interests include molluscan pathology, aquaculture, and invertebrate zoology.

JOAN LEWIS, M.S.W., A.C.S.W., L.S.W., is Chair of the Human Services Division and the Social Work Department at Alvernia College in Reading, Pennsylvania.

JENNIFER LIGHT holds an M.Phil. in history and philosophy of Science. Former Lionel de Jersey Harvard Scholar at Emmanuel College, Cambridge University, she is currently Resident Tutor in History of Science at Eliot House, Harvard University, and a Ph.D. candidate in the history of science.

ANN LINDELL, M.L.I.S., M.F.A., is Assistant Librarian at the University of Florida Architecture and Fine Arts Library, Gainesville, Florida.

ELEANOR L. LOMAX, M.L.I.S., is Reference Librarian at Florida Atlantic University Libraries in Boca Raton, Florida.

NANCY C. LONG, Ph.D., is Assistant Professor of Science at the College of General Studies, Boston University, and Research Associate in the Physiology Program, Harvard School of Public Health—the same department where Dr. Mary Ellen Avery did her research fellowship. Ms. Long's research interests are in host defense mechanisms, ways in which the body protects itself against injury and infection.

SAROJINI D. LOTLIKAR, M.S.L.S., is Assistant Professor and Catalog Librarian at the Helen Ganser Library, Millersville University of Pennsylvania.

ELIZABETH LUNT, M.S.L.S., is Access Services Librarian at Paley Library, Temple University.

THURA R. MACK, M.S.L.S., is Assistant Professor and Reference Librarian at the Hodges Library, University of Tennessee, Knoxville.

HEATHER MARTIN, M.L.I.S., M.A., is Reference and Outreach Librarian at the Paul Laurence Dunbar Library, Wright State University, Dayton, Ohio. She has written biographical sketches of Evelyn Ashford and Margaret Avery for the publication *African American Women: A Biographical Dictionary.*

LEE MCDAVID, M.L., is a librarian at the University of South Carolina in Columbia. She holds an M.A. in history.

SHIRLEY B. MCDONALD, B.S., M.L.I.S., C.L.I.S., is Librarian at Live Oak High School in Watson, Louisiana.

MARY ANN MCFARLAND, M.A.L.S., is Assistant Director for Access Services at the St. Louis University Health Sciences Center Library in St. Louis, Missouri. She is a frequent contributor to professional publications.

MARILYN R. P. MORGAN is a Ph.D. candidate in communication and rhetoric. Now doing research at Rensselaer Polytechnic Institute in Troy, New York, she previously completed internships at Oak Ridge National Laboratory and the University of Tennessee's Energy, Environment, and Resources Center. Her professional interests include computer-mediated communication, social science research methods, and the space program.

PATRICIA MURPHY is Assistant Professor of Agriculture and Life Sciences Bibliographer at Virginia Polytechnic Institute and State University in Blacksburg, Virginia.

SYLVIA NICHOLAS, M.S.L.S., is Reference Librarian at the Galter Health Sciences Library, Northwestern University.

CONNIE H. NOBLES, Ph.D. in curriculum and instruction, is Assistant Professor at Southeastern Louisiana University, where she teaches science methods incorporating women in science throughout history. She has contributed to several volumes on noted women scientists and is working on a collaborative biography with one of the sons of Leona Woods Marshall Libby, the only woman nuclear physicist involved in the Manhattan Project.

CAROL BROOKS NORRIS, M.L.S., is Associate Professor and Online Searching Librarian at Sherrod Library, East Tennessee State University, Johnson City. She has published annotated bibliographies and articles in professional publications, as well as poetry.

LESLIE O'BRIEN is Assistant Professor, Cataloging Team Leader, at Virginia Polytechnic Institute and State University in Blacksburg, Virginia.

JOAN GARRETT PACKER, M.A. in history, M.L.S., is Head of the Reference Department at Central Connecticut State University in New Britain. She has published bibliographies on Margaret Drabble and Rebecca West.

LAURIE A. POTTER, M.L.S., is Medical Reference Librarian at the Savitt Medical Library in Reno, Nevada. She is certified as a senior member of the Academy of Health Information Professionals and is an active contributor to professional publications.

ARLENE RODDA, M.L.S., is currently working at Emerson College Library in Boston, where she is a member of the Public Services Department and coordinator of the bibliographic instruction program. She hopes to devote herself full-time to writing someday, so she finds inspiration in the story of Rachel Carson.

JOY SCHABER did her undergraduate work in biology and English and holds an M.S. in ecology. She has written articles on scientific issues for the University of California Natural Reserve System and the local Davis newspaper. Currently she is working on the ecology of Giant Sequoia groves for the Sierra Nevada Ecosystem Project.

CARLA J. SHILTS, M.S.L.S., and M.S. candidate in biology, is Science Reference Librarian at the Chester Fritz Library of the University of North Dakota in Grand Forks. She is a regular contributor to professional publications.

CAROLE B. SHMURAK, Ph.D., is Associate Professor of Education at Central Connecticut State University. She holds a B.A. in chemistry, an M.A. in biochemistry, and a Ph.D. in science education. Her recent publications include articles on women in science.

MARTHA E. STONE, M.S., A.H.I.P., is Coordinator for Reference Services at Treadwell Library, Massachusetts General Hospital, Boston. She holds an M.S. in library and information science. She is a book reviewer for a variety of publications.

MARGARET SYLVIA, M.L.I.S., is Assistant Professor at St. Mary's University Academic Library in San Antonio, Texas, where she is Acquisitions and Collection Development Department Head and CD-ROM Local Area Network Systems Administrator. She has authored several articles in professional journals.

GLORIA I. TROYER, B.A. in history, has been Archival and Special Collections Librarian at the University of Guelph Library, University of Guelph, Guelph, Ontario, Canada, since 1980. She is a founding member of the Ontario Women's History Network (OWHN) and has been on its Board of Directors since 1990. She has most recently edited *Every Comfort in the Wilderness* (1994).

REBECCA LOWE WARREN, M.A., is co-author of *The Scientist within You: Experiments and Biographies of Distinguished Women in Science* and an adjunct faculty member at Maryhurst College. Her science class, "Beyond Marie Curie," highlights the contributions of women in science and mathematics. Her affiliations include the American Association of University Women.

MICHAEL WEBER, M.L.S., is Assistant Director for Technical Services at the Alvernia College Library in Reading, Pennsylvania.

MOLLY WEINBURGH, Ph.D. in educational leadership and science education, is Assistant Professor at Georgia State University, College of Education, Department of Middle/Secondary Education and Instructional Technology. She is co-author of *Tropical Rainforest: An Interdisciplinary Approach* and a frequent contributor to professional publications.

KIMERLY J. WILCOX, Ph.D., is Assistant Professor at the General College of the University of Minnesota, where she teaches introductory biology and anatomy and physiology courses. Her research interests include the genetics of epilepsy and developmental education.

IRMGARD H. WOLFE, M.L.S., is Cataloger and Preservation Librarian at Cook Memorial Library, University of Southern Mississippi, Hattiesburg.